CRC Series in
CONTEMPORARY FOOD SCIENCE

Fergus M. Clydesdale, Series Editor
University of Massachusetts, Amherst

Published Titles:

America's Foods Health Messages and Claims:
Scientific, Regulatory, and Legal Issues
James E. Tillotson

New Food Product Development: From Concept to Marketplace
Gordon W. Fuller

Food Properties Handbook
Shafiur Rahman

Aseptic Processing and Packaging of Foods:
Food Industry Perspectives
Jarius David, V. R. Carlson, and Ralph Graves

The Food Chemistry Laboratory: A Manual for Experimental Foods,
Dietetics, and Food Scientists
Connie Weaver

Handbook of Food Spoilage Yeasts
Tibor Deak and Larry R. Beauchat

Food Emulsions: Principles, Practice, and Techniques
David Julian McClements

Forthcoming Titles:

Getting the Most Out of Your Consultant: A Guide
to Selection Through Implementation
Gordon W. Fuller

Antioxidant Status, Diet, Nutrition, and Health
Andreas M. Papas

Food Shelf Life Stability
N. A. Michael Eskin and Davis S. Robinson

Bread Staling
Pavinee Chinachoti and Yael VodoVotz

FOOD EMULSIONS

Principles, Practice, and Techniques

David Julian McClements

CRC Press
Boca Raton   London   New York   Washington, D.C.

Library of Congress Cataloging-in-Publication Data

Catalog information may be obtained from the Library of Congress

No claim to original U.S. Government works
International Standard Book Number 0-8493-8008-1
Printed in the United States of America 1 2 3 4 5 6 7 8 9 0
Printed on acid-free paper

Table of Contents

Preface

As one strolls along the aisles of a supermarket, one passes a wide variety of food products, both natural and manufactured, which exist either partly or wholly as emulsions or which have been in an emulsified form sometime during their production. Common examples include milk, flavored milks, creams, salad dressings, dips, coffee whitener, ice cream, soups, sauces, mayonnaise, butter, margarine, fruit beverages, and whipped cream. Even though these products differ widely in their appearance, texture, taste, and shelf life, they all consist (or once consisted) of small droplets of one liquid dispersed in another liquid. Consequently, many of their physicochemical and sensory properties can be understood by applying the fundamental principles and techniques of *emulsion science*. It is for this reason that anyone in the food industry working with these types of products should have at least an elementary understanding of this important topic.

The primary objective of this book is to present the principles and techniques of emulsion science and show how they can be used to better understand, predict, and control the properties of a wide variety of food products. Rather than describe the specific methods and problems associated with the creation of each particular type of emulsion-based food product, I have concentrated on an explanation of the basic concepts of emulsion science, as these are applicable to all types of food emulsion. Details about the properties of particular types of food emulsion are described in the latest edition of an excellent book edited by S.E. Friberg and K. Larsson (*Food Emulsions,* 3rd edition, Marcel Dekker, New York, 1997), which should be seen as being complementary to this volume.

It is a great pleasure to acknowledge the contributions of all those who helped bring this book to fruition. Without the love and support of my best friend and partner Jayne and of my family, this book would never have been completed. I also thank all of my students and co-workers who have been a continual source of stimulating ideas and constructive criticism and my teachers for providing me with the strong academic foundations on which I have attempted to build. Finally, I thank all those at CRC Press for their help in the preparation of this book.

The Author

Dr. David Julian McClements has been an Assistant Professor in the Department of Food Science at the University of Massachusetts since 1994. He received a B.S. (Hons) in food science (1985) and a Ph.D. in "Ultrasonic Characterization of Fats and Emulsions" (1989) from the University of Leeds (United Kingdom). He then did postdoctoral research at the University of Leeds, University of California at Davis, and the University College Cork in Ireland, before joining the University of Massachusetts. Dr. McClements' research interests include ultrasonic characterization of food emulsions, food biopolymers and colloids (focusing on emulsions, gels, and micellar systems), protein functionality, and physicochemical properties of lipids.

Dr. McClements has co-authored a book entitled *Advances in Food Colloids* with Professor Eric Dickinson and co-edited a book entitled *Developments in Acoustics and Ultrasonics* with Dr. Malcolm Povey. In addition, he has published over 100 scientific articles as book chapters, encyclopedia entries, journal manuscripts, and conference proceedings. Dr. McClements recently received the Young Scientist award from the American Chemical Society's Division of Food and Agriculture in recognition of his achievements.

1 Context and Background

1.1. EMULSION SCIENCE IN THE FOOD INDUSTRY

Many natural and processed foods consist either partly or wholly as emulsions or have been in an emulsified state at some time during their production; such foods include milk, cream, butter, margarine, fruit beverages, soups, cake batters, mayonnaise, cream liqueurs, sauces, desserts, salad cream, ice cream, and coffee whitener (Friberg and Larsson 1997, Krog et al. 1983, Jaynes 1983, Dickinson and Stainsby 1982, Dickinson 1992, Swaisgood 1996). Emulsion-based food products exhibit a wide variety of different physicochemical and organoleptic characteristics, such as appearance, aroma, texture, taste, and shelf life. For example, milk is a low-viscosity white fluid, strawberry yogurt is a pink viscoelastic gel, and margarine is a yellow semisolid. This diversity is the result of the different sorts of ingredients and processing conditions used to create each type of product. The manufacture of an emulsion-based food product with specific quality attributes depends on the selection of the most appropriate raw materials (e.g., water, oil, emulsifiers, thickening agents, minerals, acids, bases, vitamins, flavors, colorants, etc.) and processing conditions (e.g., mixing, homogenization, pasteurization, sterilization, etc.).

Traditionally, the food industry largely relied on craft and tradition for the formulation of food products and the establishment of processing and storage conditions. This approach is unsuitable for the modern food industry, which must rapidly respond to changes in consumer preferences for a greater variety of cheaper, healthier, and more convenient foods (Sloan 1994, 1996; Katz 1997). In addition, the modern food industry relies increasingly on large-scale production operations to produce vast quantities of foods at relatively low cost. The development of new foods, the improvement of existing foods, and the efficient running of food-processing operations require a more systematic and rigorous approach than was used previously (Hollingsworth 1995).

Two areas which have been identified as being of particular importance to the improvement of food products are:

1. *Enhanced scientific understanding of food properties.* An improved understanding of the factors that determine the bulk physicochemical and organoleptic properties of emulsions will enable manufacturers to create low-cost high-quality food products in a more systematic and reliable fashion (Kokini et al. 1993, Rizvi et al. 1993).

2. *Development of new analytical techniques to characterize food properties.* The development and application of new analytical techniques to characterize the properties of emulsions are leading to considerable advances in research, development, and quality control (Dickinson 1995a,b; Gaonkar 1995). These techniques are used in the laboratory to enhance our understanding of the factors which determine the properties of foods and in the factory to monitor the properties of foods during processing in order to ensure that they meet the required quality specifications.

Emulsion science is a multidisciplinary subject that combines chemistry, physics, and engineering (Sherman 1968a; Becher 1957, 1983; Hiemenz 1986; Hunter 1986, 1989, 1993; Evans and Wennerstrom 1994). The aim of the emulsion scientist working in the food industry is to utilize the principles and techniques of emulsion science to enhance the quality of the food supply and the efficiency of food production. This book presents the conceptual and theoretical framework required by food scientists to understand and control the properties of emulsion-based food products.

1.2. GENERAL CHARACTERISTICS OF FOOD EMULSIONS

1.2.1. Definitions

An emulsion consists of two immiscible liquids (usually oil and water), with one of the liquids dispersed as small spherical droplets in the other (Figure 1.1). In most foods, the diameters of the droplets usually lie somewhere between 0.1 and 100 µm (Dickinson and Stainsby 1982; Dickinson 1992; Walstra 1996a,b). Emulsions can be conveniently classified according to the distribution of the oil and aqueous phases. A system which consists of oil droplets dispersed in an aqueous phase is called an *oil-in-water* or O/W emulsion (e.g., mayonnaise, milk, cream, soups, and sauces). A system which consists of water droplets dispersed in an oil phase is called a *water-in-oil* or W/O emulsion (e.g., margarine, butter, and

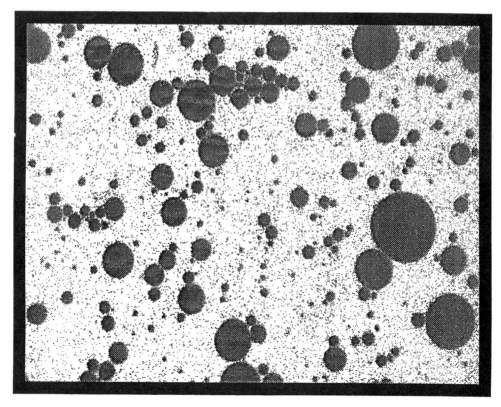

FIGURE 1.1 Microscopic image of a 20 wt% tetradecane oil-in-water emulsion stabilized by 1 wt% whey protein isolate obtained using confocal scanning fluorescence microscopy. The dark regions are the oil droplets, and the light regions are the aqueous phase.

spreads). The substance that makes up the droplets in an emulsion is referred to as the *dispersed* or *internal phase,* whereas the substance that makes up the surrounding liquid is called the *continuous* or *external phase.* It is also possible to prepare multiple emulsions of the oil-in-water-in-oil (O/W/O) or water-in-oil-in-water (W/O/W) type (Dickinson and McClements 1995). For example, a W/O/W emulsion consists of water droplets dispersed within larger oil droplets, which are themselves dispersed in an aqueous continuous phase (Evison et al. 1995). Recently, research has been carried out to create stable multiple emulsions which can be used to control the release of certain ingredients, reduce the total fat content of emulsion-based food products, or isolate one ingredient from another (Dickinson and McClements 1995).

The concentration of droplets in an emulsion is usually described in terms of the *dispersed-phase volume fraction* (φ) (Section 1.3.1). The process of converting two separate immiscible liquids into an emulsion, or of reducing the size of the droplets in a preexisting emulsion, is known as *homogenization.* In the food industry, this process is usually carried out using mechanical devices known as *homogenizers,* which subject the liquids to intense mechanical agitation (Chapter 6).

It is possible to form an emulsion by homogenizing pure oil and pure water together, but the two phases rapidly separate into a system which consists of a layer of oil (lower density) on top of a layer of water (higher density). This is because droplets tend to merge with their neighbors when they collide with them, which eventually leads to complete phase separation. The driving force for this process is the fact that the contact between oil and water molecules is energetically unfavorable (Israelachvili 1992), so that emulsions are *thermodynamically unstable* systems (Chapter 7). It is possible to form emulsions that are *kinetically stable* (metastable) for a reasonable period of time (a few days, weeks, months, or years) by including substances known as *emulsifiers* and/or *thickening agents* prior to homogenization (Chapter 4). Emulsifiers are *surface-active* molecules which absorb to the surface of freshly formed droplets during homogenization, forming a protective membrane which prevents the droplets from coming close enough together to aggregate (Chapters 6 and 7). Most emulsifiers are *amphiphilic* molecules (i.e., they have polar and nonpolar regions on the same molecule). The most common emulsifiers used in the food industry are amphiphilic proteins, small-molecule surfactants, and phospholipids (Chapter 4). Thickening agents are ingredients which are used to increase the viscosity of the continuous phase of emulsions, and they enhance emulsion stability by retarding the movement of the droplets. The most common thickening agents used in the food industry are polysaccharides (Chapter 4). A *stabilizer* is any ingredient that can be used to enhance the stability of an emulsion and may therefore be either an emulsifier or a thickening agent.

An appreciation of the difference between the thermodynamic stability of a system and its kinetic stability is crucial for an understanding of the properties of food emulsions (Dickinson 1992). Consider a system which consists of a large number of molecules that can occupy two different states: E_{low} and E_{high} (Figure 1.2). The state with the lowest free energy is the one which is thermodynamically favorable and therefore the one that the molecules are most likely to occupy. At thermodynamic equilibrium, the two states are populated according to the Boltzmann distribution (Atkins 1994):

$$\frac{\phi_{high}}{\phi_{low}} = \exp\left[-\frac{(E_{low} - E_{high})}{kT} \right] \qquad (1.1)$$

where φ is the fraction of molecules that occupies the energy level E, k is Boltzmann's constant ($k = 1.38 \times 10^{-23}$ J K^{-1}), and T is the absolute temperature. The larger the difference

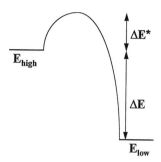

FIGURE 1.2   Difference between thermodynamic and kinetic stability. A system will remain in a thermodynamically unstable or metastable state for some time if there is a sufficiently large energy barrier preventing it from reaching the state with the lowest free energy.

between the two energy levels compared to the thermal energy of the system (kT), the greater the fraction of molecules in the lower energy state. In practice, a system may not be able to reach equilibrium during the time scale of an observation because of the presence of an energy barrier (ΔE^*) between the two states (Figure 1.2). A system in the high energy state must acquire an energy greater than ΔE^* before it can move into the low energy state. The rate at which a transformation from a high to a low energy state occurs therefore decreases as the height of the energy barrier increases. When the energy barrier is sufficiently large, the system may remain in a thermodynamically unstable state for a considerable length of time, in which case it is said to be kinetically stable or metastable (Atkins 1994). In food emulsions, there are actually a large number of intermediate metastable states between the initial emulsion and the separated phases, and there is an energy barrier associated with a transition between each of these states. Nevertheless, it is often possible to identify a single energy barrier, which is associated with a particular physicochemical process, that is the most important factor in determining the overall kinetic stability of an emulsion (Chapter 7).

1.2.2.   Mechanisms of Emulsion Instability

The term "emulsion stability" is broadly used to describe the ability of an emulsion to resist changes in its properties with time (Chapter 7). Nevertheless, there are a variety of physicochemical mechanisms which may be responsible for alterations in the properties of an emulsion, and it is crucial to be clear about which of these mechanisms are important in the system under consideration. A number of the most important physical mechanisms responsible for the instability of emulsions are shown schematically in Figure 1.3. *Creaming* and *sedimentation* are both forms of *gravitational separation*. Creaming describes the upward movement of droplets due to the fact that they have a lower density than the surrounding liquid, whereas sedimentation describes the downward movement of droplets due to the fact that they have a higher density than the surrounding liquid. *Flocculation* and *coalescence* are both types of droplet aggregation. Flocculation occurs when two or more droplets come together to form an aggregate in which the droplets retain their individual integrity, whereas coalescence is the process where two or more droplets merge together to form a single larger droplet. Extensive droplet coalescence can eventually lead to the formation of a separate layer of oil on top of a sample, which is known as "oiling off." *Phase inversion* is the process whereby an oil-in-water emulsion is converted into a water-in-oil emulsion or vice versa. The

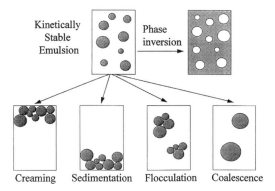

FIGURE 1.3 Food emulsions may become unstable through a variety of physical mechanisms, including creaming, sedimentation, flocculation, coalescence, and phase inversion.

factors which determine these and the other major forms of emulsion instability are discussed in Chapter 7, along with methods of controlling and monitoring them. In addition to the physical processes mentioned above, it should be noted that there are also various chemical, biochemical, and microbiological processes that occur in food emulsions which can also affect their shelf life and quality.

1.2.3. Ingredient Partitioning in Emulsions

Most food emulsions can conveniently be considered to consist of three regions which have different physicochemical properties: the interior of the droplets, the continuous phase, and the interface (Figure 1.4). The molecules in an emulsion distribute themselves among these three regions according to their concentration and polarity (Wedzicha 1988). Nonpolar molecules tend to be located primarily in the oil phase, polar molecules in the aqueous phase, and amphiphilic molecules at the interface. It should be noted that even at equilibrium, there is a continuous exchange of molecules between the different regions, which occurs at a rate that depends on the mass transport of the molecules through the system. Molecules may also move from one region to another when there is some alteration in the environmental conditions of an emulsion (e.g., a change in temperature or dilution within the mouth). The location and mass transport of the molecules within an emulsion have a significant influence on the aroma, flavor release, texture, and physicochemical stability of food products (Dickinson and Stainsby 1982, Wedzicha et al. 1991, Coupland and McClements 1996, Landy et al. 1996).

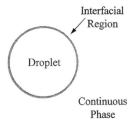

FIGURE 1.4 The ingredients in an emulsion partition themselves between the oil, water, and interfacial regions according to their concentration and interactions with the local environment.

1.2.4. Dynamic Nature of Emulsions

Many of the properties of emulsions can only be understood with reference to their dynamic nature. The formation of emulsions by homogenization is a highly dynamic process which involves the violent disruption of droplets and the rapid movement of surface-active molecules from the bulk liquids to the interfacial region (Chapter 6). Even after their formation, the droplets in an emulsion are in continual motion and frequently collide with one another because of their Brownian motion, gravity, or applied mechanical forces (Melik and Fogler 1988, Dukhin and Sjoblom 1996, Lips et al. 1993). The continual movement and interactions of droplets cause the properties of emulsions to evolve over time due to the various destabilization mechanisms mentioned in Section 1.2.2. An appreciation of the dynamic processes that occur in food emulsions is therefore extremely important for a thorough understanding of their bulk physicochemical and organoleptic properties.

1.2.5. Complexity of Food Emulsions

Most food emulsions are much more complex than the simple three-component (oil, water, and emulsifier) systems described in Section 1.2.1. The aqueous phase may contain a variety of water-soluble ingredients, including sugars, salts, acids, bases, surfactants, proteins, and carbohydrates. The oil phase usually contains a complex mixture of lipid-soluble components, such as triacylglycerols, diacylglycerols, monoacylglycerols, free fatty acids, sterols, and vitamins. The interfacial region may contain a mixture of various surface-active components, including proteins, phospholipids, surfactants, alcohols, and solid particles. In addition, these components may form various types of structural entities in the oil, water, or interfacial regions, such as fat crystals, ice crystals, protein aggregates, air bubbles, liquid crystals, and surfactant micelles. A further complicating factor is that foods are subjected to variations in their temperature, pressure, and mechanical agitation during their production, storage, and handling, which can cause significant alterations in their overall properties.

It is clear from the above discussion that food emulsions are compositionally, structurally, and dynamically complex materials and that many factors contribute to their overall properties. One of the major objectives of this book is to present the conceptual framework needed by food scientists to understand these complex systems in a more systematic and rigorous fashion. Much of our knowledge about these complex systems has come from studies of simple model systems (Section 1.5). Nevertheless, there is an increasing awareness of the need to elucidate the factors that determine the properties of actual emulsion-based food products. For this reason, many researchers are now focusing on the complex issues that need to be addressed, such as ingredient interactions, effects of processing conditions, and phase transitions (Dickinson 1992, 1995b; Dickinson and McClements 1995; Dalgleish 1996a; Hunt and Dalgleish 1994, 1995; Demetriades et al. 1997a,b).

1.3. EMULSION PROPERTIES

1.3.1. Dispersed-Phase Volume Fraction

The concentration of droplets in an emulsion is usually described in terms of the *dispersed-phase volume fraction* (ϕ), which is equal to the volume of emulsion droplets (V_D) divided by the total volume of the emulsion (V_E): $\phi = V_D/V_E$. Knowledge of the dispersed-phase volume fraction is important because the droplet concentration influences the appearance, texture, flavor, stability, and cost of emulsion-based food products. In some situations, it is more convenient to express the composition of an emulsion in terms of the dispersed-phase mass fraction (ϕ_m), which is related to the volume fraction by the following equation:

$$\phi_m = \frac{\phi\rho_2}{\rho_2\phi + (1 - \phi)\rho_1} \tag{1.2}$$

where ρ_1 and ρ_2 are the densities of the continuous and dispersed phases, respectively. When the densities of the two phases are equal, the mass fraction is equivalent to the volume fraction. The dispersed-phase volume fraction of an emulsion is often known because the concentration of the ingredients used to prepare it is carefully controlled. Nevertheless, local variations in dispersed-phase volume fraction occur within emulsions when the droplets accumulate at either the top or bottom of an emulsion due to creaming or sedimentation. In addition, the dispersed-phase volume fraction of an emulsion may vary during a food-processing operation (e.g., if a mixer or valve is not operating efficiently). Consequently, it is important to have analytical techniques to measure dispersed-phase volume fraction (Chapter 10).

1.3.2. Particle Size Distribution

Many of the most important properties of emulsion-based food products (e.g., shelf life, appearance, texture, and flavor) are determined by the size of the droplets they contain, (Dickinson and Stainsby 1982, Dickinson 1992). Consequently, it is important for food scientists to be able to reliably control, predict, measure, and report the size of the droplets in emulsions. In this section, the most important methods of reporting droplet sizes are discussed. Methods of controlling, predicting, and measuring droplet size are covered in later chapters (Chapters 6, 7, and 10).

If all the droplets in an emulsion are of the same size, the emulsion is referred to as *monodisperse,* but if there is a range of sizes present, the emulsion is referred to as *polydisperse.* The size of the droplets in a monodisperse emulsion can be completely characterized by a single number, such as the droplet diameter (d) or radius (r). Monodisperse emulsions are sometimes used for fundamental studies because the interpretation of experimental measurements is much simpler than that of polydisperse emulsions. Nevertheless, food emulsions always contain a distribution of droplet sizes, and so the specification of their droplet size is more complicated than that of monodisperse systems. Ideally, one would like to have information about the full *particle size distribution* of an emulsion (i.e., the size of each of the droplets in the system). Nevertheless, in many situations, knowledge of the average size of the droplets and the width of the distribution is sufficient (Hunter 1986).

1.3.2.1. Presenting Particle Size Data

The number of droplets in most emulsions is extremely large, and so their size can be considered to vary continuously from some minimum value to some maximum value. When presenting particle size data, it is convenient to divide this size range into a number of discrete size classes and stipulate the number of droplets that fall into each class (Hunter 1986). The resulting data can then be represented in tabular form (Table 1.1) or plotted as a histogram that shows the number of droplets in each size class (Figure 1.5). Rather than presenting the number of droplets (n_i) in each size class, it is often more informative to present the data as the *number frequency,* $f_i = n_i/N$, where N is the total number of droplets, or as the *volume frequency,* $\phi_i = v_i/V$, where v_i is the volume of the droplets in the ith size class and V is the total volume of all the droplets in the emulsion. It should be noted that the shape of a particle size distribution changes appreciably depending on whether it is presented as a number or volume frequency (Table 1.2). The volume of a droplet is propor-

TABLE 1.1
The Particle Size Distribution of an Emulsion Represented in Tabular Form

Size class (μm)	d_i (μm)	N_i	f_i (%)	ϕ_i (%)	$C(d_i)$ (%)
041–0.054	0.048	0	0.0	0.0	0.0
0.054–0.071	0.063	2	0.1	0.0	0.1
0.071–0.094	0.082	4	0.2	0.0	0.3
0.094–0.123	0.108	50	2.5	0.0	2.8
0.123–0.161	0.142	84	4.2	0.1	7.0
0.161–0.211	0.186	152	7.6	0.3	14.6
0.211–0.277	0.244	224	11.2	1.1	25.8
0.277–0.364	0.320	351	17.6	3.9	43.35
0.364–0.477	0.420	470	23.5	11.8	66.85
0.477–0.626	0.551	385	19.2	21.8	86.1
0.626–0.821	0.723	190	9.5	24.3	95.6
0.821–1.077	0.949	64	3.2	18.5	98.8
1.077–1.414	1.245	21	1.0	13.7	99.85
1.414–1.855	1.634	3	0.2	4.4	100
1.855–2.433	2.144	0	0.0	0.0	100

Note: The volume frequency is much more sensitive to larger droplets than is the number frequency.

tional to d^3, and so a volume distribution is skewed more toward the larger droplets, whereas a number distribution is skewed more toward the smaller droplets.

A particle size distribution can also be represented as a smooth curve, such as the distribution function, $F(d_i)$, or the cumulative function, $C(d_i)$ (Figure 1.5). The (number) distribution function is constructed so that the area under the curve between two droplet sizes (d_i and $d_i + \delta d_i$) is equal to the number of droplets (n_i) in that size range (i.e., $n_i = F(d_i)\delta d_i$) (Hunter 1986). This relationship can be used to convert a histogram to a distribution function or vice versa. The cumulative function represents the percentage of droplets that are smaller than d_i (Figure 1.5). The resulting curve has an S-shape which varies from 0 to 100% as the particle size increases. The particle size at which half the droplets are smaller and the other half are larger is known as the median droplet diameter (d_m).

1.3.2.2. Mean and Standard Deviation

It is often convenient to represent the size of the droplets in a polydisperse emulsion by one or two numbers, rather than stipulating the full particle size distribution (Hunter 1986). The most useful numbers are the mean diameter ($\bar{d}$), which is a measure of the central tendency of the distribution, and the standard deviation (σ), which is a measure of the width of the distribution:

$$\bar{d} = \sum n_i d_i / N \qquad \sigma = \sqrt{\left[\sum n_i (d_i - \bar{d})^2\right] / N} \qquad (1.3)$$

The above mean is also referred to as the mean length diameter (d_L) because it represents the sum of the *length* of the droplets divided by the total number of droplets. If all the droplets in a polydisperse emulsion were laid end to end, they would have the same overall length as those in a monodisperse emulsion containing an equal number of droplets of diameter d_L. It

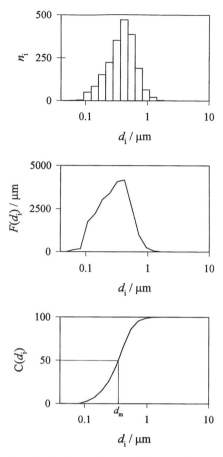

FIGURE 1.5 The particle size distribution of an emulsion can be represented by a histogram, a distribution function $F(d)$, or a cumulative function $C(d)$.

is also possible to express the mean droplet size in a number of other ways (Table 1.2). Each of these mean sizes has dimensions of length (meters), but stresses a different physical aspect of the distribution (e.g., the average length, surface area, or volume). For example, the volume–surface mean diameter is related to the surface area of droplets exposed to the continuous phase per unit volume of emulsion (A_S):

$$A_S = \frac{6\phi}{d_{VS}} \tag{1.4}$$

This relationship is particularly useful for calculating the total surface area of droplets in an emulsion from a knowledge of the mean diameter of the droplets and the dispersed-phase volume fraction. An appreciation of the various types of mean droplet diameter is also important because different experimental techniques used to measure droplet sizes are sensitive to different mean values (Orr 1988). For example, analysis of polydisperse emulsions using osmotic pressure measurements gives information about their mean length diameter, whereas light-scattering and sedimentation measurements give information about their mean surface diameter. Consequently, it is always important to be clear about which

TABLE 1.2
Different Ways of Expressing the Mean Droplet Diameter
of a Polydisperse Emulsion

Name of mean	Symbol	Definition
Length	$\bar{d}$ or d_L	$d_L = \sum n_i d_i / \sum n_i$
Surface area	d_S	$d_S = \sqrt{\sum n_i d_i^2 / \sum n_i}$
Volume	d_V	$d_V = \sqrt[3]{\sum n_i d_i^3 / \sum n_i}$
Volume–surface area	d_{VS} or d_{32}	$d_{VS} = \sum n_i d_i^3 / \sum n_i d_i^2$

mean diameter has been determined in an experiment when using or quoting droplet size data.

1.3.2.3. Mathematical Models

The particle size distribution of an emulsion can often be modeled using a mathematical theory, which is convenient because it means that the full data set can be described by a small number of parameters (Hunter 1986). If a plot of droplet frequency versus droplet size is symmetrical about the mean droplet size, the curve can be described by a normal distribution function (Figure 1.6):

$$f(d) = \frac{1}{\sigma\sqrt{2\pi}} \exp\left[\frac{-(d - \bar{d})^2}{2\sigma^2}\right] \tag{1.5}$$

where $f(d)\delta d$ is the fraction of emulsion droplets which lies within the size interval between d and $d + \delta d$. The number of droplets in each size group can be calculated from the relation $n_i = N \cdot f(d) \cdot \delta d$. Most (~68%) of the droplets fall within one standard deviation of the mean ($\bar{d} \pm \sigma$), while the vast majority (~99.7%) fall within three standard deviations ($\bar{d} \pm 3\sigma$). Only two parameters are needed to describe the particle size distribution of an emulsion that can be approximated by a normal distribution, the mean and the standard deviation.

The particle size distribution of most food emulsions is not symmetrical about the mean, but tends to extend much further at the high-droplet-size end than at the low-droplet-size end (Figure 1.6). This type of distribution can often be described by a log-normal distribution:

$$f(d) = \frac{1}{\ln \sigma_g \sqrt{2\pi}} \exp\left[\frac{-(\ln d - \ln \bar{d}_g)^2}{2 \ln^2 \sigma_g}\right] \tag{1.6}$$

where $f(d)\delta(\ln d)$ is the fraction of emulsion droplets which lies within the size interval between $\ln d$ and $\ln d + \delta(\ln d)$, and $\bar{d}_g$ and σ_g are the geometric mean and the standard deviation of the geometric mean, which are given by the following expressions:

$$\ln \bar{d}_g = \sum n_i \ln d_i / N \qquad \ln \sigma_g = \sqrt{\sum [n_i(\ln d_i - \ln \bar{d}_g)^2] / N} \tag{1.7}$$

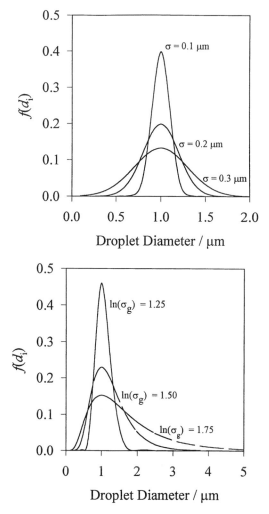

FIGURE 1.6 Comparison of emulsions that can be described by normal and log-normal droplet size distributions.

If the log-normal curve shown in Figure 1.6 was plotted as droplet frequency versus log diameter, it would be symmetrical about $\ln \bar{d}_g$.

It should be stressed that the particle size distribution of many food emulsions cannot be adequately described by the simple models given above. Bimodal distributions, which are characterized by two peaks (Figure 1.7), are often encountered in food emulsions (e.g., when extensive droplet flocculation occurs or when there is insufficient emulsifier present in an emulsion to stabilize all of the droplets formed during homogenization). For these systems, it is often better to present the data as the full particle size distribution; otherwise, considerable errors may occur if an inappropriate model is used.

1.3.3. Interfacial Properties

The droplet interface consists of a narrow region (usually a few nanometers thick) which surrounds each emulsion droplet and contains a mixture of oil, water, and emulsifier mol-

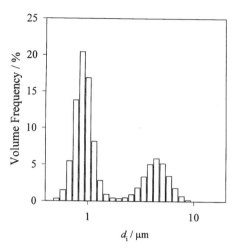

FIGURE 1.7 Example of a bimodal distribution resulting from the heat-induced flocculation of droplets in a 20 wt% corn oil-in-water emulsion stabilized by whey protein isolate. (From Demetriades, K., Coupland, J.N., and McClements, D.J., *Journal of Food Science,* **62**, 462, 1997b. With permission.)

ecules (Hunter 1986, 1989). The interfacial region only makes up a significant fraction of the total volume of an emulsion when the droplet size is less than about 1 μm (Table 1.3). Even so, it plays a major role in determining many of the most important bulk physicochemical and organoleptic properties of food emulsions. For this reason, food scientists are particularly interested in elucidating the factors which determine the composition, structure, thickness, rheology, and charge of the interfacial region. The composition and structure of the interfacial region are determined by the type and concentration of surface-active species present, as well as by the events which occur both during and after emulsion formation (Chapter 6). The thickness and rheology of the interfacial region influence the stability of emulsions to gravitational separation, coalescence, and flocculation and determine the rate at which molecules leave or enter the droplets. The major factors which determine the characteristics of the interfacial region are discussed in Chapter 5, along with experimental techniques to characterize its properties.

TABLE 1.3
Effect of Particle Size on the Physical Characteristics of 1 g of Oil Dispersed in Water in the Form of Spherical Droplets

Droplet radius (μm)	No. of droplets per gram oil (g^{-1})	Droplet surface area per gram oil (m^2 g^{-1})	% oil molecules at droplet surface
100	2.6×10^5	0.03	0.02
10	2.6×10^8	0.3	0.2
1	2.6×10^{11}	3	1.8
0.1	2.6×10^{14}	30	18

Note: Values were calculated assuming the oil had a density of 920 kg m^{-3} and the end-to-end length of the oil molecules was 6 nm.

1.3.4. Droplet Charge

The bulk physicochemical and organoleptic properties of many food emulsions are governed by the magnitude and sign of the electrical charge on the droplets (Dickinson and Stainsby 1982). The origin of this charge is normally the adsorption of emulsifier molecules that are ionized or ionizable. Surfactants have hydrophilic head groups that may be neutral, positively charged, or negatively charged (Chapter 4). Proteins may also be neutral, positively charged, or negatively charged depending on the pH of the solution compared to their isoelectric point (Chapter 4). Consequently, emulsion droplets may have an electrical charge that depends on the types of surface-active molecules present and the pH of the aqueous phase. The charge on a droplet is important because it determines the nature of its interactions with other charged species (Chapters 2 and 3) or its behavior in the presence of an electrical field (Chapter 10). Two species which have charges of opposite sign are attracted toward each other, whereas two species which have charges of similar sign are repelled (Chapters 2 and 3). All of the droplets in an emulsion are usually coated with the same type of emulsifier, and so they have the same electrical charge (if the emulsifier is ionized). When this charge is sufficiently large, the droplets are prevented from aggregating because of the electrostatic repulsion between them (Chapter 3). The properties of emulsions stabilized by ionized emulsifiers are particularly sensitive to the pH and ionic strength of the aqueous phase. If the pH of the aqueous phase is adjusted so that the emulsifier loses its charge, or if salt is added to "screen" the electrostatic interactions between the droplets, the repulsive forces may no longer be strong enough to prevent the droplets from aggregating. Droplet aggregation often leads to a large increase in emulsion viscosity (Chapter 8) and may cause the droplets to cream more rapidly (Chapter 7).

The influence of electrostatic interactions on the stability of emulsions can clearly be demonstrated if one adds a few drops of lemon juice to a glass of homogenized milk. After a few minutes, the milk changes from a low-viscosity emulsion containing isolated oil droplets to a viscous coagulum containing extensively flocculated droplets. This is because the lemon juice decreases the pH of the emulsion toward the isoelectric point of the proteins, thereby reducing the electrostatic repulsion between the droplets and causing droplet aggregation.

Electrostatic interactions also influence the interactions between emulsion droplets and other charged species, such as biopolymers, surfactants, vitamins, antioxidants, flavors, and minerals (Dickinson 1992, Landy et al. 1996, Coupland and McClements 1996, Mei et al. 1998). These interactions often have significant implications for the overall quality of an emulsion product. For example, the volatility of a flavor is reduced when it is electrostatically attracted to the surface of an emulsion droplet, which alters the flavor profile of a food (Landy et al. 1996), or the susceptibility of oil droplets to lipid oxidation depends on whether the catalyst is electrostatically attracted to the droplet surface (Mei et al. 1998). The accumulation of charged species at a droplet surface and the rate at which this accumulation takes place depend on the sign of the charge of the species relative to that of the surface, the strength of the electrostatic interaction, the concentration of the species, and the presence of any other charged species that might compete for the surface.

The above discussion highlights the importance of droplet charge in determining both the physical and chemical properties of food emulsions. It is therefore important for food scientists to be able to predict, control, and measure droplet charge. For most food emulsions, it is difficult to accurately predict droplet charge because of the complexity of their composition and the lack of suitable theories. Nevertheless, there is a fairly good understanding of the major factors which influence droplet charge (Chapter 3) and of the effect of droplet charge on the stability and rheology of emulsions (Chapters 7 and 8). In addition, a variety of experimental techniques have been developed to measure the magnitude and sign of the charge on emulsion droplets (Chapter 10).

1.3.5. Droplet Crystallinity

The physical state of the droplets in an emulsion can influence a number of its most important bulk physicochemical and organoleptic properties, including appearance, rheology, flavor, aroma, and stability (Dickinson and McClements 1995; Boode 1992; Boode and Walstra 1993a,b; Walstra 1996b). The production of margarine and butter depends on a controlled destabilization of an oil-in-water emulsion containing partly crystalline droplets. The stability of cream to shear and temperature cycling depends on the crystallization of the milk fat droplets. The rate at which milk fat droplets cream depends on their density, which is determined by the fraction of the droplet which is solidified. The cooling sensation that occurs when fat crystals melt in the mouth contributes to the characteristic mouthfeel of many food products (Walstra 1987). A knowledge of the factors that determine the crystallization and melting of emulsified substances, and of the effect that droplet-phase transitions have on the properties of emulsions, is therefore particularly important to food scientists.* In oil-in-water emulsions, we are concerned with phase transitions of emulsified fat, whereas in water-in-oil emulsions, we are concerned with phase transitions of emulsified water. In the food industry, we are primarily concerned with the crystallization and melting of emulsified fats, because these transitions occur at temperatures that are commonly encountered during the production, storage, or handling of oil-in-water emulsions and because they usually have a pronounced influence on the bulk properties of food emulsions. In contrast, phase transitions of emulsified water are less likely to occur in foods because of the high degree of supercooling required to initiate crystallization (Clausse 1985).

The percentage of total fat in a sample which is solidified at a particular temperature is known as the solid fat content (SFC). The SFC varies from 100% at low temperatures where the fat is completely solid to 0% at high temperatures where the fat is completely liquid. The precise nature of the SFC–temperature curve is an important consideration when selecting a fat for a particular food product. The shape of this curve depends on the composition of the fat, the thermal and shear history of the sample, whether the sample is heated or cooled, the heating or cooling rate, the size of the emulsion droplets, and the type of emulsifier. The melting and crystallization behavior of emulsified substances can be quite different from that of the same substance in bulk (Dickinson and McClements 1995). The crystallization of bulk fats is considered in Chapter 4, while the additional factors that influence the crystallization of emulsified fats are considered in Chapter 7. Experimental techniques that are used to provide information about the crystallization and melting of emulsion droplets are described in Chapter 10.

1.4. HIERARCHY OF EMULSION PROPERTIES

The bulk physicochemical and organoleptic properties of emulsion-based food products are ultimately determined by the concentration, dimensions, interactions, and dynamics of the various types of structural entities present within them (e.g., atoms, molecules, molecular aggregates, crystals, micelles, droplets, air bubbles, and individual phases) (Figure 1.8). The

---

* It should be noted that the continuous phase of an emulsion is also capable of melting or crystallizing, which can have a profound influence on the overall properties. For example, the characteristic texture of ice cream is partly due to the presence of ice crystals in the aqueous continuous phase, whereas the rheology of butter and margarine is determined by the existence of a network of aggregated fat crystals in the oil continuous phase.

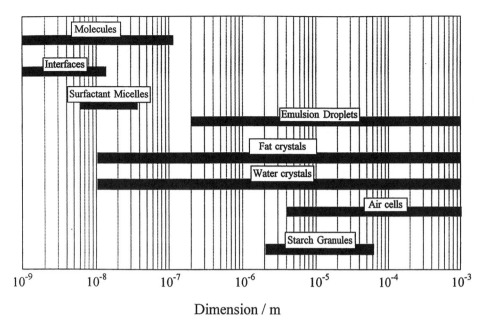

FIGURE 1.8 Typical dimensions of structural entities commonly found in emulsion-based food products.

properties of emulsions can therefore be studied at a number of different levels of structural organization (e.g., subatomic, atomic, molecular, supramolecular, colloidal, microscopic, macroscopic, and organoleptic) depending on the concerns of the investigator (Eads 1994). Subatomic particles interact with each other via strong and weak nuclear forces to form atoms. Atoms interact with each other via covalent and ionic bonds to form molecules. Atoms and molecules interact with each other via various covalent and noncovalent forces to form separate phases (which may be gas, liquid, or solid), simple solutions, molecular aggregates, and colloidal particles (Chapter 2). The bulk physicochemical and organoleptic properties of an emulsion depend on the way these structural entities interact with one another to form the emulsion droplets, interfacial region, and continuous phase. A more complete understanding of the factors that determine the properties of emulsions depends on establishing the most important processes that operate at each level of structural organization and linking the different levels together. This is an extremely ambitious and complicated task that requires many years of painstaking research. Nevertheless, the knowledge gained from such an endeavor will enable food manufacturers to design and produce higher quality foods in a more cost-effective and systematic fashion. For this reason, the connection between molecular, colloidal, and bulk physicochemical properties of food emulsions will be stressed throughout this book.

1.5. INVESTIGATION OF EMULSION PROPERTIES

Our understanding of the factors which determine the properties of food emulsions develops through a synthesis of experimentation and theory development. An investigator usually studies a particular aspect of a system by carefully designing and carrying out an experiment. The investigator then uses his or her knowledge to postulate a hypothesis to account for the observed behavior, which is then tested and refined by further experimentation. Eventually, a theory may be developed which enables one to better understand and predict the behavior of the system. As further studies are carried out, this theory evolves with time and may even be replaced by a competing theory that better accounts for the observed behavior.

Food emulsions are extremely complex systems, and many factors operate in concert to determine their overall properties. For this reason, experiments are usually carried out using simplified model systems which retain the essential features of the real system, but which ignore many of the secondary effects. For example, the emulsifying properties of proteins are often investigated by using an isolated individual protein, pure oil, and pure water (Dickinson 1992). In reality, a protein ingredient used in the food industry consists of a mixture of different proteins, sugars, salts, fats, and minerals, and the oil and aqueous phases may contain a variety of different chemical constituents (Section 1.2.5). Nevertheless, by using a well-characterized model system, it is possible to elucidate the primary factors which influence the properties of proteins in emulsions in a more quantitative fashion. Once these primary factors have been established, it is possible to increase the complexity of the model by introducing additional variables and systematically examining their influence on the overall properties. This incremental approach eventually leads to a thorough understanding of the factors that determine the properties of actual food emulsions and to the development of theories which can be used to describe and predict their behavior.

1.6.   OVERVIEW AND PHILOSOPHY

It is impossible to cover every aspect of food emulsions in a book of this size. Of necessity, one must be selective about the material presented and the style in which it is presented. Rather than reviewing the practical knowledge associated with each particular type of emulsion-based food product, the focus here will be on the fundamental principles of emulsion science as applied to food systems because these principles are generally applicable to all types of food emulsion. Even so, real food emulsions will be used as examples where possible in order to emphasize the practical importance of the fundamental approach. As mentioned earlier, particular attention will be paid to the relationship among molecular, colloidal, and bulk physicochemical properties of food emulsions, because this approach leads to the most complete understanding of their behavior.

Throughout this book, it will be necessary to introduce a number of theories which have been developed to describe the properties of emulsions. Rather than concentrating on the mathematical derivation of these theories, their physical significance will be highlighted, with a focus on their relevance to food scientists. A feeling for the major factors which determine the properties of food emulsions can often be gained by programming these theories onto a personal computer and systematically examining the role that each physical parameter plays in the equation.

2 Molecular Interactions

2.1. INTRODUCTION

Although food scientists have some control over the final properties of a product, they must work within the physical constraints set by nature (i.e., the characteristics of the individual molecules and the type of interactions that occur between them). There is an increasing awareness within the food industry that the efficient production of foods with improved quality depends on a better understanding of the molecular basis of their bulk physicochemical and organoleptic properties (Baianu 1992, Kokini et al. 1993, Eads 1994). The individual molecules within a food emulsion can interact with each other to form a variety of different structural entities (Figure 2.1). A molecule may be part of a bulk phase where it is surrounded by molecules of the same type, it may be part of a mixture where it is surrounded by molecules of a different type, it may be part of an electrolyte solution where it is surrounded by counterions and solvent molecules, it may accumulate at an interface between two phases, it may be part of a molecular aggregate dispersed in a bulk phase, it may be part of a three-dimensional network that extends throughout the system, or it may form part of a complex biological structure (Israelachvili 1992). The bulk physicochemical properties of food emulsions depend on the nature, properties, and interactions of the structures formed by the molecules. The structural organization of a particular set of molecules is largely determined by the forces that act between them and the prevailing environmental conditions (e.g., temperature and pressure). Nevertheless, foods are rarely in their most thermodynamically stable state, and therefore the structural organization of the molecules is often governed by various kinetic factors which prevent them from reaching the arrangement with the lowest free energy (Section 1.2.1). For this reason, the structural organization of the molecules in foods is largely dependent on their previous history (i.e., the temperatures, pressures, gravity, and applied mechanical forces experienced during their lifetime). To understand, predict, and control the behavior of food emulsions, it is important to be aware of the origin and nature of the forces responsible for holding the molecules together and how these forces lead to the various types of structures found in food emulsions. Only then will it be possible to create and stabilize foods that have internal structures that are known to be beneficial to food quality.

2.2. FORCES OF NATURE

There are four distinct types of force in nature: strong nuclear interactions, weak nuclear interactions, electromagnetic interactions, and gravity (Israelachvili 1992, Atkins 1994). The strong and weak nuclear forces act over extremely short distances and are chiefly responsible for holding together subatomic particles in the nucleus. As nuclear rearrangements do not normally occur in foods, these forces will not be considered further. Gravitational forces are relatively weak and act over large distances compared to other types of forces. Their strength is proportional to the product of the masses of the objects involved, and consequently they

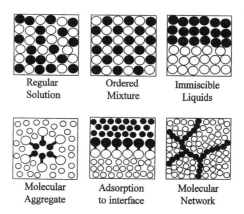

| Regular Solution | Ordered Mixture | Immiscible Liquids |

| Molecular Aggregate | Adsorption to interface | Molecular Network |

FIGURE 2.1 The molecules in food emulsions may adopt a variety of different structural arrangements depending on the nature of their interactions with their neighbors.

are insignificant at the molecular level because molecular masses are extremely small. Nevertheless, they do affect the behavior of food emulsions at the macroscopic level (e.g., sedimentation or creaming of droplets, the shape adopted by large droplets, meniscus formation, and capillary rise) (Israelachvili 1992). The forces that act at the molecular level are all electromagnetic in origin and can conveniently be divided into four types: covalent, electrostatic, van der Waals, and steric overlap (Hiemenz 1986, Israelachvili 1992, Atkins 1994). Despite acting over extremely short distances, often on the order of a few angstroms or less, intermolecular forces are ultimately responsible for the bulk physicochemical and organoleptic properties of emulsions and other food materials.

2.3. ORIGIN AND NATURE OF MOLECULAR INTERACTIONS

2.3.1. Covalent Interactions

Covalent bonds involve the sharing of outer-shell electrons between two or more atoms, so that the individual atoms lose their discrete nature (Karplus and Porter 1970, Atkins 1994). The number of electrons in the outer shell of an atom governs its *valency* (i.e., the optimum number of covalent bonds it can form with other atoms). Covalent bonds may be *saturated* or *unsaturated,* depending on the number of electrons involved. Unsaturated bonds tend to be shorter, stronger, and more rigid than saturated bonds (Israelachvili 1992). The distribution of the electrons within a covalent bond determines its *polarity.* When the electrons are shared equally among the atoms, the bond has a nonpolar character, but when the electrons are shared unequally, the bond has a polar character. The polarity of a molecule depends on the symmetry of the various covalent bonds which it contains (see Section 2.3.2.). Covalent bonds are also characterized by their *directionality* (i.e., their tendency to be directed at clearly defined angles relative to each other). The valency, saturation, polarity, strength, and directionality of covalent bonds determine the three-dimensional structure, flexibility, chemical reactivity, and physical interactions of molecules.

Chemical reactions involve the breaking and formation of covalent bonds (Atkins 1994). The bulk physicochemical and organoleptic properties of food emulsions are altered by various types of chemical and biochemical reactions that occur during their production, storage, and consumption (Coultate 1996; Fennema 1996; Fennema and Tannenbaum 1996a,b). Some of these reactions are beneficial to food quality, while others are detrimental. It is therefore important for food scientists to be aware of the various types of chemical reaction that occur in food emulsions and to establish their influence on the overall properties of the system. The chemical reactions which occur in food emulsions are similar to those that occur in any other multicomponent heterogeneous food materials (e.g., oxidation of lipids [Nawar 1996], hydrolysis of proteins or polysaccharides [Damodaran 1996, BeMiller and Whistler 1996], cross-linking of proteins [Damodaran 1996], and Maillard reactions between reducing sugars and free amino groups [BeMiller and Whistler 1996]). Nevertheless, the rates and pathways of these reactions are often influenced by the physical environment of the molecules involved (e.g., whether they are located in the oil, water, or interfacial region) (Wedzicha 1988).

Until fairly recently, emulsion scientists were principally concerned with understanding the physical changes which occur in food emulsions, rather than the chemical changes. Nevertheless, there is currently great interest in establishing the relationship between emulsion properties and the mechanisms of various chemical reactions that occur within them (Wedzicha et al. 1991, Coupland and McClements 1996, Landy et al. 1996, Huang et al. 1997).

Despite the importance of chemical reactions in emulsion quality, it should be stressed that many of the most important changes in emulsion properties are a result of alterations in the spatial distribution of the molecules, rather than the result of alterations in their chemical structure (e.g., creaming, flocculation, coalescence, and phase inversion). The spatial distribution of molecules is governed principally by their noncovalent (or physical) interactions with their neighbors (e.g., electrostatic, van der Waals, and steric overlap). It is therefore particularly important to have a good understanding of the origin and nature of these interactions.

2.3.2. Electrostatic Interactions

Electrostatic interactions occur between molecular species that possess a permanent electrical charge, such as ions and polar molecules (Murrell and Boucher 1982, Reichardt 1988, Rogers 1989). An ion is an atom or molecule that has either lost or gained one or more outer-shell electrons so that it obtains a permanent positive or negative charge (Atkins 1994) (Figure 2.2). A polar molecule has no net charge (i.e., as a whole, the molecule is neutral), but it does have an electrical *dipole* because of an uneven distribution of the charges within it. Certain atoms are able to "pull" the electrons in the covalent bonds toward them more strongly than are other atoms (Atkins 1994). As a consequence, they acquire a partial negative charge ($\delta-$), and the other atom acquires a partial positive charge ($\delta+$). If the partial charges within a molecule are distributed symmetrically, they cancel each other and the molecule has no dipole (e.g., CCl_4), but if they are distributed asymmetrically, the molecule will have a dipole

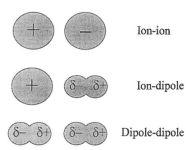

FIGURE 2.2   Schematic representation of the most important types of intermolecular electrostatic interactions that arise between molecules.

(Israelachvili 1992). For example, the chlorine atom in HCl pulls the electrons in the covalent bond more strongly than the hydrogen atom, and so a dipole is formed: $H^{\delta+}Cl^{\delta-}$. The strength of a dipole is characterized by the *dipole moment* $\mu = ql$, where l is the distance between two charges q^+ and q^-. The greater the magnitude of the partial charges, or the farther they are apart, the greater the dipole moment of a molecule.

The interaction between two molecular species is characterized by an *intermolecular pair potential, w(s)*, which is the energy required to bring two molecules from an infinite distance apart to a separation s (Israelachvili 1992). There are a number of different types of electrostatic interactions that can occur between permanently charged molecular species (ion–ion, ion–dipole, and dipole–dipole), but they can all be described by a similar equation (Hiemenz 1986):

$$w_E(s) = \frac{Q_1 Q_2}{4\pi\varepsilon_0\varepsilon_R s^n} \tag{2.1}$$

where Q_1 and Q_2 are the effective charges on the two species, ε_0 is the dielectric constant of a vacuum (8.85×10^{-12} C^2 J^{-1} m^{-1}), ε_R is the relative dielectric constant of the intervening medium, s is the center-to-center distance between the charges, and n is an integer that depends on the nature of the interaction. For ions, the value of Q is determined by their valency (z) and electrical charge (e) (1.602×10^{-19} C), whereas for dipoles, it is determined by their dipole moment μ and orientation (Table 2.1). Numerical calculations of the intermolecular pair potential for representative ion–ion, ion–dipole, and dipole–dipole interactions are illustrated in Figure 2.3a.

Examination of Equation 2.1 and Figure 2.3a provides a number of valuable insights into the nature of intermolecular electrostatic interactions and the factors which influence them:

1. They may be either attractive or repulsive depending on the sign of the charges. If the charges have similar signs, $w_E(s)$ is positive and the interaction is repulsive, but if they have opposite signs, $w_E(s)$ is negative and the interaction is attractive.
2. Their strength depends on the magnitudes of the charges involved (Q_1 and Q_2). Thus, ion–ion interactions are stronger than ion–dipole interactions, which are in turn stronger than dipole–dipole interactions. In addition, the strength of interactions involving ions increases as their valency increases, whereas the strength of interactions involving polar species increases as their dipole moment increases.

TABLE 2.1
Parameters Needed to Calculate the Interaction Pair Potential for Ion–Ion, Ion–Dipole, and Dipole–Dipole Electrostatic Interactions Using Equation 2.1 (see also Figure 2.3a)

Interaction type	Example	$Q_1 Q_2$	n
Ion–ion	Na^+ Cl^-	$(z_1 e)(z_2 e)$	1
Ion–dipole	Na^+ H_2O	$(z_1 e)\mu_2 \cos\phi$	2
Dipole–dipole	H_2O H_2O	$\mu_1\mu_2 f(\phi)$	3

Note: z is the valence, μ is the dipole moment, e is the electronic charge, and ϕ is the angle between the charges.

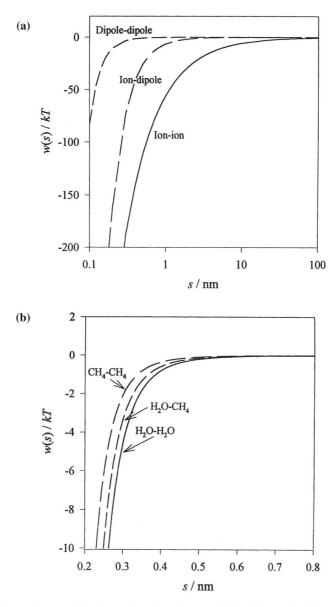

FIGURE 2.3 Dependence of the intermolecular pair potential on intermolecular separation for (a) electrostatic, (b) van der Waals, and (c) steric overlap interactions.

3. Their strength increases as the center-to-center separation of the charged species decreases. Thus, interactions between small ions or molecules (which can get close together) are stronger than those between large ions or molecules of the same charge.

4. The range of ion–ion ($1/s$) interactions is longer than that of ion–dipole interactions ($1/s^2$), which is longer than that of dipole–dipole interactions ($1/s^3$).

5. Their strength depends on the nature of the material separating the charges (via ε_R): the higher the relative dielectric constant, the weaker the interaction (Table 2.2). Electrostatic interactions between two charged species in water ($\varepsilon_R = 80$) are

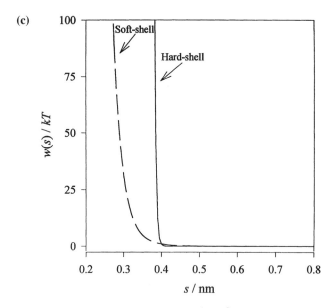

FIGURE 2.3 (continued)

therefore much weaker than those between the same species in oil ($\varepsilon_R = 2$), which accounts for the much higher solubility of salts in water than in nonpolar solvents (Israelachvili 1992).

6. Their strength depends on the orientation of any dipoles involved, being strongest when partial charges of opposite sign are brought close together. When the electrostatic interaction between a dipole and another charged species is much stronger than the thermal energy (Section 2.5), the dipole becomes permanently aligned so as to maximize the strength of the attraction. This alignment of dipoles is responsible for the high degree of structural organization of molecules in bulk water and the ordering of water molecules around ions in aqueous solutions (Chapter 4).

The ionization of many biological molecules depends on the pH of the surrounding aqueous phase, and so electrostatic interactions involving these molecules are particularly sensitive to pH (Nakamura 1996; Damodaran 1989, 1994, 1996, 1997). For example, a protein may exist as an individual molecule in solution when the pH is sufficiently far from its isoelectric point because of strong electrostatic repulsion between the protein molecules, but it may precipitate when the pH of the solution is close to its isoelectric point because the electrostatic repulsion between the molecules is no longer strong enough to prevent them from aggregating (Kinsella 1982, Kinsella and Whitehead 1989, Damodaran 1996). The strength of electrostatic interactions between molecules suspended in an aqueous solution is also sensitive to the type and concentration of electrolyte present (Bergethon and Simons 1990, de Wit and van Kessel 1996). A charged molecule tends to be surrounded by oppositely charged ions (counterions), which effectively "screen" (reduce) the electrostatic interaction between other molecules of the same type (Chapter 3).

Electrostatic interactions play an extremely important role in determining the overall properties of food emulsions because many of the major constituents of these products are either ionic or dipolar (e.g., water, sugars, salts, proteins, polysaccharides, surfactants, acids, and bases) (Fennema 1996a). The unique physiochemical properties of water are governed

TABLE 2.2
Compilation of Molecular Properties of Some Common Liquids and Solutes Needed to Calculate Intermolecular Interactions

Static Relative Dielectric Constants (ε_R)

Water	78.5	Chloroform	4.8
Ethylene glycol	40.7	Edible oils	2.5
Methanol	32.6	Carbon tetrachloride	2.2
Ethanol	24.3	Liquid paraffin	2.2
Acetone	20.7	Dodecane	2.0
Propanol	20.2	Hexane	1.9
Acetic acid	6.2	Air	1.0

Molecular Diameters, Polarizabilities, and Dipole Moments

Molecule type	σ (nm)	$\alpha/4\pi\varepsilon_0$ ($\times 10^{-30}$ m^3)	μ (D^a)
H_2O	0.28	1.48	1.85
CH_4	0.40	2.60	0
HCl	0.36	2.63	1.08
CH_3Cl	0.43	4.56	1.87
CCl_4	0.55	10.5	0
NH_3	0.36	2.26	1.47
Methanol	0.42	3.2	1.69
Ethanol	b	5.2	1.69
Acetone	b	6.4	2.85
Benzene	0.53	10.4	0

[a] $D = 3.336 \times 10^{-30}$ C m.

[b] Cannot be treated as spheres.

Taken from Israelachvili 1992 and Buffler 1995.

by relatively strong dipole–dipole interactions which cause the water molecules to become highly organized (Chapter 4). The accumulation and organization of water molecules around solutes are determined by various types of dipole–dipole, dipole–ion, and ion–ion interactions (Chapter 4). The "screening" of electrostatic interactions between charged emulsion droplets is due to the attraction of counterions to the surface of the droplets (Chapters 3 and 7). The conformation and interactions of biopolymers in aqueous solution are governed by electrostatic interactions between the charged groups and the surrounding molecules (Chapter 4). These examples highlight the importance of understanding the origin and nature of electrostatic interactions in food emulsions.

2.3.3. van der Waals Interactions

van der Waals forces act between all types of molecular species, whether they are ionic, polar, or nonpolar (Hiemenz 1986, Israelachvili 1992). They are conveniently divided into three separate contributions, which all rely on the polarization of molecules (Figure 2.4):

1. *Dispersion forces.* These forces arise from the interaction between an instantaneous dipole and a dipole induced in a neighboring molecule by the presence of the instantaneous dipole. The electrons in a molecule are continually moving around the nucleus. At any given instant in time, there is an uneven distribution of

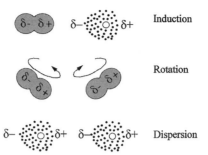

FIGURE 2.4 Schematic representation of van der Waals intermolecular interactions which involve either the electronic or orientational polarization of molecules.

the negatively charged electrons around the positively charged nucleus, and so an *instantaneous* dipole is formed. This instantaneous dipole generates an electrical field which induces a dipole in a neighboring molecule. Consequently, there is an instantaneous attractive force between the two dipoles. On average, the attraction between the molecules is therefore finite, even though the average net charge on the molecules involved is zero.

2. *Induction forces.* These forces arise from the interaction between a permanent dipole and a dipole induced in a neighboring molecule by the presence of the permanent dipole. A permanent dipole causes an alteration in the distribution of electrons of a neighboring molecule, which leads to the formation of an induced dipole. The interaction between the permanent dipole and the induced dipole leads to an attractive force between the molecules.

3. *Orientation forces.* These forces arise from the interaction between two permanent dipoles that are continuously rotating. On average, these rotating dipoles have no net charge, but there is still a weak attractive force between them because the movement of one dipole induces some correlation in the movement of a neighboring dipole. When the interaction between the two dipoles is strong enough to cause them to be permanently aligned, this contribution is replaced by the electrostatic dipole–dipole interaction described in the previous section.

As will be seen in the next chapter, an understanding of the origin of these three contributions to the van der Waals interaction has important consequences for predicting the stability of emulsion droplets to aggregation.

The overall intermolecular pair potential due to van der Waals interactions is given by:

$$w_{\mathrm{VDW}}(s) = \frac{-(C_{\mathrm{disp}} + C_{\mathrm{ind}} + C_{\mathrm{orient}})}{(4\pi\varepsilon_0\varepsilon_R)^2 s^6} \tag{2.2}$$

where C_{disp}, C_{ind}, and C_{orient} are constants which depend on the dispersion, induction, and orientation contributions, respectively (Hiemenz 1986). Their magnitude depends on the dipole moment (for permanent dipoles) and the polarizability (for induced dipoles) of the molecules involved in the interaction (Table 2.2). The polarizability is a measure of the strength of the dipole induced in a molecule when it is in the presence of an electrical field: the larger the polarizability, the easier it is to induce a dipole in a molecule. For most biological molecules, the dominant contribution to the van der Waals interaction is the dispersion force, with the important exception of water, where the major contribution is from the orientation force (Israelachvili 1992). Examination of Equation 2.2 and Figure 2.3b provides some useful physical insights into the factors that influence the van der Waals interactions between molecules:

1. The value of C_{disp}, C_{ind}, and C_{orient} is always positive, which means that the overall interaction potential between two molecules is always negative (attractive).
2. The interaction is relatively short range, decreasing rapidly with intermolecular separation ($1/s^6$).
3. The attraction between molecules decreases as the dielectric constant of the intervening medium increases, which highlights the electromagnetic origin of van der Waals interactions.
4. The magnitude of the interaction increases as the polarizability and dipole moment of the molecules involved increase.

Although van der Waals interactions act between all types of molecular species, they are considerably weaker than electrostatic interactions (Figure 2.3 and Table 2.3). For this reason, they are most important in determining interactions between nonpolar molecules, where electrostatic interactions do not make a significant contribution. Indeed, the structure and physicochemical properties of organic liquids are largely governed by the van der Waals interactions between the molecules (Israelachvili 1992).

All types of van der Waals interaction involve either the electronic or orientational polarization of molecules and have a $1/s^6$ dependence on intermolecular separation (Hiemenz 1986). Another type of interaction that depends on molecular polarization but which does not have a $1/s^6$ dependence on intermolecular separation is *ion polarization* (Israelachvili 1992). Although this type of interaction is not strictly a van der Waals interaction, it is convenient to consider it in this section because it also involves polarization. A positively charged ion causes the electrons in a neighboring molecule to be pulled toward it, thus inducing a dipole whose δ^- pole faces toward the ion. Similarly, a negatively charged ion causes the electrons in a neighboring molecule to be repelled away from it, thus inducing a dipole whose δ^+ pole faces toward the ion. Thus there is an attractive force between the ion and the induced dipole because of electronic polarization. For polar molecules, there may be an additional contribution due to orientational polarization. When the interaction between an ion and a dipole is not strong enough to cause the dipole to become permanently aligned, the dipole continuously rotates because of its thermal energy (Section 2.5). On average, there is no net charge on a rotating dipole because of its continuous rotation, but in the presence of an ion there is a net attraction between the ion and the dipole because the low-energy orientations are preferred (Israelachvili 1992). When the interaction between the ion and the dipole is strong enough to cause the dipole to be permanently aligned, this contribution should be replaced by the electrostatic ion–dipole interaction described in the previous section.

The intermolecular pair potential for ion polarization is given by:

$$w_{IP}(s) = \frac{-(ze)^2}{2(4\pi\varepsilon_0\varepsilon_R)^2 s^4}\left(\alpha_0 + \frac{\mu^2}{3kT}\right) \tag{2.3}$$

where the α_0 term is the contribution from the electronic polarizability of the molecule and the $\mu^2/3kT$ term is the contribution from the orientational polarizability. For nonpolar molecules, only the electronic polarization term contributes to this interaction, but for polar molecules, both electronic and orientational polarization contribute. This type of interaction is significantly stronger than the van der Waals interactions mentioned above and should therefore be included in any calculation of the interaction energy between molecules involving ions.

2.3.4. Steric Overlap Interactions

When two atoms or molecules come so close together that their electron clouds overlap, there is an extremely large repulsive force generated between them (Figure 2.3c). This steric overlap force is very short range and increases rapidly when the separation between the two molecules becomes less than the sum of their radii ($\sigma = r_1 + r_2$). A number of empirical equations have been derived to describe the dependence of the steric overlap intermolecular pair potential, $w_{steric}(s)$, on molecular separation (Israelachvili 1992). The "hard-shell" model assumes that the repulsive interaction is zero when the separation is greater than σ but infinitely large when it is less than σ:

$$w_{steric}(s) = \left(\frac{\sigma}{s} \right)^{\infty} \tag{2.4}$$

In reality, molecules are slightly compressible, and so the increase in the steric overlap repulsion is not as dramatic as indicated by Equation 2.4. The slight compressibility of molecules is accounted for by a "soft-shell" model, such as the power-law model:

$$w_{steric}(s) = \left(\frac{\sigma}{s} \right)^{12} \tag{2.5}$$

At separations greater than σ, the steric overlap repulsion is negligible, but at separations less than this value, there is a steep increase in the interaction pair potential, which means that the molecules strongly repel one another. The strong repulsion that arises from steric overlap determines the effective size of atoms and molecules and how closely they can come together. It therefore has a strong influence on the packing of molecules in liquids and solids.

2.4. OVERALL INTERMOLECULAR PAIR POTENTIAL

We are now in a position to calculate the overall interaction between a pair of molecules. Assuming that no chemical reactions occur between the molecules, the overall intermolecular pair potential is the sum of the various physical interactions mentioned above:

$$w(s) = w_E(s) + w_{VDV}(s) + w_{steric}(s) \tag{2.6}$$

The magnitude of each of the individual contributions to the overall interaction potential is strongest at close separations and decreases as the molecules move apart. Nevertheless, the overall intermolecular pair potential has a more complex dependence on separation, which may be attractive at some separations and repulsive at others, because it is the sum of a number of interactions which each have different magnitudes, ranges, and signs.

To highlight some of the most important features of intermolecular interactions, it is useful to consider the interaction of a pair of spherical nonpolar molecules (i.e., no electrostatic interactions). The overall intermolecular pair potential for this type of system is given

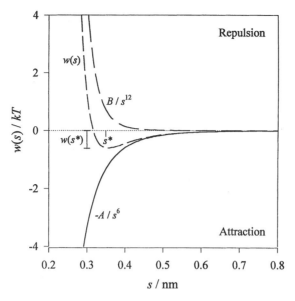

FIGURE 2.5 Intermolecular pair potential for a pair of spherical nonpolar molecules. The curves were calculated assuming typical values for the constants: $A = 10^{-77}$ J m^6 and $B = 10^{-134}$ J m^{12}.

by an expression known as the Lennard–Jones potential (Bergethon and Simons 1990, Baianu 1992):

$$w(s) = \frac{-A}{s^6} + \frac{B}{s^{12}} \tag{2.7}$$

where the A term represents the contribution from the van der Waals interactions (Equation 2.2) and the B term represents the contribution from the steric overlap interaction (Equation 2.5). The dependence of the intermolecular pair potential on separation is illustrated in Figure 2.5. The van der Waals interactions are attractive at all separations, whereas the steric overlap interactions are repulsive. At large separations, $w(s)$ is so small that there is no effective interaction between the molecules. As the molecules are brought closer together, the pair potential becomes increasingly attractive (negative) because the van der Waals interactions dominate. Eventually, the molecules get so close together that their electron clouds overlap and the pair potential becomes strongly repulsive (positive) because steric overlap interactions dominate. Consequently, there is a minimum in the overall intermolecular pair potential at some intermediate separation (s^*). Two molecules will tend to remain associated in this potential energy minimum in the absence of any disruptive influences (such as thermal energy or applied external forces), with a "bond length" of s^* and a "bond strength" of $w(s^*)$.

It is often more convenient to describe the interaction between a pair of molecules in terms of *forces* rather than *potential energies* (Israelachvili 1992). The force acting between two molecules can simply be calculated from the intermolecular pair potential using the following relationship: $F(s) = -dw(s)/ds$. The minimum in the potential energy curve therefore occurs at a separation where the net force acting between the molecules is zero (i.e., the attractive and repulsive forces exactly balance). If the molecules move closer together, they experience a repulsive force, and if they move farther apart, they experience an attractive force.

2.5. BOND STRENGTHS AND THE ROLE OF THERMAL ENERGY

The molecules in a substance are in continual motion (translational, rotational, and vibrational) because of their thermal energy (kT) (Israelachvili 1992, Atkins 1994). The thermally induced movement of molecules has a disorganizing influence, which opposes the formation of intermolecular bonds. For this reason, the strength of intermolecular interactions is usually judged relative to the thermal energy: $kT \approx 4.1 \times 10^{-24}$ kJ per bond or $RT \approx 2.5$ kJ mol^{-1}. If the bond strength is sufficiently greater than kT, the molecules will remain together, but if it is sufficiently smaller, they will tend to move apart. At intermediate bond strengths, the molecules spend part of their time together and part of their time apart (i.e., bonds are rapidly breaking and reforming).

The bond strengths of a number of important types of intermolecular interaction are summarized in Table 2.3. In a vacuum, the strength of these bonds decreases in the following order: ion–ion, covalent > ion–dipole > dipole–dipole > van der Waals. With the exception

TABLE 2.3
Approximate Bond Strengths for Some of the Most Important Types of Molecular Interactions That Occur in Foods at Room Temperature

Type of interaction	In vacuum		In water	
	$w(s^*)$ (kJ mol^{-1})	$w(s^*)$ (RT)	$w(s^*)$ (kJ mol^{-1})	$w(s^*)$ (RT)
Covalent bonds				
C–O	340	140		
C–C	360	140		
C–H	430	170		
O–H	460	180		
C=C	600	240		
C≡N	870	350		
Electrostatic ion–ion				
Na$^+$ Cl$^-$	500	200	6.3	2.5
Mg^{2+} Cl$^-$	1100	460	14.1	5.7
Al^{3+} Cl$^-$	1800	730	22.5	9.1
Ion–dipole				
Na$^+$ H$_2$O	97	39	1.2	0.5
Mg^{2+} H$_2$O	255	103	3.2	1.3
Al^{3+} H$_2$O	445	180	5.6	2.3
Dipole–dipole				
H$_2$O H$_2$O	38	15	0.5	0.2
Ion polarization				
Na$^+$ CH$_4$	24	10		
van der Waals				
CH$_4$ CH$_4$	1.5	0.60		
C$_6$H$_{14}$ C$_6$H$_{14}$	7.4	3.0		
C$_{12}$H$_{26}$ C$_{12}$H$_{26}$	14.3	5.7		
C$_{18}$H$_{38}$ C$_{18}$H$_{38}$	21.2	6.1		
CH$_4$ H$_2$O	2.6	0.7		
H$_2$O H$_2$O	17.3	6.9		

Note: All dipole interactions assuming that the molecules are aligned so they get maximum attraction. van der Waals forces calculated from Israelachvili (1992) assuming that $w(s^*)$ is approximately equal to the cohesive energy over 6.

of methane (a small nonpolar molecule), the bonds between the molecules shown in Table 2.3 are sufficiently strong (compared to the thermal energy) to hold them together in a liquid or solid. It must be stressed that the strength of the electrostatic and van der Waals interactions between molecules decreases appreciably when they are surrounded by a solvent rather than a vacuum, especially when the solvent has a high dielectric constant. The strength of the electrostatic interaction between molecules dispersed in water is about 80 times less than that in a vacuum because of the high relative dielectric constant of water ($\varepsilon_R \approx 80$). This largely accounts for the high water solubility of many ionic crystals (Israelachvili 1992). The strength of ion–dipole interactions between molecules dispersed in water is usually sufficiently large (compared to the thermal energy) to cause water molecules in the immediate vicinity of an ion to be attracted to its surface and to become aligned, especially when the ion is small and highly charged (Chapter 4). Even some types of dipole–dipole interaction are sufficiently strong in water to cause a high degree of structural organization of the molecules (e.g., the tetrahedral structure formed in bulk water) (Chapter 4).

The strength of van der Waals interactions is also reduced when the molecules involved are surrounded by a solvent (Israelachvili 1992). At large separations between the molecules, the van der Waals interaction between oil molecules is reduced by a factor of about 20 when they are surrounded by water rather than a vacuum (Israelachvili 1992). At present, there is no theory that can accurately account for the reduction in van der Waals interactions between molecules which are in close contact within a solvent. Nevertheless, we can reasonably postulate that van der Waals bonds in a solvent are very weak and that only for relatively large molecules will these interactions be strong enough to hold the molecules together. This accounts for the fact that the volatility of molecules increases as their molecular weight or polarity decreases (Israelachvili 1992).

2.6. THE STRUCTURAL ORGANIZATION OF MOLECULES IN LIQUIDS

2.6.1. Thermodynamics of Mixing

In food emulsions, we are usually concerned with the interactions of large numbers of molecules in a liquid, rather than between a pair of isolated molecules in a vacuum. We must therefore consider the interaction of a molecule with its neighbors and how these interactions determine the overall organization of the molecules within a liquid (Murrell and Boucher 1982, Murrell and Jenkins 1994, Evans and Wennerstrom 1994). The behavior of large numbers of molecules at equilibrium can be described by statistical thermodynamics (Sears and Salinger 1975, Atkins 1994). A molecular ensemble tends to organize itself so that the molecules are in an arrangement which minimizes the free energy of the system. The free energy of a molecular ensemble is governed by both enthalpy and entropy contributions (Bergethon and Simons 1990). The enthalpy contributions are determined by the molecular interaction energies discussed above, while the entropy contributions are determined by the tendency of a system to adopt its most disordered state.

Consider a hypothetical system that consists of a collection of two different types of equally sized spherical molecules, **A** and **B** (Figure 2.6). The free energy change that occurs when these molecules are mixed is given by:

$$\Delta G_{\text{mix}} = \Delta E_{\text{mix}} - T\Delta S_{\text{mix}} \tag{2.8}$$

where ΔE_{mix} and ΔS_{mix} are the differences in the molecular interaction energy and entropy of the mixed and unmixed states, respectively. Practically, we may be interested in whether the

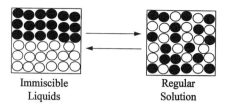

Immiscible Liquids → Regular Solution

FIGURE 2.6 System in which two types of molecules may be completely miscible or form a regular solution depending on the strength of the interactions between them and the entropy of mixing.

resulting system consists of two immiscible liquids or is a mixture where the molecules are more or less intermingled (Figure 2.6). Thermodynamics tells us that if ΔG_{mix} is positive, mixing is unfavorable and the molecules tend to exist as two separate phases (i.e., they are immiscible); if ΔG_{mix} is negative, mixing is favorable and the molecules tend to be intermingled with each other (i.e., they are miscible); and if $\Delta G_{mix} \approx 0$, the molecules are partly miscible and partly immiscible. For simplicity, we assume that if the two types of molecules do intermingle with each other, they form a regular solution (i.e., a completely random arrangement of the molecules) (Figure 2.6 right) rather than an ordered solution, in which the type **A** molecules are preferentially surrounded by type **B** molecules or vice versa. In practice, this means that the attractive forces between the two different types of molecules are not much stronger than the thermal energy of the system (Atkins 1994, Evans and Wennerstrom 1994). This argument is therefore only applicable to mixtures that contain nonpolar or slightly polar molecules, where strong ion–ion or ion–dipole interactions do not occur. Despite the simplicity of this model system, we can still gain considerable insight into the behavior of more complex systems that are relevant to food emulsions. In the following sections, we separately consider the contributions of the interaction energy and the entropy to the overall free energy change that occurs on mixing.

2.6.2. Potential Energy Change on Mixing

An expression for ΔE_{mix} can be derived by calculating the total interaction energy of the molecules before and after mixing (Israelachvili 1992, Evans and Wennerstrom 1994). For both the mixed and the unmixed system, the total interaction energy is determined by summing the contribution of each of the different types of bond:

$$E = n_{AA}w_{AA} + n_{BB}w_{BB} + n_{AB}w_{AB} \tag{2.9}$$

where n_{AA}, n_{BB}, and n_{AB} are the total number of bonds, and w_{AA}, w_{BB}, and w_{AB} are the intermolecular pair potentials at equilibrium separation that correspond to interactions between **A–A**, **B–B**, and **A–B** molecules, respectively. The total number of each type of bond formed is calculated from the number of molecules present in the system, the coordination number of the individual molecules (i.e., the number of molecules in direct contact with them), and their spatial arrangement. For example, many of the **A–A** and **B–B** interactions that occur in the unmixed system are replaced by **A–B** interactions in the mixed system. The difference in the total interaction energy between the mixed and unmixed states is then calculated: $\Delta E_{mix} = E_{mix} - E_{unmixed}$. This type of analysis leads to the following equation (Evans and Wennerstrom 1994).

$$\Delta E_{mix} = nX_A X_B w \tag{2.10}$$

where n is the total number of moles, w is the *effective interaction parameter,* and X_A and X_B are the mole fractions of molecules of type **A** and **B**, respectively. The effective interaction parameter is a measure of the compatibility of the molecules in a mixture and is related to the intermolecular pair potential between isolated molecules by the expression

$$w = zN_A[w_{AB} - \tfrac{1}{2}(w_{AA} + w_{BB})] \tag{2.11}$$

where z is the coordination number of a molecule and N_A is Avogadro's number. The effective interaction parameter determines whether the transfer of a molecule from a liquid where it is surrounded by similar molecules to one in which it is partly surrounded by dissimilar molecules is favorable (w is negative), unfavorable (w is positive), or indifferent ($w = 0$). It should be stressed that even though there may be attractive forces between all the molecules involved (i.e., w_{AA}, w_{BB}, and w_{AB} may all be negative), the overall interaction potential can be either negative (favorable to mixing) or positive (unfavorable to mixing) depending on the *relative* magnitude of the interactions. If the strength of the interaction between two different types of molecules (w_{AB}) is greater (more negative) than the average strength between similar molecules ($w_{AB} < [w_{AA} + w_{BB}]/2$), then w is negative, which favors the intermingling of the different types of molecules. On the other hand, if the strength of the interaction between two different types of molecules is weaker (less negative) than the average strength between similar molecules ($w_{AB} > [w_{AA} + w_{BB}]/2$), then w is positive, which favors phase separation. If the strength of the interaction between different types of molecules is the same as the average strength between similar molecules ($w_{AB} = [w_{AA} + w_{BB}]/2$), then the system has no preference for any particular arrangement of the molecules within the system. In summary, the change in the overall interaction energy may either favor or oppose mixing, depending on the relative magnitudes of the intermolecular pair potentials.

2.6.3. Entropy Change on Mixing

An expression for ΔS_{mix} is obtained from simple statistical considerations (Israelachvili 1992, Evans and Wennerstrom 1994). The entropy of a system depends on the number of different ways the molecules can be arranged. For an immiscible system, there is only one possible arrangement of the two different types of molecules (i.e., zero entropy), but for a regular solution, there are a huge number of different possible arrangements (i.e., high entropy). A statistical analysis of this situation leads to the derivation of the following equation for the entropy of mixing:

$$\Delta S_{mix} = -nR(X_A \ln X_A + X_B \ln X_B) \tag{2.12}$$

ΔS_{mix} is always positive because X_A and X_B are both between zero and one (so that the natural logarithm terms are negative), which reflects the fact that there is always an increase in entropy after mixing. For regular solutions, the entropy contribution ($-T\Delta S_{mix}$) always decreases the free energy of mixing (i.e., favors the intermingling of the molecules). It should be stressed that for more complex systems, there may be additional contributions to the entropy due to the presence of some order within the mixed state (e.g., organization of water molecules around a solute molecule) (Chapter 4).

2.6.4. Free Energy Change on Mixing

For a regular solution, the free energy change on mixing depends on the combined contributions of the interaction energies and the entropy:

$$\Delta G_{mix} = n[X_A X_B w + RT(X_A \ln X_A + X_B \ln X_B)] \qquad (2.13)$$

We are now in a position to investigate the relationship between the strength of the interactions between molecules and their structural organization. The dependence of the free energy of mixing on the effective interaction parameter and the composition of a system consisting of two different types of molecules is illustrated in Figure 2.7. The molecules are completely miscible when the free energy of mixing is negative and large compared to the thermal energy, are partly miscible when $\Delta G_{mix} \approx 0$, and are completely immiscible when the free energy of mixing is positive and large compared to the thermal energy. Figure 2.7 indicates that mixing occurs even when the effective interaction parameter is zero, because of the contribution of the entropy of mixing term. This accounts for the miscibility of liquids in which the interactions between the two types of molecules are fairly similar (e.g., two nonpolar oils). Two liquids are completely immiscible when the effective interaction parameter is large and positive. The above approach enables us to use thermodynamic considerations to relate bulk physicochemical properties of liquids (such as immiscibility) to molecular properties (such as the effective interaction parameter and the coordination number).

2.6.5.   The Properties of More Complex Systems

The derivation of Equation 2.13 depends on making a number of simplifying assumptions about the properties of the system that are not normally valid in practice (e.g., that the molecules are spherical, that they all have the same size and coordination number, and that there is no ordering of the molecules within the mixture) (Israelachvili 1992). It is possible

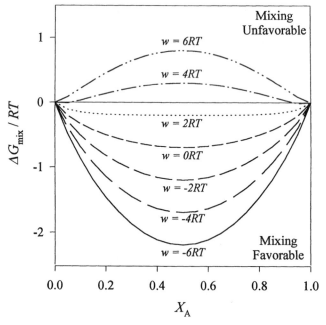

FIGURE 2.7 Dependence of the free energy of mixing (calculated using Equation 2.13) on the composition and effective interaction parameter of a binary liquid. When ΔG_{mix} is much less than $-RT$, the system tends to be mixed; otherwise, it will be partly or wholly immiscible.

to incorporate some of these features into the above theory, but a more elaborate mathematical analysis is required. Food molecules come in all sorts of different sizes, shapes, and flexibilities. They may be nonpolar, polar, or amphiphilic; they may have specific binding sites; or they may have to be in a certain orientation before they can interact with their neighbors. In addition, a considerable degree of structural organization of the molecules within a solvent often occurs when a solute is introduced. The variety of molecular characteristics exhibited by food molecules accounts for the great diversity of structures that are formed in food emulsions, such as bulk liquids, regular solutions, organized solutions, micelles, molecular networks, and immiscible liquids (Figure 2.1).

Another problem with the thermodynamic approach is that food systems are rarely at thermodynamic equilibrium because of the presence of various kinetic energy barriers that prevent the system from reaching its lowest energy state. This approach cannot therefore tell us whether or not two liquids will exist as an emulsion, because an emulsion is a thermodynamically unstable system. Nevertheless, it can tell us whether two liquids are *capable* of forming an emulsion (i.e., whether they are immiscible or miscible). Despite the obvious limitations of the simple thermodynamic approach, it does highlight some of the most important features of molecular organization, especially the importance of considering both interaction energies and entropy effects.

2.7. MOLECULAR INTERACTIONS AND CONFORMATION

So far, we have only considered the way that molecular interactions influence the spatial distribution of molecules in a system. Molecular interactions can also determine the three-dimensional conformation and flexibility of individual molecules (Lehninger et al. 1993, Atkins 1994, Gelin 1994). Small molecules, such as H_2O and CH_4, normally exist in a single conformation which is determined by the relatively strong covalent bonds that hold the atoms together (Karplus and Porter 1970, Atkins 1994). On the other hand, many larger molecules can exist in a number of different conformations because of the possibility of rotation around saturated covalent bonds (e.g., proteins and polysaccharides) (Baianu 1992, Bergethon and Simons 1990, Lehninger et al. 1993, Fennema 1996a). A macromolecule will tend to adopt the conformation that has the lowest free energy under the prevailing environmental conditions (Alber 1989). The conformational free energy of a molecule is determined by the interaction energies and entropy of the system that contains it (Dill 1990). The molecular interactions may be between different parts of the same molecule (intramolecular) or between the molecule and its neighbors (intermolecular). Similarly, the entropy is determined by the number of conformations that the molecule can adopt, as well as by any changes in the entropy caused by interactions with its neighbors (e.g., restriction of their translational or rotational motion) (Alber 1989, Dill 1990).

To highlight the importance of molecular interactions and entropy in determining the conformation of molecules in solution, it is useful to examine a specific example. Consider a hydrophilic biopolymer molecule in an aqueous solution that can exist in either a helical or a random-coil conformation depending on the environmental conditions (Figure 2.8). Many types of food biopolymers are capable of undergoing this type of transformation, including the protein gelatin (Walstra 1996b) and the polysaccharide xanthan (BeMiller and Whistler 1996). The free energy associated with the transition (helix $\Leftrightarrow$ coil) between these two different conformations is given by:

$$\Delta G_{h \to c} = \Delta E_{h \to c} - T\Delta S_{h \to c} \qquad (2.14)$$

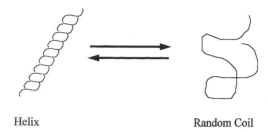

Helix   Random Coil

FIGURE 2.8   The conformation of a molecule in solution is governed by a balance of interaction energies and entropic effects. A helical molecule unfolds when it is heated above a certain temperature because the random-coil conformation is entropically more favorable than the helical conformation.

where $\Delta G_{h \to c}$, $\Delta E_{h \to c}$, and $\Delta S_{h \to c}$ are the free energy, interaction energy, and entropy changes associated with the helix-to-coil transformation. If $\Delta G_{h \to c}$ is negative, the random-coil conformation is favored; if $\Delta G_{h \to c}$ is positive, the helix conformation is favored; and if $\Delta G_{h \to c} \approx 0$, the molecule spends part of its time in each of the conformations. A helical conformation often allows a molecule to maximize the number of energetically favorable intermolecular and intramolecular interactions while minimizing the number of energetically unfavorable ones (Bergethon and Simons 1990, Dickinson and McClements 1995). Nevertheless, it has a much lower entropy than the random-coil state because the molecule can only exist in a single conformation, whereas in the random-coil state the molecule can exist in a large number of different conformations that have similar low energies. At low temperatures, the interaction energy term dominates the entropy term and so the molecule tends to exist as a helix, but as the temperature is raised, the entropy term ($-T\Delta S_{h \to c}$) becomes increasingly important until eventually it dominates and the molecule unfolds. The temperature at which the helix-to-coil transformation takes place is referred to as the transition temperature ($T_{h \to c}$), which occurs when $\Delta G_{h \to c} = 0$. Similar arguments can be used to account for the unfolding of globular proteins when they are heated above a particular temperature, although the relative contribution of the various types of interaction energy is different (Dickinson and McClements 1995). It must be stressed that many food molecules are unable to adopt their thermodynamically most stable conformation because of the presence of various kinetic energy barriers (Section 1.2.1). When an energy barrier is much greater than the thermal energy of the system, a molecule may be "trapped" in a metastable state indefinitely.

The flexibility of molecules in solution is also governed by both thermodynamic and kinetic factors. Thermodynamically, a flexible molecule must be able to exist in a number of conformations that have fairly similar ($\pm kT$) low free energies. Kinetically, the energy barriers that separate these energy states must be small compared to the thermal energy of the system. When both of these criteria are met, a molecule will rapidly move between a number of different configurations and therefore be highly flexible. If the free energy difference between the conformations is large compared to the thermal energy, the molecule will tend to exist predominantly in the minimum free energy state (unless it is locked into a metastable state by the presence of a large kinetic energy barrier).

Knowledge of the conformation and flexibility of a macromolecule under a particular set of environmental conditions is particularly important in understanding and predicting the behavior of many ingredients in food emulsions. The conformation and flexibility of a molecule determine its chemical reactivity, catalytic activity, intermolecular interactions, and functional properties (e.g., solubility, dispersability, water-holding capacity, gelation, foaming, and emulsification) (Damodaran 1994, 1996, 1997).

2.8. HIGHER ORDER INTERACTIONS

When one consults the literature dealing with molecular interactions in foods and other biological systems, one often comes across the terms "hydrogen bonding" and "hydrophobic interactions" (Bergethon and Simons 1990, Baianu 1992, Fennema 1996a). In reality, these terms are a shorthand way of describing certain combinations of interactions which occur between specific chemical groups commonly found in food molecules. Both of these *higher order* interactions consist of contributions from various types of interaction energy (van der Waals, electrostatic, and steric overlap), as well as some entropy effects. It is useful to highlight the general features of hydrogen bonds and hydrophobic interactions in this section, before discussing their importance in determining the properties of individual food components later (Chapter 4).

2.8.1. Hydrogen Bonds

Hydrogen bonds play a crucial role in determining the functional properties of many of the most important molecules present in food emulsions, including water, proteins, lipids, carbohydrates, surfactants, and minerals (Chapter 4). They are formed between a lone pair of electrons on an electronegative atom (such as oxygen) and a hydrogen atom on a neighboring group (i.e., $O–H^{\delta+} \ldots O^{\delta-}$) (Baker and Hubbard 1984, Baianu 1990, Bergethon and Simons 1990, Lehninger et al. 1993). The major contribution to hydrogen bonds is electrostatic (dipole–dipole), but van der Waals forces and steric repulsion also make a significant contribution (Dill 1990). Typically, they have bond strengths between about 10 and 40 kJ mol^{-1} and lengths of about 0.18 nm (Israelachvili 1992). The actual strength of a particular hydrogen bond depends on the electronegativity and orientation of the donor and acceptor groups (Baker and Hubbard 1984). Hydrogen bonds are stronger than most other examples of dipole–dipole interaction because hydrogen atoms have a strong tendency to become positively polarized and because they have a small radius. In fact, hydrogen bonds are so strong that they cause appreciable alignment of the molecules involved. The strength and directional character of hydrogen bonds are responsible for many of the unique properties of water (Chapter 4).

2.8.2. Hydrophobic Interactions

Hydrophobic interactions also play a major role in determining the behavior of many important ingredients in food emulsions, particularly lipids, surfactants, and proteins (Nakai and Li-Chan 1988). They manifest themselves as a strong attractive force that acts between nonpolar groups separated by water (Ben-Naim 1980, Tanford 1980, Israelachvili 1992). Nevertheless, the actual origin of hydrophobic interactions is the ability of water molecules to form relatively strong hydrogen bonds with their nearest neighbors, whereas nonpolar molecules can only form relatively weak van der Waals bonds (Israelachvili 1992). When a nonpolar molecule is introduced into liquid water, it causes the water molecules in its immediate vicinity to rearrange themselves, which changes both the interaction energy and entropy of the system (Chapter 4). It turns out that these changes are thermodynamically unfavorable, and so the system attempts to minimize contact between water and nonpolar groups, which appears as an attractive force between the nonpolar groups (Ben-Naim 1980, Evans and

Wennerstrom 1994). It is this effect that is largely responsible for the immiscibility of oil and water, the adsorption of surfactant molecules to an interface, the aggregation of protein molecules, and the formation of surfactant micelles, and it is therefore particularly important for food scientists to have a good understanding of its origin and the factors which influence it (Nakai and Li-Chan 1988).

2.9. COMPUTER MODELING OF LIQUID PROPERTIES

Our understanding of the way that molecules organize themselves in a liquid can be greatly enhanced by the use of computer modeling techniques (Murrell and Jenkins 1994, Gelin 1994). Computer simulations of molecular properties have provided a number of valuable insights that are relevant to a better understanding of the behavior of food emulsions, including the miscibility/immiscibility of liquids, the formation of surfactant micelles, the adsorption of emulsifiers at an interface, the transport of nonpolar molecules through an aqueous phase, and the conformation and flexibility of biopolymers in solution (van Gunsteren 1988; Brady 1989; Ludescher 1990; Kumosinski et al. 1991a,b; Dickinson and McClements 1995; Esselink et al. 1994; Gelin 1994). The first step in a molecular simulation is to define the characteristics of the molecules involved (size, shape, flexibility, and polarity) and the nature of the intermolecular pair potentials that act between them (Gelin 1994).* A collection of these molecules is arbitrarily distributed within a "box" that represents a certain region of space, and the change in the structural organization of the molecules is then monitored as they are allowed to interact with each other. Depending on the simulation technique used, one can obtain information about the evolution of the structure with time and/or about the equilibrium structure of the molecular ensemble. The two most commonly used computer simulation techniques are the Monte Carlo approach and the molecular dynamics approach (Murrell and Boucher 1982, Murrell and Jenkins 1994).

2.9.1. Monte Carlo Techniques

This technique is named after Monte Carlo, a town in southern France which is famous for its gambling and casinos. The reason for this peculiar name is the fact that the movement of the molecules in the "box" is largely determined by a random selection process, just as the winner in a roulette game is selected. Initially, one starts with an arbitrary arrangement of the molecules in the box. The overall interaction energy is then calculated from a knowledge of the positions of all the molecules and their intermolecular pair potentials. One of the molecules is then randomly selected and moved to a new location and the overall interaction energy is recalculated. If the energy decreases, the move is definitely allowed, but if it increases, the probability of the move being allowed depends on the magnitude of the energy change compared to the thermal energy. When the increase in energy is much greater than RT, the move is highly unlikely and will probably be rejected, but if it is on the same order as RT, it is much more likely to be accepted. This procedure is continued until there is no further change in the average energy of the system after moving a molecule, which is taken to be the minimum potential energy state of the system. Entropy effects are accounted for by continuously monitoring the fluctuations in the overall interaction energy after successive molecules have been moved once the system has reached equilibrium. At thermodynamic equilibrium, the probability of finding a molecular ensemble in a particular potential energy

---

* It should be noted that it is usually necessary to make a number of simplifying assumptions about the properties and interactions of the molecules in order to create computer programs which can be solved in a reasonable time period.

level (E) is proportional to the Boltzmann factor ($P \propto e^{-E/RT}$). Thus the free energy of the system is found by calculating the average value of $e^{-E/RT}$ over all the possible states of the system:

$$G = -RT \ln\langle e^{-E/RT} \rangle \tag{2.15}$$

where $<>$ represents the average over all possible states of the system. The Monte Carlo technique therefore provides information about the equilibrium properties of a system, rather than about the evolution in its properties with time.

2.9.2. Molecular Dynamics Techniques

This technique is named for the fact that it relies on monitoring the movement of molecules with time. Initially, one starts with an arbitrary arrangement of the molecules in the box. The computer then calculates the force which acts on each of the molecules as a result of its interactions with the surrounding molecules. Newton's equations of motion are then used to determine the direction and speed at which each of the molecules moves within a time interval that is short compared to the average time between molecular collisions (typically 10^{-15} to 10^{-14} s). By carrying out the computation over a large number of successive time intervals, it is possible to monitor the evolution of the system with time. At present, it typically takes about 1 s of computer time to simulate 1 ps (10^{-12} s) of real time (Israelachvili 1992). Events at the molecular level typically occur over time scales ranging from a few nanoseconds (e.g., molecular rotation) to a few microseconds (e.g., collisions of colloidal particles), which corresponds to computer times ranging from a few minutes to a few days or weeks (Israelachvili 1992). As computer technology advances, these computation times will inevitably decrease, but they are still likely to be too long to allow computer modeling to be routinely used to study many important molecular processes. A molecular dynamics simulation should lead to the same final state as a Monte Carlo simulation if it is allowed to proceed long enough to reach equilibrium. The free energy of the system is determined by the same method as for Monte Carlo simulations (i.e., by taking into account the fraction of molecules that occupies each energy state once the system has reached equilibrium) (Equation 2.15).

In practice, molecular dynamics simulations are more difficult to set up and take much longer to reach equilibrium than Monte Carlo simulations. For this reason, Monte Carlo simulations are more practical if a researcher is only interested in equilibrium properties, but molecular dynamics simulations are used when information about both the kinetics and thermodynamics of a system is required. Molecular dynamics techniques are particularly suitable for studying nonequilibrium processes, such as mass transport, fluid flow, adsorption kinetics, and solubilization processes (Dickinson and McClements 1995). It is clear from the above discussion that each technique has its own advantages and disadvantages and that both techniques can be used to provide useful insights into the molecular basis for the bulk physiochemical properties of food emulsions.

3 Colloidal Interactions

3.1. INTRODUCTION

Food emulsions are microheterogeneous materials that contain a variety of different structural entities which range in size, shape, and physicochemical properties, including atoms, molecules, molecular aggregates, micelles, emulsion droplets, crystals, and air cells (Dickinson and Stainsby 1982, Dickinson 1992). Many of these structural entities have at least one dimension that falls within the colloidal size range (i.e., between a few nanometers and a few micrometers) (see Figure 1.8). The characteristics of these colloidal particles, and their interactions with each other, are responsible for many of the most important physicochemical and organoleptic properties of food emulsions. The ability of food scientists to understand, predict, and control the properties of food emulsions therefore depends on a knowledge of the interactions that arise between colloidal particles. In this chapter, we examine the origin and nature of the most important types of colloidal interaction, while in later chapters we consider the relationship between these interactions and the stability, rheology, and appearance of food emulsions (Chapters 7 to 9). This is one of the most exciting and rewarding areas of research currently being pursued in emulsion science, and our understanding is rapidly advancing as a result of recent developments in computer modeling, experimental techniques, and the concerted endeavors of individuals working in a variety of scientific disciplines, including mathematics, physics, chemistry, biology, and food science.

The interaction between a pair of colloidal particles is the result of interactions between all of the molecules within them, as well as those within the intervening medium (Hunter 1986, Israelachvili 1992). For this reason, many of the interactions between colloidal particles appear at first glance to be similar to those between molecules (e.g., van der Waals, electrostatic, and steric) (Chapter 2). Nevertheless, the characteristics of these colloidal interactions are often different from their molecular counterparts, because of additional features that arise due to the relatively large size of colloidal particles compared to individual molecules. The major emphasis of this chapter will be on interactions between emulsion droplets, although the same principles can be applied to the various other types of colloidal particles that are commonly found in foods.

3.2. COLLOIDAL INTERACTIONS AND DROPLET AGGREGATION

Colloidal interactions govern whether emulsion droplets aggregate or remain as separate entities, as well as determine the characteristics of any aggregates formed (e.g., their size, shape, porosity, and deformability) (Dickinson 1992, Dickinson and McClements 1995, Bijsterbosch et al. 1995). Many of the bulk physicochemical and organoleptic properties of food emulsions are determined by the degree of droplet aggregation and the characteristics of the aggregates (Chapters 7 to 9). It is therefore extremely important for food scientists to understand the relationship among colloidal interactions, droplet aggregation, and bulk properties.

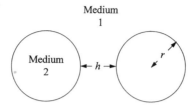

FIGURE 3.1   Emulsion droplets of radius r separated by a surface-to-surface separation h.

In Chapter 2, the interaction between two isolated molecules was described in terms of an *intermolecular* pair potential. In a similar fashion, the interactions between two emulsion droplets can be described in terms of an *interdroplet* pair potential. The interdroplet pair potential, $w(h)$, is the energy required to bring two emulsion droplets from an infinite distance apart to a surface-to-surface separation of h (Figure 3.1). Before examining specific types of interactions between emulsion droplets, it is useful to examine the features of colloidal interactions in a more general fashion.

Consider a system which consists of two emulsion droplets of radius r at a surface-to-surface separation h (Figure 3.1). For convenience, we will assume that only two types of interactions occur between the droplets, one attractive and one repulsive:

$$w(h) = w_{\text{attractive}}(h) + w_{\text{repulsive}}(h) \tag{3.1}$$

The overall interaction between the droplets depends on the relative magnitude and range of the attractive and repulsive interactions. A number of different types of behavior can be distinguished depending on the nature of the interactions involved (Figure 3.2):

1.   ***Attractive interactions dominate at all separations.*** If the attractive interactions are greater than the repulsive interactions at all separations, then the overall interaction is always attractive (Figure 3.2A), which means that the droplets will tend to aggregate (provided the strength of the interaction is greater than the disorganizing influence of the thermal energy).

2.   ***Repulsive interactions dominate at all separations.*** If the repulsive interactions are greater than the attractive interactions at all separations, then the overall interaction is always repulsive (Figure 3.2B), which means that the droplets tend to remain as individual entities.

3.   ***Attractive interactions dominate at large separations, but repulsive interactions dominate at short separations.*** At very large droplet separations, there is no effective interaction between the droplets. As the droplets move closer together, the attractive interaction initially dominates, but at closer separations the repulsive interaction dominates (Figure 3.2C). At some intermediate surface-to-surface separation, there is a minimum in the interdroplet interaction potential (h_{min}). The depth of this minimum, $w(h_{\text{min}})$, is a measure of the strength of the interaction between the droplets, while the position of the minimum (h_{min}) corresponds to the most likely separation of the droplets. Droplets aggregate when the strength of the interaction is large compared to the thermal energy, $|w(h_{\text{min}})| \gg kT$; remain as separate entities when the strength of the interaction is much smaller than the thermal energy, $|w(h_{\text{min}})| \ll kT$; and spend some time together and some time apart at intermediate interaction strengths, $|w(h_{\text{min}})| \approx kT$. When droplets fall into a deep

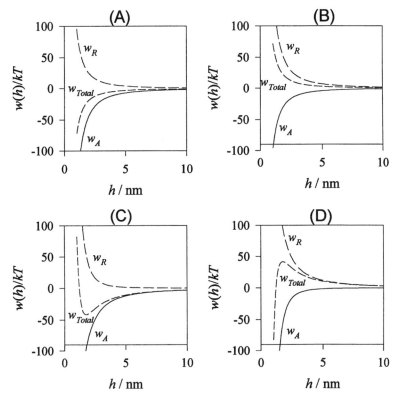

FIGURE 3.2   The interaction of a pair of emulsion droplets depends on the relative magnitude and range of attractive and repulsive interactions.

potential energy minimum, they are said to be *strongly flocculated* or *coagulated* because a large amount of energy is required to pull them apart again. When they fall into a shallow minimum, they are said to be *weakly flocculated* because they are fairly easy to pull apart. The fact that there is an extremely large repulsion between the droplets at close separations prevents them from coming close enough together to coalesce.

4.   ***Repulsive interactions dominate at large separations, but attractive interactions dominate at short separations.***   At very large droplet separations, there is no effective interaction between the droplets. As the droplets move closer together, the repulsive interaction initially dominates, but at closer separations the attractive interaction dominates (Figure 3.2D). At some intermediate surface-to-surface separation (h_{max}), there is an *energy barrier* which the droplets must overcome before they can move any closer together. If the height of this energy barrier is large compared to the thermal energy of the system, $w(h_{max}) \gg kT$, the droplets are effectively prevented from coming close together and will therefore remain as separate entities. If the height of the energy barrier is small compared to the thermal energy, $w(h_{max}) \ll kT$, the droplets easily have enough thermal energy to "jump" over it, and they rapidly fall into the deep minimum that exists at close separations. At intermediate values, $w(h_{max}) \approx kT$, the droplets still tend to aggregate, but this process occurs slowly because only a fraction of droplet–droplet

collisions has sufficient energy to "jump" over the energy barrier. The fact that there is an extremely strong attraction between the droplets at close separations is likely to cause them to coalesce (i.e., merge together).

Despite the simplicity of the above model (Equation 3.1), we have already gained a number of valuable insights into the role that colloidal interactions play in determining whether emulsion droplets are likely to be unaggregated, flocculated, or coalesced. In particular, the importance of the sign, magnitude, and range of the colloidal interactions has become apparent. As would be expected, the colloidal interactions that arise between the droplets in real food emulsions are much more complex than those considered above (Dickinson 1992). First, there are a number of different types of repulsive and attractive interaction that contribute to the overall interaction potential, each with a different sign, magnitude, and range. Second, food emulsions contain a huge number of droplets and other colloidal particles that have different sizes, shapes, and properties. Third, the liquid that surrounds the droplets may be compositionally complex, containing various types of ions and molecules. Droplet–droplet interactions in real food emulsions are therefore influenced by the presence of the neighboring droplets, as well as by the precise nature of the surrounding liquid. For these reasons, it is difficult to accurately account for colloidal interactions in real food emulsions because of the mathematical complexity of describing interactions between huge numbers of molecules, ions, and particles (Dickinson 1992). Nevertheless, considerable insight into the factors which determine the properties of food emulsions can be obtained by examining the interaction between a pair of droplets. In addition, our progress toward understanding complex food systems depends on first understanding the properties of simpler model systems. These model systems can then be incrementally increased in complexity and accuracy as advances are made in our knowledge.

In the following sections, the origin and nature of the major types of colloidal interaction which arise between emulsion droplets are reviewed. In Section 3.11, we then consider ways in which these individual interactions combine with each other to determine the overall interdroplet pair potential and thus the stability of emulsion droplets to aggregation. A knowledge of the contribution that each of the individual colloidal interactions makes to the overall interaction enables one to identify the most effective means of controlling the stability of a given system to aggregation.

3.3. VAN DER WAALS INTERACTIONS

3.3.1. Origin of van der Waals Interactions

Intermolecular van der Waals interactions arise because of the attraction between molecules that have been electronically or orientationally polarized (Section 2.5). In addition to acting between individual molecules, van der Waals interactions also act between macroscopic bodies that contain large numbers of molecules, such as emulsion droplets (Hiemenz 1986). The van der Waals interactions between macroscopic bodies can be calculated using two different mathematical approaches (Hunter 1986, Derjaguin et al. 1987, Israelachvili 1992). In the *microscopic* approach, the van der Waals interaction between a pair of droplets is calculated by carrying out a pairwise summation of the interaction energies of all the molecules in one of the droplets with all of the molecules in the other droplet. Calculations made using this approach rely on a knowledge of the properties of the individual molecules, such as polarizabilities, dipole moments, and electronic energy levels. In the *macroscopic* approach, the droplets and surrounding medium are treated as continuous liquids which interact with each other because of the fluctuating electromagnetic fields generated by the movement of the electrons within them. Calculations made using this approach rely on a knowledge of the bulk physicochemical properties of the liquids, such as dielectric constants, refractive indices, and absorption frequencies. Under certain circumstances, both theoretical approaches give similar predictions of the van der Waals interaction between macroscopic bodies. In general, however, the macroscopic approach is usually the most suitable for describing interactions between emulsion droplets because it automatically takes into account the effects of retardation and the liquid surrounding the droplets (Hunter 1986).

3.3.2. Interdroplet Pair Potential

The van der Waals interdroplet pair potential, $w_{VDW}(h)$, of two emulsion droplets of equal radius r separated by a surface-to-surface distance h is given by the following expression (Figure 3.1):

$$w_{VDW}(h) = \frac{-A_{121}}{6}\left[\left(\frac{2r^2}{h^2 + 4rh}\right) + \left(\frac{2r^2}{h^2 + 4rh + 4r^2}\right) + \ln\left(\frac{h^2 + 4rh}{h^2 + 4rh + 4r^2}\right)\right]$$

(3.2)

where A_{121} is the *Hamaker function* for emulsion droplets (medium 1) separated by a liquid (medium 2). The value of the Hamaker function can be calculated using either the microscopic or macroscopic approach mentioned above (Mahanty and Ninham 1976). At close separations ($h \ll r$), the above equation can be simplified considerably:

$$w_{VDW}(h) = \frac{-A_{121}r}{12h}$$

(3.3)

This equation indicates that van der Waals interactions between colloidal particles ($w \propto 1/h$) are much longer range than those between molecules ($w \propto 1/s^6$), which has important consequences for determining the stability of food emulsions. Equations for calculating the van der Waals interaction between spheres of unequal radius are given by Hiemenz (1986).

3.3.3. Hamaker Function

In general, an accurate calculation of the Hamaker function of a pair of emulsion droplets is a complicated task (Mahanty and Ninham 1976, Hunter 1986, Israelachvili 1992). A knowledge of the optical properties (dielectric permittivities) of the oil, water, and interfacial phases over a wide range of frequencies is required, and this information is not readily available for most substances (Hunter 1986, Roth and Lenhoff 1996). In addition, the full theory must be solved numerically using a digital computer (Pailthorpe and Russel 1982). Nevertheless, approximate expressions for the Hamaker function have been derived, which can be calculated using data that can easily be found in the literature (Israelachvili 1992):

$$A_{121} = A_{v=0} + A_{v>0}$$

(3.4)

where

$$A_{v=0} = \frac{3}{4}kT\sum_{s=1}^{\infty}\frac{1}{s^3}\left(\frac{\varepsilon_1 - \varepsilon_2}{\varepsilon_1 + \varepsilon_2}\right)^{2s} \qquad A_{v>0} = \frac{3hv_e}{16\sqrt{2}}\frac{\left(n_1^2 - n_2^2\right)^2}{\left(n_1^2 + n_2^2\right)^{3/2}}$$

Here, ε is the static relative dielectric constant, n is the refractive index, v_e is the major electronic absorption frequency in the ultraviolet region of the electromagnetic spectrum (which is assumed to be equal for both phases), h is Planck's constant, and the subscripts 1 and 2 refer to the continuous phase and droplets, respectively. Equation 3.4 indicates that the Hamaker function of two similar droplets is always positive, which means that $w_{VDW}(h)$ is always negative, so that the van der Waals interaction is always attractive. It should be noted, however, that the interaction between two colloidal particles containing different materials may be either attractive or repulsive, depending on the relative physical properties of the particles and intervening medium (Israelachvili 1992, Milling et al. 1996).

In Equation 3.4, the Hamaker function is divided into two contributions: a zero-frequency component ($A_{v=0}$) and a frequency-dependent component ($A_{v>0}$). The overall interdroplet pair potential is therefore given by:

$$w_{VDW}(h) = w_{v=0}(h) + w_{v>0}(h) \qquad (3.5)$$

where $w_{v=0}(h)$ and $w_{v>0}(h)$ are determined by inserting the expressions for $A_{v=0}$ and $A_{v>0}$ into Equation 3.2 or 3.3. The zero-frequency component is due to orientation and induction contributions to the van der Waals interaction, whereas the frequency-dependent component is due to the dispersion contribution (Section 2.3.3). The separation of the Hamaker function into these two components is particularly useful for understanding the influence of electrostatic screening and retardation on van der Waals interactions (see below).

For food emulsions, the Hamaker function is typically about 0.75×10^{-20} J, with about 42% of this coming from the zero-frequency contribution and 58% from the frequency-dependent contribution. The physicochemical properties needed to calculate Hamaker functions for ingredients typically found in food emulsions are summarized in Table 3.1. In practice, the magnitude of the Hamaker function depends on droplet separation and is considerably overestimated by Equation 3.4 because of the effects of electrostatic screening, retardation, and interfacial layers (see below).

3.3.4. Electrostatic Screening

The zero-frequency component of the Hamaker function ($A_{v=0}$) is electrostatic in origin because it depends on interactions that involve permanent dipoles (Section 2.3.3). Consequently, this part of the van der Waals interaction is "screened" (reduced) when droplets are

TABLE 3.1
Physicochemical Properties Needed to Calculate the Nonretarded
Hamaker Function (Equation 3.4) for Some Materials Commonly
Found in Food Emulsions

Medium	Static relative dielectric constant (ε_R)	Refractive index (n)	Absorption frequency ($v_e/10^{15}$ s^{-1})
Water	80	1.333	3.0
Oil	2	1.433	2.9
Pure protein	5	1.56	2.9
x% protein in water	$5x + 80(1 - x)$	$1.56x + 1.333(1 - x)$	2.9
Pure Tween 20		1.468	2.9

Data compiled from Israelachvili (1992), Wei et al. (1994), and Hato et al. (1996).

suspended in an electrolyte solution because of the accumulation of counterions around the droplets (Section 3.4.2). Electrostatic screening causes the zero-frequency component to decrease with increasing droplet separation and with increasing electrolyte concentration (Mahanty and Ninham 1976; Marra 1986; Israelachvili 1992; Mishchuk et al. 1995, 1996). At high electrolyte concentrations, the zero-frequency contribution decays rapidly with distance and makes a negligible contribution to the overall interaction energy at distances greater than a few nanometers (Figure 3.3). On the other hand, the frequency-dependent component ($A_{v>0}$) is unaffected by electrostatic screening because the ions in the electrolyte solution are so large that they do not have time to move in response to the rapidly fluctuating dipoles (Israelachvili 1992). Consequently, the van der Waals interaction may decrease by as much

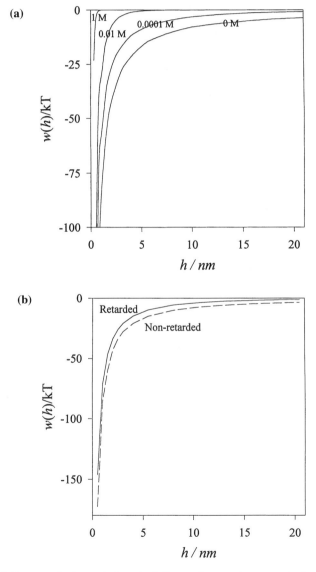

FIGURE 3.3 Influence of electrostatic retardation and screening on van der Waals interactions between oil droplets suspended in water (see Table 3.1 for the properties of the phases used in the calculations). The curves in Figure 3.3a are for different electrolyte concentrations.

as 42% in oil-in-water emulsions at high-ionic-strength solutions because the zero-frequency component is completely screened.

3.3.5. Retardation

The strength of the van der Waals interaction between emulsion droplets is reduced because of a phenomenon known as *retardation* (Israelachvili 1992). The origin of retardation is the finite time taken for an electromagnetic field to travel from one droplet to another and back (Mahanty and Ninham 1976). The frequency-dependent contribution to the van der Waals interaction ($w_{v>0}$) is the result of a transient dipole in one droplet inducing a dipole in another droplet, which then interacts with the first dipole (Section 2.3.3). The strength of the resulting attractive force is reduced if the time taken for the electromagnetic field to travel between the droplets is comparable to the lifetime of a transient dipole, because then the orientation of the first dipole will have changed by the time the field from the second dipole arrives (Israelachvili 1992). This effect becomes appreciable at dipole separations greater than a few nanometers and results in a decrease in the frequency-dependent ($A_{v>0}$) contribution to the Hamaker function with droplet separation. The zero-frequency contribution ($A_{v=0}$) is unaffected by retardation because it is electrostatic in origin (Mahanty and Ninham 1976). Consequently, the contribution of the $A_{v>0}$ term becomes increasingly small as the separation between the droplets increases. For example, the retarded value of $w_{v>0}(h)$ between two emulsion droplets at a separation of 20 nm is only 33% of the nonretarded value (Figure 3.3). Any accurate prediction of the van der Waals interaction between droplets should therefore include retardation effects. A number of authors have developed relatively simple correction functions which can be used to account for retardation effects (Schenkel and Kitchner 1960; Gregory 1969, 1981; Anandarajah and Chen 1995; Chen and Anandarajah 1996), although the most accurate method is to solve the full theory numerically (Mahanty and Ninham 1976, Pailthorpe and Russel 1982).

3.3.6. Influence of an Interfacial Layer

So far, we have assumed that the van der Waals interaction occurs between two homogeneous spheres separated by an intervening medium (Figure 3.1). In reality, emulsion droplets are normally surrounded by a thin layer of emulsifier molecules, and this interfacial layer has different physicochemical properties (ε_R, n, and v_e) than either the oil or water phases (Figure 3.4). The molecules nearest the surface of a particle make the greatest contribution to the overall van der Waals interaction, and so the presence of an interfacial layer can have a large effect on the interactions between emulsion droplets, especially at close separations (Vold 1961, Israelachvili 1992, Parsegian 1993).

  The influence of an adsorbed layer on the van der Waals interactions between emulsion droplets has been considered by Vold (1961):

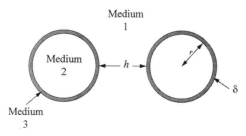

FIGURE 3.4 The droplets in food emulsions are normally surrounded by an adsorbed emulsifier layer, which modifies their van der Waals interactions.

$$w_{VDW}(h) =$$

$$-\frac{1}{12}\left[A_{131}H\left(\frac{h}{2(r+\partial)},1\right) + A_{232}H\left(\frac{h+2\partial}{2r},1\right) + 2A_{132}H\left(\frac{h+\partial}{2r},\frac{r+\partial}{r}\right)\right]$$

(3.6)

where the subscripts 1, 2, and 3 refer to the continuous phase, droplet, and emulsifier layer, respectively; h is the surface-to-surface separation between the *outer* regions of the adsorbed layers; δ is the thickness of the adsorbed layer; and $H(x,y)$ is a function given by:

$$H(x,y) = \frac{y}{x^2 + xy + x} + \frac{y}{x^2 + xy + x + y} + 2\ln\left(\frac{x^2 + xy + x}{x^2 + xy + x + y}\right)$$

The dependence of the (nonretarded and nonscreened) van der Waals interaction between two emulsion droplets on the thickness and composition of an interfacial layer consisting of a mixture of protein and water was calculated using Equation 3.6 and the physical properties listed in Table 3.1 (Figure 3.5). In the absence of the interfacial layer, the attraction between the droplets was about $-110\ kT$ at a separation of 1 nm. Figure 3.5 clearly indicates that the interfacial layer causes a significant alteration in the strength of the interactions between the droplets, leading to either an increase or decrease in the strength of the attraction, depending on the protein concentration. At high protein concentrations (>60%), the attraction is greater than that between two bare emulsion droplets, whereas at low protein concentrations (<60%) it is smaller.

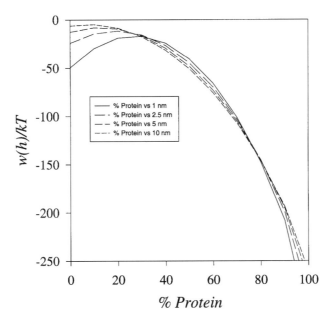

FIGURE 3.5 Influence of the composition of an interfacial layer, consisting of water and protein, on the van der Waals interactions between emulsion droplets. The interdroplet pair potential is reported at an outer surface-to-surface separation of 1 nm for 1-μm droplets. The physical properties of the oil, water, and interfacial layer used in the calculations are reported in Table 3.1

3.3.7. General Features of van der Waals Interactions

1. The interaction between two oil droplets (or between two water droplets) is always attractive.
2. The strength of the interaction decreases with droplet separation, and the interaction is fairly long range ($w \propto 1/h$).
3. The interaction becomes stronger as the droplet size increases.
4. The strength of the interaction depends on the physical properties of the droplets and the surrounding liquid (through the Hamaker function).
5. The strength of the interaction depends on the thickness and composition of the adsorbed emulsifier layer.
6. The strength of the interaction decreases as the concentration of electrolyte in an oil-in-water emulsion increases because of electrostatic screening.

van der Waals interactions act between all types of colloidal particles, and therefore they must always be considered when calculating the overall interaction potential between emulsion droplets (Hiemenz 1986, Israelachvili 1992). Nevertheless, it must be stressed that an accurate calculation of their magnitude and range is extremely difficult, because of the lack of physicochemical data required to perform the calculations and because of the need to simultaneously account for the effects of screening, retardation, and interfacial layers (Hunter 1986). The fact that van der Waals interactions are relatively strong and long range, and that they are always attractive, suggests that emulsion droplets would tend to associate with each other. In practice, many food emulsions are stable to droplet aggregation, which indicates the existence of repulsive interactions that are strong enough to overcome the van der Waals attraction. Some of the most important types of these repulsive interactions, including electrostatic, polymeric steric, hydration, and thermal fluctuation interactions, are discussed in the following sections.

3.4. ELECTROSTATIC INTERACTIONS

3.4.1. Origins of Surface Charge

The droplets in many food emulsions have electrically charged surfaces because of the adsorption of emulsifiers which are either ionic or capable of being ionized (e.g., proteins, polysaccharides, and surfactants) (Chapter 4). All food proteins have acidic ($-COOH \rightarrow COO^- + H^+$) and basic ($NH_2 + H^+ \rightarrow NH_3^+$) groups whose degree of ionization depends on the pH and ionic strength of the surrounding aqueous phase (Charalambous and Doxastakis 1989, Damodaran 1996, Magdassi 1996, Magdassi and Kamyshny 1996). Some surface-active polysaccharides, such as modified starch and gum arabic, also have acidic groups which may be ionized (BeMiller and Whistler 1996). Ionic surfactants may be either positively or negatively charged depending on the nature of their hydrophilic head group (Linfield 1976, Myers 1988, Richmond 1990). The magnitude and sign of the electrical charge on an emulsion droplet therefore depend on the type of emulsifier used to stabilize it, the concentration of the emulsifier at the interface, and the prevailing environmental conditions (e.g., pH, temperature, and ionic strength). All the droplets in an emulsion are usually stabilized by the same type of emulsifier and therefore have the same electrical charge. The electrostatic interaction between similarly charged droplets is repulsive, and so electrostatic interactions play a major role in preventing droplets from coming close enough together to aggregate.

It is convenient to divide the different types of ions which can influence surface charge into three categories (Hunter 1986, 1989):

1. ***Potential-determining ions.*** This type of ion is responsible for the association–dissociation of charged groups (e.g., –COOH → COO⁻ + H⁺). In food emulsions, the most important potential-determining ions are H⁺ and OH⁻, because they govern the degree of ionization of acidic and basic groups on many proteins and polysaccharides. The influence of potential-determining ions on surface charge is therefore determined principally through the pH of the surrounding solution.

2. ***Indifferent electrolyte ions.*** This type of ion accumulates around charged groups because of electrostatic interactions (e.g., Na⁺ ions may accumulate around a negatively charged –COO⁻ group). These ions reduce the strength of the electrical field around a charged group principally due to electrostatic screening (Section 3.4.2), rather than causing association–dissociation of the charged group. At high ionic strengths, some "indifferent" electrolyte ions can actually alter the degree of ionization of charged groups. They do this either by altering the dissociation constant of the surface groups (i.e., their pK value) or by acting as potential-determining ions that compete with the H⁺ or OH⁻ ions (e.g., –COO⁻ + Na⁺ → –COO⁻Na⁺). The influence of indifferent electrolyte ions on surface charge is therefore determined principally by the ionic strength of the surrounding solution.

3. ***Adsorbed ions.*** Surface charge can also be altered by the adsorption of surface-active ions. In food emulsions, the most important types of surface-active ions are ionic emulsifiers, including many surfactants, proteins, and polysaccharides (Chapter 4). The contribution of adsorbed ions to surface charge is governed mainly by the type and concentration of emulsifiers present in the system and their relative affinities for the droplet surface.

Emulsion scientists are interested in understanding the role that each of these different types of ions play in determining the electrical charge on emulsion droplets, because the magnitude of this charge determines the stability of many food emulsions to aggregation and therefore has a pronounced influence on their appearance, taste, texture, and stability (Chapters 7 to 9).

3.4.2. Ion Distribution Near a Charged Surface

An understanding of the origin and nature of electrostatic interactions between emulsion droplets relies on an appreciation of the way that the various types of ions are organized close to a charged surface. Consider a charged surface which is in contact with an electrolyte solution (Figure 3.6). Ions of opposite charge to the surface (*counterions*) are attracted toward it, whereas ions of similar charge (*co-ions*) are repelled from it. Nevertheless, the tendency for ions to be organized in the vicinity of a charged surface is opposed by the disorganizing influence of the thermal energy (Evans and Wennerstrom 1994). Consequently, the concentration of counterions is greatest at the charged surface and decreases as one moves away from the surface until it reaches the bulk counterion concentration, whereas the concentration of co-ions is smallest at the charged surface and increases as one moves away from the surface until it reaches the bulk co-ion concentration (Figure 3.6). The concentration of counterions near a charged surface is always greater than the concentration of co-ions, and so a charged surface can be considered to be surrounded by a cloud of counterions. Nevertheless, the overall system must be electrically neutral, and so the charge on the surface must be completely balanced by the excess charge of the counterions in the electrolyte solution. The distribution of ions close to a charged surface is referred to as the *electrical double layer,* because it is convenient to assume that the system consists of two oppositely charged layers:

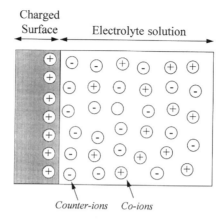

FIGURE 3.6   The organization of ions near a charged surface is governed by two opposing tendencies: (1) electrostatic interactions which favor accumulation of counterions near a surface and (2) thermal energy which favors a random distribution of the ions.

the surface and the surrounding liquid (Kitakara and Watanabe 1984; Hunter 1986, 1989; Hiemenz 1986).

It is important to establish the relationship between the characteristics of a charged surface, the properties of the solution in contact with it, and the distribution of ions in its immediate vicinity, because this information is needed to calculate the strength of electrostatic interactions between emulsion droplets. A surface is usually characterized by its surface charge density (σ) and its surface potential (Ψ_0). The surface charge density is the amount of electrical charge per unit surface area, whereas the surface potential is the amount of energy required to increase the surface charge density from zero to σ. These values depend on the type and concentration of emulsifier present at a surface, as well as the nature of the electrolyte solution (e.g., pH, ionic strength, and temperature). The important characteristics of an electrolyte solution are its dielectric constant and the concentration and valency of the ions it contains.

A mathematical relationship, known as the Poisson–Boltzmann equation, has been derived to relate the electrical potential in the vicinity of a charged surface to the concentration and type of ions present in the adjacent electrolyte solution (Evans and Wennerstrom 1994):

$$\frac{d^2\psi(x)}{dx^2} = -\frac{e}{\varepsilon_0\varepsilon_R} \sum_i z_i n_{0i} \exp\left(\frac{-z_i e\psi(x)}{kT}\right) \tag{3.7}$$

where n_{0i} is the concentration of ionic species of type i in the bulk electrolyte solution (in molecules per cubic meter), z_i is their valency, e is the electrical charge of a single proton, ε_0 is the dielectric constant of a vacuum, ε_R is the relative dielectric constant of the solution, and $\psi(x)$ is the electrical potential at a distance x from the charged surface. This equation is of central importance to emulsion science because it is the basis for the calculation of electrostatic interactions between emulsion droplets. Nevertheless, its widespread application has been limited because it does not have an explicit analytical solution (Hunter 1986). When accurate calculations are required, it is necessary to solve Equation 3.7 numerically using a digital computer (Carnie et al. 1994). For certain systems, it is possible to derive much simpler analytic formulas which can be used to calculate the electrical potential near a surface by making certain simplifying assumptions (Evans and Wennerstrom 1994, Sader et al. 1995).

If it is assumed that the electrostatic attraction between the charged surface and the counterions is relatively weak compared to the thermal energy (i.e., $z_i e \psi_0 < kT$, which means that ψ_0 must be less than about 25 mV at room temperature in water), then a simple expression, known as the Debye–Huckel approximation, can be used to calculate the dependence of the electrical potential on distance from the surface (Hunter 1986, Hiemenz 1986):

$$\Psi(x) = \Psi_0 \exp(-\kappa x) \tag{3.8}$$

This equation indicates that the electrical potential decreases exponentially with distance from the surface at a decay rate which is determined by the parameter κ^{-1}, which is known as the *Debye screening length*. The Debye screening length is a measure of the "thickness" of the electrical double layer and it is related to the properties of the electrolyte solution by the following equation:

$$\kappa^{-1} = \sqrt{\frac{\varepsilon_0 \varepsilon_R kT}{e^2 \sum n_{0i} z_i^2}} \tag{3.9}$$

For aqueous solutions at room temperatures, $\kappa^{-1} \approx 0.304/\sqrt{I}$ nm, where I is the ionic strength expressed in moles per liter (Israelachvili 1992). For example, the Debye screening lengths for NaCl solutions with different ionic strengths are 0.3 nm for a 1 M solution, 0.96 nm for a 100 mM solution, 3 nm for a 10 mM solution, 9.6 nm for a 1 mM solution, and 30.4 nm for a 0.1 mM solution.

The Debye screening length is an extremely important characteristic of an electrolyte solution because it determines how rapidly the electrical potential decreases with distance from the surface. Physically, κ^{-1} corresponds to the distance from the charged surface where the electrical potential has fallen to $1/e$ of its value at the surface. This distance is particularly sensitive to the concentration and valency of the ions in an electrolyte solution. As the ion concentration or valency increases, κ^{-1} becomes smaller, and therefore the electrical potential decreases more rapidly with distance (Figure 3.7). The physical explanation for this phenomenon is that the neutralization of the surface charge occurs at shorter distances when the concentration of opposite charge in the surrounding solution increases.

A charged surface can be considered to be surrounded by a "cloud" of counterions with a thickness equal to the Debye screening length and which depends strongly on the ion concentration and valency (Hunter 1986). As will be seen in later chapters, the screening of electrostatic interactions by electrolytes has important consequences for the stability and rheology of many food emulsions (Chapters 7 and 8).

The Poisson–Boltzmann theory assumes an electrolyte solution is a continuum that contains ions which are infinitesimally small. It therefore allows ions to accumulate at an unphysically large concentration near to a charged surface (Evans and Wennerstrom 1994, Kitakara and Watanabe 1984). In reality, ions have a finite size and shape, and this limits the number of them which can be present in the first layer of molecules that are in direct contact with the surface (Derjaguin et al. 1987, Derjaguin 1989, Israelachvili 1992). This assumption is not particularly limiting for systems in which there is a weak interaction between the ions and the charged surface. Nevertheless, it becomes increasingly unrealistic as the strength of the electrostatic interactions between a charged surface and the surround-

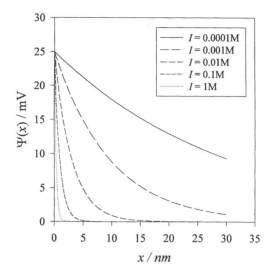

FIGURE 3.7   Influence of ionic strength on the electric field near a charged surface. The electrical double layer shrinks as the electrolyte concentration increases or the ion valency increases.

ing ions increases relative to the thermal energy (Evans and Wennerstrom 1994). In these cases, the Poisson–Boltzmann theory must be modified to take into account the finite size of the ions in the electrolyte solution. It has proved convenient to divide the counterion distribution near a highly charged surface into two regions: an inner and an outer region (Figure 3.8).

3.4.2.1.   Inner Region

In the inner region, the attraction between the counterions and charged surface is strong and therefore they are relatively immobile, whereas in the outer region, the attraction is much weaker and therefore the counterions are more mobile (Evans and Wennerstrom 1994). The thickness of the inner region (δ) is approximately equal to the *radius* of the hydrated counterions, rather than their diameter, because the effective charge of an ion is located at its center (Hiemenz 1986). The inner region is sometimes referred to as the *Stern layer*, while the boundary between the inner and outer regions is referred to as the *Stern plane* (Figure 3.8), after Otto Stern, the scientist who first proposed this concept (Hiemenz 1986, Derjaguin 1989). The electrical potential at the Stern plane (Ψ_δ) is different from that at the surface (Ψ_0) because of the presence of the counterions in the Stern layer. For monovalent indifferent electrolyte counterions, Ψ_δ is less than Ψ_0, because the surface charge is partly neutralized by the charge on the counterions (Figure 3.9a). The extent of this decrease depends on the number and packing of the counterions within the Stern plane (Derjaguin 1989). The same behavior is observed for multivalent indifferent electrolyte counterions at low concentrations, but at higher concentrations the surface may adsorb such a large number of oppositely charged multivalent counterions that its charge is actually reversed, so that Ψ_δ has an opposite sign to Ψ_0 (Figure 3.9b). If a charged surface adsorbs surface-active co-ions (e.g., ionic emulsifiers), it is even possible for Ψ_δ to be larger than Ψ_0 (Figure 3.9c). An increase in surface charge may occur when the hydrophobic attraction between the nonpolar tail of a surfactant and a surface is greater than any electrostatic repulsive interactions. Thus the magnitude of the electrical potential at the Stern plane depends on the precise nature and concentration of the ions present in the system (Derjaguin 1989).

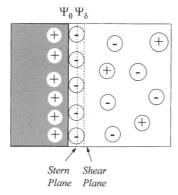

$\Psi_0\,\Psi_\delta$

Stern   Shear
Plane   Plane

FIGURE 3.8 When the electrostatic attraction between a charged surface and the surrounding counterions is relatively strong compared to the thermal energy, it is convenient to divide the electrolyte solution into an inner and an outer region.

A number of theories have been developed to take into account the effect of the finite size and limited packing of ions in the Stern layer on the relationship between Ψ_δ and Ψ_0. One of the most widely used is the *Stern isotherm*, which assumes that there are only a finite number of binding sites at the surface and that once these are filled, the surface becomes saturated and cannot adsorb any more ions:

$$\theta = \frac{K\chi}{1 + K\chi} \tag{3.10}$$

where θ is the fraction of occupied surface sites, K is the adsorption equilibrium constant, and χ is the mole fraction of the ion in the bulk phase. The adsorption equilibrium constant depends on the strength of the interaction between the ion and the surface compared to the thermal energy (Hiemenz 1986):

$$K \approx \exp\!\left(\frac{\Delta G_{ads}}{kT}\right) \tag{3.11}$$

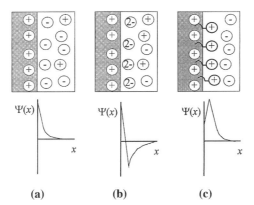

(a)   (b)   (c)

FIGURE 3.9 The electrical potential at the Stern plane may be lower, higher, or a different sign than that at the surface, depending on the strength of the interaction and the type of ions adsorbed.

where ΔG_{ads} is the free energy associated with the adsorption of an ion to the interface: ΔG_{ads} $= ze\Psi_\delta + \phi$. The $ze\Psi_\delta$ term is due to the electrostatic attraction between the ion and the surface, while the ϕ term accounts for any specific binding effects. These specific binding effects could be due to hydrophobic interactions (e.g., when an emulsifier adsorbs) or chemical interactions (e.g., $-COO^- + Na^+ \rightarrow -COO^-Na^+$). The fraction of surface sites which are occupied increases as the free energy of adsorption of an ion increases. The electrical potential at the Stern layer can be related to the electrical potential at the charged surface using the following equation (Hiemenz 1986):

$$\Psi_\delta = \Psi_0 - \frac{\delta\theta\sigma^*}{\varepsilon_\delta\varepsilon_0} \tag{3.12}$$

where σ^* is the surface charge density when the surface is completely saturated with ions and ε_δ is the relative dielectric constant of the Stern layer. Equation 3.12 indicates that the difference between the potential at the surface and that at the Stern plane depends on the fraction of surface sites that are occupied. In principle, this equation can be used to calculate the change in the electrical potential of a surface due to ion adsorption. In practice, this equation is difficult to use because of a lack of knowledge about the values of δ, ϕ, and ε_δ in the Stern layer (Hiemenz 1986, Derjaguin 1989). These parameters are unique for every ion–surface combination and are difficult to measure experimentally. For this reason, it is usually more convenient to experimentally measure the electrical potential at the Stern plane, rather than attempting to predict it theoretically (Hunter 1986).

Experiments have shown that Ψ_δ is closely related to the electrical potential at the *shear plane* (Hunter 1986). When a liquid flows past a charged surface, it "pulls" those counterions which are only weakly attached to the surface along with it, but leaves those ions that are strongly attached in place (i.e., those ions in the Stern plane). The shear plane is defined as the distance from the charged surface below which the counterions remain strongly attached and is approximately equal to the diameter of the hydrated ions (Figure 3.8). The electrical potential at the shear plane is referred to as the *zeta potential* (ζ) and can be measured using various types of electrokinetic techniques (Chapter 10).

3.4.2.2. Outer Region

In the outer region, the electrostatic interaction between the surface and the counterions is usually fairly weak (because the ions in the Stern layer partially screen the surface charge), and so the variation in electrical potential with distance can be described by Equation 3.8, by replacing Ψ_0 with Ψ_δ:

$$\Psi(x) = \Psi_\delta \exp(-\kappa x) \tag{3.13}$$

where x is now taken to be the distance from the shear plane, rather than from the charged surface. The dependence of the electrical potential on distance from the shear plane can then be calculated once Ψ_δ is known.

3.4.3. Electrostatic Interactions Between Charged Droplets

In food emulsions, we are interested in the strength of the electrostatic interactions between droplets (Dickinson and Stainsby 1982, Dickinson 1992). An isolated charged droplet is surrounded by a cloud of counterions with a thickness determined by the Debye screening

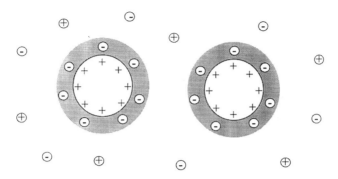

FIGURE 3.10 Distribution of ions around electrically charged emulsion droplets.

length (Figure 3.10). When two similarly charged droplets approach each other, their counterion clouds overlap, and this gives rise to a repulsive interaction (Evans and Wennerstrom 1994). There are two major contributions to this electrostatic interaction: (1) an *enthalpic* contribution associated with the change in the strength of the attractive and repulsive electrostatic interactions between the various charged species involved and (2) an *entropic* contribution associated with the confinement of the counterions between the droplets to a smaller volume. The entropic contribution is strongly repulsive, whereas the enthalpic contribution is weakly attractive, and therefore the overall interaction is repulsive (Evans and Wennerstrom 1994). The fact that the major contribution to the electrostatic interaction is entropic means that it increases in strength with increasing temperature.

Theoretical equations based on the Poisson–Boltzmann theory have been derived to relate the electrostatic interdroplet pair potential to the physical characteristics of the emulsion droplets and the intervening electrolyte solution (Hiemenz 1986, Hunter 1986, Carnie et al. 1994, Okshima 1994, Sader et al. 1995). The full equations cannot be solved analytically, but they can be solved numerically on a computer (Carnie et al. 1994) or by making simplifying assumptions that lead to relatively simple equations that are applicable under certain conditions (Sader et al. 1995).

If it is assumed that there is a relatively low surface potential ($\Psi_\delta < 25$ mV) and that the Debye screening length and surface-to-surface separation are much less than the droplet size (i.e., $\kappa^{-1} < r/10$ and $h < r/10$), then fairly simple expressions for the electrostatic interdroplet pair potential between two similar droplets can be derived (Hunter 1986).

At constant surface potential:

$$w_{\text{electrostatic}}^{\psi}(h) = 2\pi\varepsilon_0\varepsilon_R r\psi_\delta^2 \ln[1 + \exp(-\kappa h)] \tag{3.14}$$

At constant surface charge:

$$w_{\text{electrostatic}}^{\sigma}(h) = -2\pi\varepsilon_0\varepsilon_R r\psi_\delta^2 \ln[1 - \exp(-\kappa h)] \tag{3.15}$$

The smallest droplets in most food emulsions are about 0.1 μm in radius, which means that these equations are likely to be applicable at droplet separations less than about 10 nm and at electrolyte concentrations greater than about 1 mM. Whether the electrostatic interaction between two droplets takes place under conditions of constant surface potential or constant surface charge depends on the ability of the surface groups to regulate their charge (Israelachvili 1992).

So far, it has been assumed that the charge on the droplets is evenly spread out over the whole of the surface. In practice, droplets may have surfaces which have some regions which are negatively charged, some regions which are positively charged, and some regions which are neutral. The heterogeneous distribution of the charges on a droplet may influence their electrostatic interactions (Holt and Chan 1997). Thus, two droplets (or molecules) which have no net charge may still be electrostatically attracted to each other if they have patches of positive and negative charge.

3.4.3.1. Charge Regulation

As two similarly charged emulsion droplets move closer together, the interaction between them becomes increasingly repulsive. Certain systems are capable of reducing the magnitude of this increase by undergoing structural rearrangements, which is referred to as charge regulation. For example, the surface charge may be regulated by adsorption–desorption of ionic emulsifiers (Yaminsky et al. 1996a,b) or by association–dissociation of charged groups (Hunter 1986, 1989). Depending on the physical characteristics of a system, it is possible to discern three different situations which may occur when two droplets approach each other (Reiner and Radke 1993):

1. ***Constant surface charge.*** As the droplets move closer together, the number of charges per unit surface area remains constant (i.e., no adsorption–desorption or association–dissociation of ions occurs). In this case, the electrostatic repulsion between the surfaces is at the maximum possible value because the surfaces are fully charged.

2. ***Constant surface potential.*** As the droplets move closer together, the number of charges per unit surface area decreases (e.g., by an adsorption–desorption or association–dissociation mechanism). In this case, the electrostatic repulsion between the surfaces is at the minimum possible value because the surface charge is reduced.

3. ***Charge regulation.*** In reality, the electrostatic repulsion usually falls somewhere between the two extremes mentioned above because of charge regulation. The number of charges per unit surface area depends on the characteristics of the adsorption–desorption or association–dissociation mechanisms (e.g., the surface activity of an ionic emulsifier or the surface pK value of an ionizable group). These processes take a finite time to occur, and therefore the surface charge density may also depend on the speed at which the droplets come together (Israelachvili 1992, Israelachvili and Berman 1995).

The variation of the interdroplet pair potential with separation is shown for two similarly charged droplets in Figure 3.11. There is a strong repulsive interaction between the droplets at close separations, which decreases as the droplets move farther apart. This repulsive interaction is often sufficiently strong and long range to prevent droplets from aggregating. At relatively large droplet separations, Equations 3.14 and 3.15 give approximately the same predictions for the electrostatic interaction, but at closer separations, the assumption of constant charge predicts a significantly higher repulsion than the assumption of constant potential (Figure 3.11). In practice, the interdroplet pair potential always lies somewhere between these two extremes and depends on the precise nature of the system.

The magnitude and range of the electrostatic repulsion between two droplets decrease as the ionic strength of the solution separating them increases because of electrostatic screening (i.e., the accumulation of counterions around the surfaces) (Figure 3.12). This has important

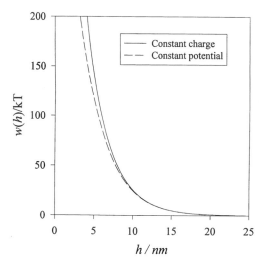

FIGURE 3.11 Comparison of electrostatic interaction between a pair of emulsion droplets under conditions of constant surface charge and constant surface potential.

consequences for the texture and stability of many food emulsions and explains the susceptibility of protein-stabilized emulsions to flocculation when the electrolyte concentration is increased above a critical level (Demetriades et al. 1997a).

3.4.3.2. Effect of Electrolyte on Surface Potential

When the electrostatic interaction between a charged surface and the counterions is relatively weak, the surface charge density is simply related to the surface potential: $\sigma = \varepsilon_R \varepsilon_0 \kappa \Psi_\delta$. This equation indicates that the electrical properties of a surface are altered by the presence of electrolytes in the aqueous phase and has important consequences for the calculation of the

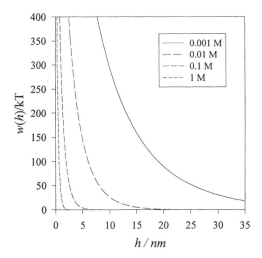

FIGURE 3.12 Electrolyte reduces the magnitude and range of the electrostatic repulsion between emulsion droplets due to electrostatic screening.

electrostatic interdroplet pair potential. If the surface charge density remains constant when salt is added to the aqueous phase, then the surface potential decreases (because less energy is needed to bring a charge from infinity to the droplet surface through an electrolyte solution). Conversely, if the electrical potential remains constant as the salt concentration is increased, this means that the surface charge density must decrease. In practice, both σ and Ψ_δ tend to change simultaneously. In food emulsions, one can usually assume that the surface charge density is independent of ionic strength at low to moderate electrolyte concentrations, and so one must take into account the variation in Ψ_δ with ionic strength when calculating the electrostatic repulsion.

3.4.4. Ion Bridging

Ion bridging is another type of colloidal interaction which involves electrostatic interactions (Ducker and Pashley 1992). It occurs when a polyvalent ion simultaneously binds to the surface of two emulsion droplets that have an opposite charge to the ion (Figure 3.13). These polyvalent ions may be low-molecular-weight species, such as Ca^{2+}, Mg^{2+}, or Al^{3+} (Dickinson et al. 1992; Agboola and Dalgleish 1995, 1996), or high-molecular-weight biopolymers, such as polysaccharides or proteins. The tendency for ion bridging to occur depends on the strength of the ion bridge that holds the droplets together compared to the electrostatic repulsion between the similarly charged droplets. For this reason, large polyvalent species, such as ionic polysaccharides, are often most effective at forming ion bridges because they are able to act as a bridge between the droplets without allowing them to get too close together. The ability of polyvalent ions to form ion bridges is superimposed on their ability to reduce the electrostatic repulsion between droplets through charge screening.

3.4.5. General Features of Electrostatic Interactions

1. Electrostatic interactions may be either attractive or repulsive depending on the sign of the charges on the droplets. The interaction is repulsive when droplets have similar charges (which is usually the case), but is attractive when they have opposite charges.
2. The strength of the interaction decreases with droplet separation and may be either long or short range depending on the ionic strength of the electrolyte solution surrounding the droplets. The interaction becomes increasingly short range as the ionic strength increases because of electrostatic screening.
3. The strength of the interaction is proportional to the size of the emulsion droplets.
4. The strength of the interaction depends on the electrical characteristics of the droplet surfaces (e.g., the number of emulsifier molecules adsorbed per unit surface area, the number of ionizable groups per emulsifier molecule, and the concentration of any potential-determining ions in the aqueous phase, for example, H^+ or OH^-).

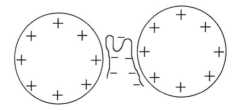

FIGURE 3.13   Polyvalent ions are capable of forming ion bridges between emulsion droplets.

5. The interaction becomes more difficult to predict when association–dissociation of ionizable groups or adsorption–desorption of ionic emulsifiers occurs, especially at close droplet separations.
6. Ion bridging effects have to be taken into account when polyvalent ions are involved.

In this section, we have seen that under certain conditions repulsive electrostatic interactions may be relatively strong and long range compared to attractive van der Waals interactions (compare Figures 3.3 and 3.11). This suggests that they may be strong enough to prevent droplets from aggregating in certain systems. Indeed, it is widely recognized that electrostatic stabilization plays an important role in determining the aggregation of droplets in many food emulsions and particularly those stabilized by proteins (Friberg and Larsson 1997, Dickinson and Stainsby 1982, Dickinson 1992). It should also be recognized that electrostatic interactions influence various other properties of food emulsions, such as the partitioning of ingredients and the rates of chemical reactions. For example, the partitioning of an ionizable volatile flavor compound, such as butyric acid, between the head space and bulk of an emulsion is influenced by electrostatic interactions (Guyot et al. 1996). The negatively charged flavor component is attracted to positively charged droplets, which causes its volatility to be decreased, thus reducing the aroma. Lipid oxidation in food emulsions is often catalyzed by polyvalent ions, such as Fe^{3+}, that are normally present in the aqueous phase. The rate of iron-catalyzed lipid oxidation in oil-in-water emulsions has been shown to increase when the droplets have a negative charge because the Fe^{3+} catalyst and oil molecules are brought into close contact (Coupland and McClements 1996, Mei et al. 1998). A knowledge of the factors which determine the magnitude and range of electrostatic interactions is therefore extremely important to food scientists.

3.5. POLYMERIC STERIC INTERACTIONS

3.5.1. Polymeric Emulsifiers

In Section 3.3, we saw that van der Waals interactions always operate between emulsion droplets and that these interactions are strong enough to cause droplets to aggregate, unless there is a sufficiently strong repulsive interaction to prevent them from coming close together. When emulsion droplets are surrounded by a layer of electrically charged emulsifier molecules, they may be stabilized against aggregation by electrostatic repulsion (Section 3.4). Nevertheless, many food emulsions are stable to droplet aggregation, despite being surrounded by a layer of emulsifier molecules that has no electrical charge, which indicates that other types of repulsive interaction also play an important role in stabilizing these systems (Dickinson and Stainsby 1982, Dickinson 1992). The most important of these is *polymeric steric stabilization,* which is due to the presence of polymeric emulsifiers at the oil–water interface. To be effective at preventing droplet aggregation, steric interactions must be comparable, both in magnitude and range (but opposite in sign), to van der Waals and any other attractive interactions.

Most emulsifiers used in the food industry are either partly or entirely polymeric: proteins are polymers of amino acids, polysaccharides are polymers of sugars, and many nonionic surfactants have polar head groups which are polymers of oxyethylene (Chapter 4). The conformation of a polymeric emulsifier at an interface depends on the number, type, and sequence of monomers along its backbone (Dalgleish 1989, 1995, 1996a,b; Damodaran 1989, 1990, 1996; Dickinson and McClements 1995). An improved understanding of the relationship between the interfacial conformation of polymeric emulsifiers and their ability to stabi-

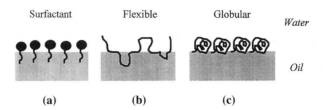

FIGURE 3.14   Orientation of some polymeric emulsifiers at an oil–water interface: (a) small-molecule surfactants, (b) flexible biopolymers, and (c) globular biopolymers.

lize emulsions is particularly important for food scientists because it enables them to create and select ingredients in a more systematic fashion.

An emulsifier tends to adopt an interfacial conformation which minimizes the free energy of the system (Damodaran 1989). The conformational free energy is determined by various enthalpic (intermolecular interaction energies) and entropic (configurational entropy and solvation) contributions. In practice, the major factor which determines the interfacial conformation of food emulsifiers is the hydrophobic effect (i.e., the tendency for a molecule to adopt an arrangement which minimizes the number of unfavorable contacts between polar and nonpolar groups) (Damodaran 1989, Dickinson and McClements 1995). Thus, small-molecule surfactants adopt a conformation in which the hydrocarbon tails protrude into the oil phase, while the hydrophilic head groups protrude into the aqueous phase (Figure 3.14a). Flexible biopolymers, such as casein or modified starches, exist as a series of loops, tails, and trains at an interface (Figure 3.14b). The predominantly hydrophilic or hydrophobic segments which protrude into the aqueous or oil phases (respectively) are referred to as *loops* or *tails*, whereas the predominantly neutral segments that lie flat against the interface are referred to as *trains* (Dickinson 1992). Immediately after adsorption, compact biopolymers (such as globular proteins) tend to adopt an orientation which maximizes the contact area between any hydrophobic patches on their surface and the oil phase (Damodaran 1989). After adsorption, the protein molecules may change their conformation in response to their new environment (Dickinson and McClements 1995; Corredig and Dalgleish 1995; Dalgleish 1996a,b). It must be stressed that many biopolymers take an appreciable time to undergo conformational changes, and therefore the interface may not be at thermodynamic equilibrium (which is assumed in many theories describing their behavior).

3.5.2.   Interdroplet Pair Potential

Polymeric steric interactions arise when emulsion droplets get so close together that the emulsifier layers overlap (Figure 3.15). This type of interaction can be conveniently divided into two contributions (Hiemenz 1986, Hunter 1986):

$$w_{\text{steric}}(h) = w_{\text{elastic}}(h) + w_{\text{mix}}(h) \tag{3.16}$$

The elastic contribution is due to the compression of the interfacial membrane, whereas the mixing contribution is due to the intermingling of the polymer chains (Figure 3.15).

3.5.2.1.   Mixing Contribution

If it is assumed that the polymer molecules in the layers interpenetrate each other without the layers being compressed (Figure 3.15a), then the interaction is entirely due to mixing of the

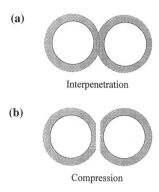

(a)

Interpenetration

(b)

Compression

FIGURE 3.15   Steric interactions between emulsion droplets can be divided into an *elastic* contribution which involves compression of the polymer layers and a *mixing* contribution which involves interpenetration of the polymer chains.

polymers. The theories describing polymeric steric interactions are much less well developed than those describing electrostatic or van der Waals interactions. The major reason for this is that polymeric steric interactions are particularly sensitive to the precise structure, orientation, packing, and interactions of the polymer molecules at the interface (Hunter 1986, Claesson et al. 1995). These parameters vary from system to system and are difficult to account for theoretically or to measure experimentally. Mathematical theories have been developed for a number of simple well-defined systems, and it is informative to examine these because they provide some useful insights into more complex systems (Hunter 1986). For example, the following equation has been derived to account for the mixing contribution when the polymer molecules are permanently attached to the droplet surface and there is a constant number of polymer chains per unit surface area (Hunter 1986):

$$w_{\mathrm{mix}}(h) = 4\pi r k T m^2 N_{\mathrm{A}} \frac{\bar{v}_P^2}{V_S} \left(\frac{1}{2} - \chi \right)\left(1 - \frac{1}{2}\frac{h}{\delta} \right)^2 \tag{3.17}$$

where m is the mass of polymer chains per unit area, δ is the thickness of the adsorbed layer, N_{A} is Avogadro's number, χ is the Flory–Huggins parameter, $\bar{v}_P$ is the partial specific volume of the polymer chains, and $\bar{V}_S$ is the molar volume of the solvent. The Flory–Huggins parameter depends on the relative magnitude of the solvent–solvent, solvent–segment, and segment–segment interactions and is a measure of the *quality* of a solvent. It is related to the effective interaction parameter (w) which was introduced in Chapter 2 to characterize the compatibility of molecules in mixtures: $\chi = w/RT$. In a good solvent ($\chi < 0.5$), the polymer molecules prefer to be surrounded by solvent molecules. In a poor solvent ($\chi > 0.5$), the polymer molecules prefer to be surrounded by each other. In an indifferent (theta) solvent ($\chi = 0.5$), the polymer molecules have no preference for either solvent or polymer molecules. In the original Flory–Huggins theory, it was assumed that χ was entirely due to enthalpic contributions associated with the molecular interactions. In practice, it is more convenient to assume that χ also contains entropic contributions since interactions involving changes in the structural organization of the solvent can then be accounted for (e.g., hydrophobic interactions) (Evans and Wennerstrom 1994). Whether the mixing contribution is attractive or repulsive depends on the quality of the solvent. In a good solvent, the increase in concentration of polymer molecules in the interpenetration zone is thermodynamically unfavorable

(w_{mix} positive) because it reduces the number of polymer–solvent contacts and therefore leads to a repulsive interaction between the droplets. Conversely, in a poor solvent, it is thermodynamically favorable (w_{mix} negative) because it increases the number of polymer–polymer contacts and therefore leads to an attractive interaction between the droplets. In an indifferent solvent, the polymer molecules have no preference as to whether they are surrounded by solvent or by other polymer molecules, and therefore the mixing contribution is zero. Thus, by altering solvent quality, it is possible to change the mixing contribution from attractive to repulsive or vice versa. In food emulsions, this could be done by varying temperature or by adding alcohol or electrolyte to the aqueous phase.

3.5.2.2. Elastic Contribution

If it is assumed that the polymer layers surrounding the emulsion droplets are compressed without any interpenetration of the polymer molecules (Figure 3.15b), then the interaction is entirely elastic. When the layers are compressed, a smaller volume is available to the polymer molecules and therefore their configurational entropy is reduced, which is energetically unfavorable, and so this type of interaction is always repulsive ($w_{elastic}$ positive).

The magnitude of the elastic contribution can be calculated from a statistical analysis of the number of configurations the polymer chains can adopt before and after the layers are compressed (Hiemenz 1986, Dickinson 1992):

$$w_{elastic}(h) = 2kTv \ln \frac{\Omega(\infty)}{\Omega(h)} \tag{3.18}$$

where $\Omega(h)$ and $\Omega(\infty)$ are the number of configurations available to the chains at a separation of h (compressed) and of infinity (uncompressed), respectively. In practice, it is difficult to calculate the number of conformations that a polymer molecule can adopt, and therefore more empirical methods have been developed to account for the elastic interaction. One of the most convenient to use was derived by Jackel (1964):

$$w_{elastic}(h) = 0.77E\left(\frac{1}{2}\delta - \frac{1}{2}h\right)^{5/2}(r + \delta) \qquad (h < \delta)$$
$$w_{elastic}(h) = 0 \qquad\qquad\qquad\qquad\qquad\qquad (h \geq \delta) \tag{3.19}$$

where E is the elastic modulus of the adsorbed layer. This equation indicates that there is a negligible interaction between the droplets when the separation is greater than the thickness of the emulsifier layers, but that there is a steep increase in the repulsive interaction energy when the droplets approach closer than this distance (Figure 3.16). As a first approximation, it may be possible to use measurements of the elastic modulus of macroscopic solutions and gels with polymer concentrations similar to those found in the interfacial region in Equation 3.19. Alternatively, the elastic modulus of an interfacial layer could be measured directly using various types of surface force apparatus (Claesson et al. 1995).

3.5.3. Distance Dependence of Polymeric Steric Interactions

Steric interactions between emulsion droplets are conveniently divided into three regimes, according to the separation of the surface of the bare droplets (h) relative to the thickness of the polymer layers (δ):

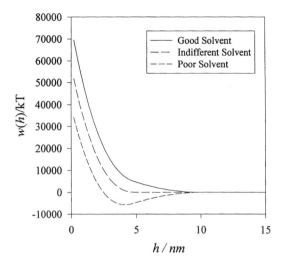

FIGURE 3.16 Interdroplet pair potential due to steric polymeric interactions. At intermediate separations, the steric polymeric interaction can be either attractive or repulsive depending on the quality of the solvent because of the mixing contribution, but at short separations it is strongly repulsive because of the elastic contribution.

1. ***Zero interaction regime ($h \geq 2\delta$).*** At sufficiently large droplet separations, the polymer layers do not overlap with each other and the steric interaction between the droplets is zero.
2. ***Interpenetration regime ($\delta \leq h < 2\delta$).*** When the droplets are sufficiently close together for the polymer layer on one droplet to interpenetrate that on another droplet, without significantly compressing it, the major contribution to the steric interaction is the mixing contribution (w_{mix}), which may be either positive or negative depending on the quality of the solvent
3. ***Interpenetration and compression regime ($h < \delta$).*** When the droplets get so close together that the polymer layers start to compress each other, the overall steric interaction is a combination of elastic and mixing contributions, although the strongly repulsive elastic component usually dominates, and so the overall interaction is repulsive.

It should be stressed that the length of the interpenetration and elastic regions actually depends on the precise nature of the polymer molecules present at the interface (Claesson et al. 1995). Flexible biopolymer molecules will have relatively large interpenetration regions, whereas compact globular proteins will have relatively small ones. Consequently, the choice of δ as the distance where the elastic contribution first contributes to the interaction is arbitrary, and different values will be more appropriate for some systems. As mentioned earlier, the only way these values can accurately be established for a particular system is by measuring the force between two polymer-coated surfaces as they are brought closer together (Israelachvili 1992, Claesson et al. 1995, 1996).

3.5.4. Optimum Characteristics of Polymeric Emulsifier

To be effective at providing steric stabilization, a polymeric emulsifier must have certain physicochemical characteristics (Hunter 1986, Dickinson 1992). First, it must have some segments which bind strongly to the droplet surface (to anchor the polymer to the surface) and other segments which protrude a significant distance into the surrounding liquid (to prevent the droplets from coming close together). This means that the emulsifier must be amphiphilic, with some hydrophobic segments which protrude into the oil phase and some hydrophilic segments which protrude into the aqueous phase (Figure 3.14). The binding to the interface must be strong enough to prevent the emulsifier from desorbing from the droplet surface as the droplets approach one another. Second, the continuous phase surrounding the droplets must be a sufficiently good solvent for the segments which protrude into it, so that the mixing contribution to the overall interaction energy is repulsive (w_{mix} positive). Third, the polymeric-repulsive interaction must act over a distance that is comparable to the range of the attractive van der Waals interactions. Thus uncharged biopolymers that form thick interfacial layers, such as modified starch, are much more effective at stabilizing emulsions against flocculation than biopolymers that form thin layers, such as globular proteins at their isoelectric point.* Finally, the surface must be covered by a sufficiently high concentration of polymer. If too little polymer is present, a single polymer may adsorb to the surface of two different emulsion droplets, forming a bridge which causes the droplets to flocculate. In addition, some of the nonpolar regions will be exposed to the aqueous phase, which leads to a hydrophobic attraction between the droplets (see Section 3.7). Many polymeric emulsifiers are charged, and therefore they stabilize emulsion droplets against aggregation through a combination of electrostatic and steric repulsion (Claesson et al. 1995).

3.5.5. General Features of Polymeric Steric Stabilization

1. Interactions are always strongly repulsive at short separations ($h < \delta$), but may be either attractive or repulsive at intermediate separations ($\delta < h < 2\delta$) depending on the quality of the solvent (Figure 3.16).
2. The range of the interaction increases with the thickness of the adsorbed layer.
3. The strength of the interaction increases with droplet size.
4. The strength of the interaction depends on the molecular architecture of the adsorbed polymeric emulsifier layer and therefore varies considerably from system to system and is difficult to predict from first principles.

Polymer steric interactions are one of the most common and important stabilizing mechanisms in food emulsions. Unlike electrostatic interactions, they occur in almost every type of food emulsion because most emulsifiers are polymeric. Some emulsions are stabilized almost entirely by polymeric steric stabilization, whereas others are stabilized by a combination of steric and electrostatic stabilization. Food scientists must often decide which is the most appropriate emulsifier for a particular application, and so it is useful to compare the differences between steric and electrostatic stabilization (Table 3.2). The principal difference is their sensitivity to pH and ionic strength (Hunter 1986). The electrostatic repulsion between emulsion droplets is dramatically decreased when the electrical charge on the droplet surfaces is reduced (e.g., by altering the pH) or screened (e.g., by increasing the concentration of electrolyte in the aqueous phase). In contrast, steric repulsion is fairly insensitive to both electrolyte concentration and pH.** Another major difference is the fact that the electrostatic repulsion is usually weaker than the van der Waals attraction at short distances, whereas the

---

 * Thick biopolymer layers may also help stabilize droplets against aggregation by reducing the magnitude of the attractive van der Waals interaction (Section 3.3.6).
** It should be stressed that the polymeric steric interaction may be affected by pH and ionic strength if the polymer molecules are charged, because this will alter the thickness of the interfacial layer and the interaction of the polymer chains.

TABLE 3.2
**Comparison of the Advantages and Disadvantages of Polymeric
and Electrostatic Stabilization Mechanisms in Food Emulsions**

Polymeric steric stabilization	Electrostatic stabilization
1. Insensitive to pH	pH dependent — aggregation tends to occur near the isoelectric point of biopolymers
2. Insensitive to electrolyte	Aggregation tends to occur at high electrolyte concentrations
3. Large amounts of emulsifier needed to cover droplet surface	Small amounts of emulsifier needed to cover droplet surface
4. Weak flocculation (easily reversible)	Strong flocculation (often irreversible)
5. Good freeze–thaw stability	Poor freeze–thaw stability

Adapted from Hunter 1986.

polymeric steric stabilization is stronger (Hunter 1986). This means that droplets stabilized by electrostatic repulsion are always thermodynamically unstable with respect to aggregation (because the aggregated state has a lower free energy than the unaggregated state), whereas sterically stabilized systems may be thermodynamically stable if the thickness of the adsorbed layer extends a sufficient distance into the continuous phase.

From a practical standpoint, another important difference is the fact that considerably more emulsifier is usually required to provide steric stabilization (because a thick interfacial layer is required) than to provide electrostatic stabilization. Thus, >5% modified starch is required to stabilize a 20 wt% oil-in-water emulsion containing 1-μm droplets, whereas <0.5% whey protein is required to stabilize the same system. The amount of emulsifier required to prepare an emulsion is often an important financial consideration when formulating a food product.

3.6. DEPLETION INTERACTIONS

3.6.1. Origin of Depletion Interactions

Many food emulsions contain small colloidal particles that are dispersed in the continuous phase which surrounds the droplets (Figure 3.17). These colloidal particles may be surfactant

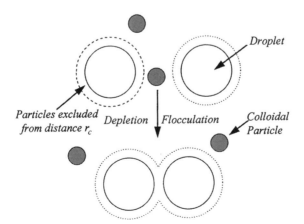

FIGURE 3.17   An attractive depletion interaction arises between emulsion droplets when they are surrounded by small nonadsorbing colloidal particles.

micelles formed when the free surfactant concentration exceeds some critical value (Aronson 1992, Bibette 1991, McClements 1994), individual polymer molecules (Sperry 1982, Seebergh and Berg 1994, Dickinson et al. 1995, Smith and Williams 1995, Jenkins and Snowden 1996), or aggregated polymers (Dickinson and Golding 1997a,b). The presence of these colloidal particles causes an attractive interaction between the droplets which is often large enough to promote emulsion instability (Jenkins and Snowden 1996). The origin of this interaction is the exclusion of colloidal particles from a narrow region surrounding each droplet (Figure 3.17). This region extends a distance approximately equal to the radius (r_c) of a colloidal particle away from the droplet surface. The concentration of colloidal particles in this *deple-tion zone* is effectively zero, while it is finite in the surrounding continuous phase. As a consequence, there is an osmotic potential difference which favors the movement of solvent molecules from the depletion zone into the bulk liquid, so as to dilute the colloidal particles and thus reduce the concentration gradient. The only way this process can be achieved is by two droplets aggregating and thereby reducing the volume of the depletion zone, which manifests itself as an attractive force between the droplets (Figure 3.17). Thus there is an osmotic driving force that favors droplet aggregation and which increases as the concentration of colloidal particles in the aqueous phase increases.

3.6.2. Interdroplet Pair Potential

When the separation between two droplets is small compared to their size ($h \ll r_d$), the interdroplet pair potential due to exclusion of the colloidal particles from the depletion zone is given by the following expression (Sperry 1982):

$$w_{\text{depletion}}(h) =$$

$$-\frac{2}{3}\pi r^3 P_{\text{OSM}}\left[2\left(1 + \frac{r_c}{r}\right)^3 + \left(1 + \frac{h}{2r}\right)^3 - 3\left(1 + \frac{r_c}{r}\right)^2\left(1 + \frac{h}{2r}\right)\right] \qquad (3.20)$$

where P_{OSM} is the osmotic pressure arising from the exclusion of the colloidal particles and r_c is the radius of the colloidal particles. The osmotic pressure difference is given by the following equation (Hiemenz 1986):

$$P_{\text{OSM}} = \frac{CRT}{M}\left(1 + \frac{N_A Cv}{2M}\right) \qquad (3.21)$$

Here, C, M, and v are the concentration, molecular weight, and volume of the colloidal particles, and N_A is Avogadro's number.

For homogeneous spherical particles, the equation for the interdroplet pair potential can be written more conveniently in the following form:

$$w_{\text{depletion}}(h) =$$

$$-\frac{kT}{2}\left(\frac{r}{r_c}\right)^3 \phi_c(1 + 2\phi_c)\left[2\left(1 + \frac{r_c}{r}\right)^3 + \left(1 + \frac{h}{2r}\right)^3 - 3\left(1 + \frac{r_c}{r}\right)^2\left(1 + \frac{h}{2r}\right)\right]$$

$$(3.22)$$

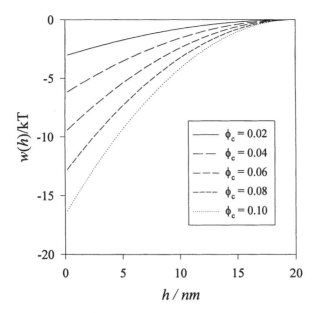

FIGURE 3.18   Influence of the volume fraction (ϕ_c) of colloidal particles on the attractive depletion interaction between emulsion droplets ($r_d = 1$ μm, $r_c = 10$ nm).

where ϕ_c is the volume fraction of the colloidal particles. The dependence of the interdroplet pair potential due to depletion effects on the droplet separation is shown in Figure 3.18. The interaction potential is zero at droplet separations greater than the diameter of the colloidal particles and decreases to a constant value when the droplets come into close contact.

Depletion interactions rely on the colloidal particles not interacting strongly with the surface of the emulsion droplets. Otherwise, the bound particles would have to be displaced as the droplets moved closer together, which would require the input of energy and therefore be repulsive.

3.6.3.   General Features of Depletion Interactions

1.   The strength of the interaction increases as the concentration of colloidal particles in the continuous phase increases (at constant r_c).
2.   The strength of the interaction increases as the size of the droplets in an emulsion increases.
3.   The strength of the interaction increases as the size of the colloidal particles decreases (at constant ϕ_c), because osmotic effects are influenced more by the number of particles involved (which increases with decreasing r_c) than by the volume of the depletion zone (which decreases with decreasing r_c).
4.   The range of the interaction increases as the size of the colloidal particles increases.

Equation 3.22 suggests that the strength of the depletion interaction is independent of pH and ionic strength. Nevertheless, these parameters may indirectly influence the depletion interaction by altering the effective size of the colloidal particles and the depletion zone. For example, changing the number of charges on a biopolymer molecule by altering the pH can either increase or decrease its effective size (Launay et al. 1986, Rha and Pradipasena 1986). Increasing the number of similarly charged groups usually causes a biopolymer to become more extended because of electrostatic repulsion between the groups. On the other hand, decreasing the number of similarly charged groups or having a mixture of positively and negatively charged groups usually causes a biopolymer to reduce its effective size. Altering the ionic strength of an aqueous solution also causes changes in the effective size of biopolymer molecules (e.g., adding salt to a highly charged biopolymer molecule screens the electrostatic repulsion between charged groups and therefore causes a decrease in biopolymer size) (Launay et al. 1986, Rha and Pradipasena 1986). Thus, if the colloidal particles are ionic, one would expect the strength of the depletion interaction to depend on pH and ionic strength, but if they are nonionic, one would expect them to be fairly insensitive to these parameters.

3.7. HYDROPHOBIC INTERACTIONS

Compared to the other major forms of colloidal interaction, the contribution of hydrophobic interactions to emulsion stability has largely been ignored by emulsion scientists. Nevertheless, this type of interaction is of great importance in many types of foods and has recently been shown to promote droplet flocculation in protein-stabilized emulsions (Monahan et al. 1996, Demetriades et al. 1997b). Hydrophobic interactions are important when the surfaces of the droplets have some nonpolar character, either because they are not completely covered by emulsifier (e.g., during homogenization or at low emulsifier concentrations) or because the emulsifier has some hydrophobic regions exposed to the aqueous phase (e.g., adsorbed proteins). The origin of the hydrophobic interaction is the ability of water molecules to form relatively strong hydrogens bonds with each other, but not with nonpolar molecules (Chapter 4). Consequently, the interaction between nonpolar substances and water is thermodynamically unfavorable, which means that a system will attempt to minimize the contact area between these substances by causing them to associate (Tanford 1980, Israelachvili 1992, Israelachvili and Wennerstrom 1996, Alaimo and Kumosinski 1997). This process manifests itself as a relatively strong attractive force between hydrophobic substances dispersed in water and is responsible for many important phenomena which occur in food emulsions, such as protein conformation, micelle formation, adsorption of surfactants, and the low water solubility of nonpolar compounds (Chapter 4).

One of the major reasons that hydrophobic interactions were ignored in the past is that there were no theories available to predict their magnitude and range. The complex nature of their origin, which depends on changes in the interactions and structural organization of a large number of water molecules in the vicinity of the nonpolar groups, means that it is extremely difficult to develop mathematical theories from first principles (Israelachvili 1992, Paulaitis et al. 1996). Nevertheless, recent advances in the development of sensitive instruments for measuring the forces between macroscopic bodies have enabled researchers to develop empirical equations to describe the magnitude and range of hydrophobic interactions (Israelachvili and Pashley 1984, Pashley et al. 1985, Claesson 1987, Claesson and Christenson 1988, Rabinovich and Derjaguin 1988). These experiments have shown that the hydrophobic interaction between nonpolar surfaces is relatively strong and long range and that it decays exponentially with surface-to-surface separation. Considerable progress in understanding the nature of hydrophobic interactions has also been achieved using computer simulations (Paulaitis et al. 1996).

The interdroplet pair potential between two emulsion droplets with hydrophobic surfaces separated by water is given by (Israelachvili and Pashley 1984):

$$w_{\text{hydrophobic}}(h) = -2\pi r \gamma_i \phi \lambda_0 e^{-h/\lambda_0} \tag{3.23}$$

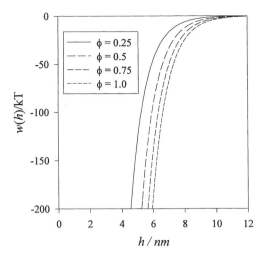

FIGURE 3.19   An attractive hydrophobic interaction arises between emulsion droplets when their surfaces have some hydrophobic character, where ϕ is approximately equal to the fraction of the droplet surface which is nonpolar.

where γ_i is the interfacial tension between the nonpolar groups and water (typically between 10 to 50 mJ m^{-2} for food oils), ϕ is a parameter that varies between 0 and 1 which takes into account the fact that only part of the droplet surface is hydrophobic, and λ_0 is the decay length of the interaction (typically between 1 to 2 nm) (Israelachvili 1992). This equation indicates that the magnitude of the hydrophobic interaction increases as the surfaces become more hydrophobic (i.e., ϕ tends toward unity). Experiments have shown that for bare nonpolar surfaces, the hydrophobic attraction is stronger than the van der Waals attraction up to separations of 80 nm (Israelachvili 1992).

When hydrophobic surfaces are covered by amphiphilic molecules, such as small-molecule surfactants or biopolymers, the hydrophobic interaction between them is effectively screened and the overall attraction is mainly due to van der Waals interactions (Israelachvili 1992). Nevertheless, hydrophobic interactions are significant when the surface has some hydrophobic character (e.g., if the surface is not completely saturated with emulsifier molecules, if it is bent to expose the oil molecules below [Israelachvili 1992], or if the emulsifier molecules have some hydrophobic regions exposed to the aqueous phase [Demetriades et al. 1997b]). Experiments have shown that the hydrophobic interaction is not directly proportional to the number of nonpolar groups at a surface, because the alteration in water structure imposed by nonpolar groups is disrupted by the presence of any neighboring polar groups (Israelachvili 1992). Thus it is not possible to assume that ϕ is simply equal to the fraction of nonpolar sites at a surface. As a consequence, it is difficult to accurately predict their magnitude from first principles.

Hydrophobic interactions become increasingly strong as the temperature is raised (Israelachvili 1992). Thus, hydrophobic interactions between emulsion droplets become more important at higher temperatures. Because the strength of hydrophobic interactions depends on the magnitude of the interfacial tension, any change in the properties of the solvent which increases the interfacial tension will increase the hydrophobic attraction. The addition of small amounts of alcohol to the aqueous phase of an emulsion lowers γ_i and therefore reduces the hydrophobic attraction between nonpolar groups. Electrolytes which alter the structural arrangement of water molecules also influence the magnitude of the hydrophobic effect when they are present at sufficiently high concentrations (Christenson et al. 1990). Structure breakers tend to enhance hydrophobic interactions, whereas structure promoters tend to reduce them (Chapter 5). Variations in pH have little direct effect on the strength of hydrophobic interactions, unless there are accompanying alterations in the structure of the water or the interfacial tension (Israelachvili and Pashley 1984).

3.8. HYDRATION INTERACTIONS

Hydration interactions arise from the structuring of water molecules around dipolar and ionic groups (in contrast to hydrophobic interactions, which arise from the structuring of water around nonpolar groups). Most food emulsifiers naturally have dipolar or ionic groups that are hydrated (e.g., $-OH$, $-COO^-$, and $-NH_3^+$), and some are also capable of binding hydrated ions (e.g., $-COO^- + Na^+ \rightarrow -COO^- Na^+$). As two droplets approach each other, the bonds between the polar groups and the water molecules in their immediate vicinity must be disrupted, which results in a repulsive interaction (Figure 3.20) (Besseling 1997). The magnitude and range of the hydration interaction therefore depend on the number and strength of the bonds formed between the polar groups and the water molecules: the greater the degree of hydration, the more repulsive and long range the interaction (Israelachvili 1992). Just as with hydrophobic interactions, it is difficult to develop theories from first principles to describe this type of interaction because of the complex nature of its origin and its dependence on the specific type of ions and polar groups present. Nevertheless, experimental measurements of the forces between two liquid surfaces have shown that hydration interactions are fairly short-range repulsive forces that decay exponentially with surface-to-surface separation (Claesson 1987, Israelachvili 1992):

$$w_{hydration}(h) = Ar\lambda_0 e^{-h/\lambda_0} \tag{3.24}$$

where A is a constant which depends on the degree of hydration of the surface (typically between 3 and 30 mJ m^{-2}) and λ_0 is the characteristic decay length of the interaction (typically

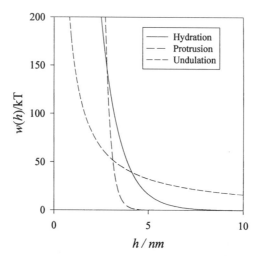

FIGURE 3.20 Short-range repulsive interactions arise between emulsion droplets when they come into close contact due to hydration, protrusion, and undulation of interfacial layers.

between 0.6 and 1.1 nm) (Israelachvili 1992). The greater the degree of hydration of a surface group, the larger the values of A and λ_0. In practice, it is often difficult to isolate the contribution of the hydration forces from other short-range interactions that are associated with mobile interfacial layers at small separations (such as steric and thermal fluctuation interactions) (Figure 3.20), and so there is still much controversy about their origin and nature. Nevertheless, it is widely accepted that they make an important contribution to the overall interaction energy in many systems.

At high electrolyte concentrations, it is possible for ionic surface groups to specifically bind hydrated ions to their surfaces (Hunter 1986, 1989; Miklavic and Ninham 1990). Some of these ions have large amounts of water associated with them and can therefore provide strong repulsive hydration interactions. Specific binding depends on the radius and valency of the ion involved, because these parameters determine the degree of ion hydration. Ions that have small radii and high valencies tend to bind less strongly because they are surrounded by a relatively thick layer of tightly "bound" water molecules and some of these must be removed before the ion can be adsorbed (Israelachvili 1992). As a general rule, the adsorbability of ions from water can be described by a lyotropic series: $I^- > Br^- > Cl^- > F^-$ for monovalent ions, and $K^+ > Na^+ > Li^+$ for monovalent cations (in order of decreasing adsorbability). On the other hand, once an ion is bound to a surface, the strength of the repulsive hydration interaction between the emulsion droplets increases with degree of ion hydration because more energy is needed to dehydrate the ion as the two droplets approach each other. Therefore, the ions that adsorb the least strongly are those that provide the greatest hydration repulsion. Thus it is possible to control the interaction between droplets by altering the type and concentration of ions present in the aqueous phase.

Hydration interactions are often strong enough to prevent droplets from aggregating (Israelachvili 1992). Thus, oil-in-water emulsions that should contain enough electrolyte to cause droplet flocculation through electrostatic screening have been found to be stable because of specific binding of ions (Israelachvili 1992). This effect is dependent on the pH of the aqueous phase because the electrolyte ions have to compete with the H^+ or OH^- ions in the water (Miklavic and Ninham 1990). For example, at relatively high pH and electrolyte concentrations (>10 mM), it has been observed that Na^+ ions can adsorb to negatively charged surface groups and prevent droplets from aggregating through hydration repulsion, but when the pH of the solution is decreased, the droplets aggregate because the high concentration of H^+ ions displaces the Na^+ ions from the droplet surface (Israelachvili 1992). Nonionic emulsifiers are less sensitive to pH and ionic strength, and they do not usually bind highly hydrated ions. The magnitude of the hydration interaction decreases with increasing temperature because polar groups become progressively dehydrated as the temperature is raised (Israelachvili 1992). In summary, the importance of hydration interactions in a particular system depends on the nature of the hydrophilic groups on the droplet surfaces, as well as on the type and concentration of ions present in the aqueous phase.

3.9.   THERMAL FLUCTUATION INTERACTIONS

The interfacial region which separates the oil and aqueous phases of an emulsion is often highly dynamic (Israelachvili 1992). In particular, interfaces that are comprised of small-molecule surfactants tend to exhibit undulations because their bending energy is relatively small compared to the thermal energy of the system (Figure 3.21a). In addition, the surfactant molecules may be continually twisting and turning, as well as moving in and out of the interfacial region (Figure 3.21b). When two dynamic interfaces move close to each other, they experience a number of repulsive *thermal fluctuation* interactions which are entropic in

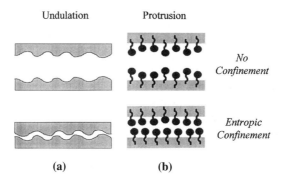

FIGURE 3.21   Interfaces that are comprised of small-molecule surfactants are susceptible to protrusion and undulation interactions.

origin (Figure 3.20) (Israelachvili 1992). In emulsions, the two most important of these are protrusion and undulation interactions.

3.9.1.   Protrusion Interactions

Protrusion interactions are short-range repulsive interactions which arise when two surfaces are brought so close together that the movement of the surfactant molecules in and out of the interface of one droplet is restricted by the presence of another droplet, which is entropically unfavorable. The magnitude of this repulsive interaction depends on the distance that the surfactant molecules are able to protrude from the interface, which is governed by their molecular structure. The interdroplet pair potential due to protrusion interactions is given by the following expression (Israelachvili 1992):

$$w_{\text{protrusion}}(h) \approx 3\pi\Gamma rkT\lambda_0 e^{-h/\lambda_0} \tag{3.25}$$

where Γ is the number of surfactant molecules (or head groups) per unit surface area and λ_0 is the characteristic decay length of the interaction (typically between 0.07 to 0.6 nm), which depends on the distance the surfactant can protrude from the surface.

3.9.2.   Undulation Interactions

Undulation interactions are short-range repulsive interactions which arise when the wave-like undulations of the interfacial region surrounding one emulsion droplet are restricted by the presence of another emulsion droplet, which is entropically unfavorable. The magnitude and range of this repulsive interaction increase as the amplitude of the oscillations increases. The interdroplet pair potential due to undulation interactions is given by the following expression (Israelachvili 1992):

$$w_{\text{undulation}}(h) \approx \frac{\pi r(kT)^2}{4k_b h} \tag{3.26}$$

where k_b is the bending modulus of the interfacial layer, which typically has values of between about 0.2 and 20 × 10⁻²⁰ J depending on the surfactant type. The magnitude of the bending modulus is related to the molecular geometry of the surfactant molecules (Chapter

4) and tends to be higher for surfactants that have two nonpolar chains than those that have only one.

Thermal fluctuation interactions are much more important for small-molecule surfactants which form flexible interfacial layers than for biopolymers which form rigid interfacial layers. Both types of interaction tend to increase with temperature because the interfaces become more mobile. Nevertheless, this effect may be counteracted by increasing dehydration of any polar groups with increasing temperature. The fact that these interactions are extremely short range means that they are unlikely to make any significant contribution to the flocculation of emulsion droplets. Nevertheless, they may be important in preventing droplet coalescence because of the relatively large repulsive interaction which arises at close droplet separations. The strength of this interaction is governed by the structure and dynamics of the interfacial layer and therefore varies considerably from system to system (Israelachvili 1992).

3.10. HYDRODYNAMIC INTERACTIONS AND NONEQUILIBRIUM EFFECTS

So far, it has been assumed that the interactions between droplets occur under equilibrium conditions. In practice, the droplets in an emulsion are in continual motion, which influences their interactions in a number of ways (Evans and Wennerstrom 1994). First, the system may not have time to reach equilibrium when two droplets rapidly approach each other because molecular rearrangements (such as adsorption–desorption of emulsifiers, ionization/deionization of charged groups, and conformational changes of biopolymers) take a finite time to occur (Israelachvili 1992, Israelachvili and Berman 1995). As a consequence, the colloidal interactions between droplets may be significantly different from those observed under equilibrium conditions. These nonequilibrium effects depend on the precise nature of the system and are therefore difficult to account for theoretically. Second, the movement of a droplet causes an alteration in the flow profile of the surrounding liquid, which can be "felt" by another droplet (Dukhin and Sjoblom 1996). As two droplets move closer together, the continuous phase must be squeezed out from the narrow gap separating them against the friction of the droplet surfaces. This effect manifests itself as a decrease in the effective diffusion coefficient of the emulsion droplets, $D(s) = D_0 G(s)$, where D_0 is the diffusion coefficient of a single droplet and $G(s)$ is a correction factor which depends on the dimensionless separation between the droplets: $s = (h + 2r)/r$ (Davis et al. 1989). Mathematical expressions for $G(s)$ have been derived from a consideration of the forces that act on particles as they approach each other in a viscous liquid (Davis et al. 1989, Zhang and Davis 1991, Dukhin and Sjoblom 1996). For rigid spherical particles, the hydrodynamic correction factor can be approximated by the following expression (Dickinson and Stainsby 1982):

$$G(s) = \frac{6\tilde{d}^2 + 4\tilde{d}}{6\tilde{d}^2 + 13\tilde{d}^2 + 2} \tag{3.27}$$

where $\tilde{d} = (s - 2)$. The value of $G(s)$ varies from 0 when the particles are in close contact ($h = 0$, $s = 2$) to unity when they are far apart and therefore have no influence on each other ($h,s \rightarrow \infty$). Thus, as particles approach each other, their speed gets progressively slower, and therefore they would not aggregate unless there was a sufficiently strong attractive colloidal interaction to overcome the repulsive hydrodynamic interaction. Equation 3.27 must be modified for emulsions to take into account the fact that there is less resistance to the movement of the continuous phase out of the gap between the droplets when their surfaces have some fluid-like characteristics (Davis et al. 1989, Zhang and Davis 1991). Thus the hydrodynamic resistance to the approach of fluid droplets is less than that for solid droplets. Hydrodynamic interactions are particularly important for determining the stability of droplets to flocculation and coalescence in emulsion systems (Chapter 7).

3.11. TOTAL INTERACTION POTENTIAL

The overall interdroplet pair potential is the sum of the various attractive and repulsive contributions:*

$$w_{total}(h) = w_{VDW}(h) + w_{electrostatic}(h) + w_{steric}(h) + w_{depletion}(h)$$
$$+ w_{hydrophobic}(h) + w_{hydration}(h) + w_{thermal}(h) \tag{3.28}$$

Not all of these interactions play an important role in every type of food emulsion, and it is often possible to identify two or three interactions which dominate the overall interaction. For this reason, it is informative to examine the characteristics of certain combinations of colloidal interaction which are particularly important in food emulsions. A summary of the characteristics of the various types of interaction is given in Table 3.3.

3.11.1. van der Waals and Electrostatic

The classical approach to describing the interactions between charge-stabilized emulsion droplets is the DLVO theory, which is named after the four scientists who first proposed it: Derjaguin, Landau, Verwey, and Overbeek (Hiemenz 1986, Hunter 1986, Derjaguin et al. 1987, Derjaguin 1989). This theory assumes that the overall interaction between a pair of emulsion droplets is the result of van der Waals attractive and electrostatic repulsive interactions:

$$w(h) = w_{VDW}(h) + w_{electrostatic}(h) \tag{3.29}$$

The dependence of the van der Waals and electrostatic interactions on separation for two electrically charged oil droplets dispersed in water is illustrated in Figure 3.22. The van der Waals interaction potential is negative (attractive) at all separations, whereas the electrostatic interaction potential is positive (repulsive). The overall interdroplet pair potential has a more complex dependence on separation because it is the sum of two opposing interactions, and it may therefore be attractive at some separations and repulsive at others. When the two droplets are separated by a large distance, there is no effective interaction between them. As they move closer together, the van der Waals attraction dominates initially and there is a shallow minimum in the profile, which is referred to as the *secondary minimum*, $w(h^o_{2min})$. When the depth of this minimum is large compared to the thermal energy, $w(h^o_{2min}) \gg kT$, the droplets tend to be flocculated, but if it is small compared to the thermal energy, the droplets tend to remain unaggregated. At closer separations, the repulsive electrostatic interaction dominates and there is an *energy barrier*, $w(h_{max})$, which must be overcome before the droplets can come any closer. If this energy barrier is sufficiently large compared to the thermal energy, $w(h_{max}) \gg kT$, it will prevent the droplets from falling into the deep

---

* In reality, it is not always appropriate to simply sum the contribution from all of the separate interactions because some of them are coupled (Ninham and Yaminsky 1997). Nevertheless, this approach gives a good first approximation.

TABLE 3.3

Summary of the Characteristics of the Various Types of Colloidal Interaction Between Emulsion Droplets

Interaction	Sign	Strength	Range	Affected by pH	Effect of ionic strength	Effect of temperature
van der Waals	A	S	LR	No	Reduced	Decreases
Electrostatic	R	W → S	SR → LR	Yes	Reduced	Increases
Steric						
Mixing	A or R	W → S	SR	SD	SD	SD
Elastic	R	S	SR	SD	SD	Increases
Depletion	A	W → S	SR	SD	SD	Increases
Hydrophobic	A	S	LR	No	Yes	Increases
Hydration	R	S	SR → LR	Indirectly	Indirectly	Decreases
Thermal fluctuation	R	S	SR	Indirectly	No	Increases

Note: A = attractive, R = repulsive, S = strong, W = weak, SR = short range (<10 nm), LR = long range (>10 nm), and SD = system dependent.

primary minimum that exists at closer separations. On the other hand, if it is not large compared to the thermal energy, then the droplets tend to fall into the primary minimum, $w(h^{\circ}_{1min})$, which would lead to droplet coalescence. An example of this type of system would be two electrically charged oil droplets suspended in pure water in the absence of any adsorbed emulsifier.

Electrostatically stabilized emulsions are particularly sensitive to the ionic strength and pH of the aqueous phase (Figure 3.23). At low electrolyte concentrations, there may be a sufficiently high energy barrier to prevent the droplets from coming close enough together to aggregate into the primary minimum (Figure 3.23). As the ion concentration is increased, the screening of the electrostatic interaction becomes more effective (Section 3.4.3), which reduces the height of the energy barrier. Above a certain electrolyte concentration, often

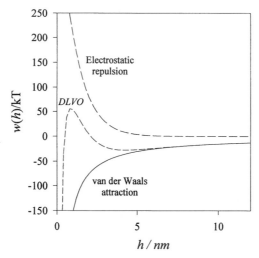

FIGURE 3.22 Typical interdroplet pair potential for an electrostatically stabilized emulsion, where the only interactions are electrostatic repulsion and van der Waals attractions (i.e., DLVO theory).

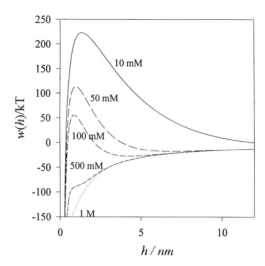

FIGURE 3.23   The stability of emulsion droplets to aggregation decreases as the ionic strength of the intervening medium increases because of electrostatic screening effects.

referred to as the *critical aggregation concentration* or *CAC*, the energy barrier is no longer high enough to prevent the droplets from falling into the deep primary minimum, and so the droplets tend to aggregate. This accounts for the susceptibility of many electrostatically stabilized food emulsions to droplet aggregation when salt is added to the aqueous phase (Hunt and Dalgleish 1994, 1995; Demetriades et al. 1997a). The electrical charge of many food emulsifiers is sensitive to the pH of the aqueous phase. For example, the droplet charge of protein-stabilized emulsions decreases as the pH tends toward the isoelectric point of the proteins, which reduces the magnitude of the electrostatic repulsion between the droplets. This accounts for the tendency of protein-stabilized emulsions to become aggregated when their pH is adjusted to the protein isoelectric point (Demetriades et al. 1997a).

3.11.2.   van der Waals, Electrostatic, and Steric

One of the major limitations of the DLVO theory is that it does not take into account the short-range repulsive interactions which arise between droplets when they come close together (i.e., polymeric steric, hydration, and thermal fluctuation) (Israelachvili 1992, Evans and Wennerstrom 1994). Consequently, it predicts that two liquid droplets would coalesce at close separations because of the large attraction between them. In practice, many of the emulsifiers used in food emulsions are capable of stabilizing the droplets through a combination of electrostatic and short-range repulsive interactions. The DLVO theory can therefore be extended by including the effects of these short-range repulsive interactions. For convenience, we will assume that only the polymeric steric repulsion needs to be included:

$$w(h) = w_{\text{VDW}}(h) + w_{\text{electrostatic}}(h) + w_{\text{steric}}(h) \tag{3.30}$$

The dependence of the interdroplet pair potential on droplet separation for the same system as that shown in Figure 3.22 but with an additional polymeric steric contribution is shown in Figure 3.24. The major difference between this system and the one with no steric interactions is that the droplets cannot get close enough together to coalesce. If the droplets do have sufficient energy to overcome the energy barrier, they fall into a deep primary minimum but cannot get any closer because of the extremely strong short-range repulsive interactions. In

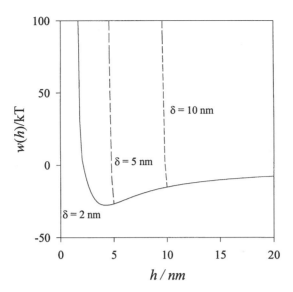

FIGURE 3.24 The presence of an adsorbed layer of emulsifier molecules modifies the interaction between electrostatically stabilized droplets ($r = 1$ μm, $I = 0.1$ M). As the thickness of the interfacial layer increases, the depth of the energy minimum decreases and it moves farther away from the droplet surfaces.

this case, the droplets would tend to form strong flocs, rather than become coalesced. It should be noted that the adsorbed emulsifier layer would also influence the magnitude of the van der Waals interaction by an amount that depends on its thickness and composition (Section 3.3.6). Inclusion of the polymeric steric interactions in the DLVO theory makes it much more realistic because most food emulsifiers are polymeric.

A practical example of a real system which approximates this type of behavior would be a protein-stabilized oil-in-water emulsion (Demetriades et al. 1997a). The droplets are stable to aggregation when they are highly charged (i.e., when the pH is significantly lower or higher than the isoelectric point) and the Debye length is sufficiently long (i.e., low ionic strengths). When the pH is adjusted close to the isoelectric point of the proteins or the ionic strength is increased, the droplets flocculate because the van der Waals attraction dominates the electrostatic repulsion, but they do not coalesce because of the strong steric repulsion which operates at short separations.

3.11.3. van der Waals and Steric

Many food emulsifiers have no electrical charge, but they are still capable of stabilizing emulsions against aggregation because of the various short-range repulsive interactions mentioned earlier. Emulsifiers of this type include nonionic surfactants (such as Tweens and Spans) and some biopolymers (such as modified starch). For convenience, we will consider a system in which only van der Waals attraction and polymeric steric interactions occur:

$$w(h) = w_{VDW}(h) + w_{steric}(h) \tag{3.31}$$

The dependence of the overall interdroplet pair potential on droplet separation for emulsions with interfacial layers of different thickness is shown in Figure 3.25. It is assumed that

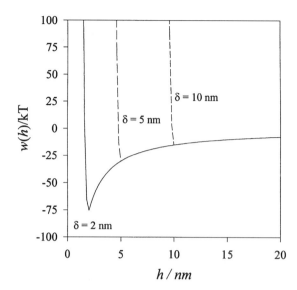

FIGURE 3.25 An adsorbed layer of emulsifier molecules at the surface of the droplets causes a short-range repulsive interaction which prevents droplets from coalescing. As the thickness of the interfacial layer increases, the depth of the energy minimum decreases and it moves farther away from the droplet surfaces.

the continuous phase is an indifferent quality solvent for the polymer, so that the mixing contribution to the polymeric steric interaction is zero (Section 3.5.2.1). At wide separations, the overall interaction between the droplets is negligible. As the droplets move closer together, the attractive van der Waals interaction begins to dominate and so there is a net attraction between the droplets. However, once the droplets get so close together that their interfacial layers overlap, then the repulsive polymeric steric interaction dominates and there is a net repulsion between the droplets. At a particular separation, there is a minima in the interdroplet pair potential, and this is the location where the droplets tend to reside. If the depth of this minima is large compared to the thermal energy of the system, the droplets remain aggregated; otherwise, they move apart. The depth and position of the minimum depend on the thickness and properties of the interfacial layer surrounding the droplets. As the thickness of the adsorbed layer increases, the repulsive interaction between droplets becomes more significant at larger separations, and consequently the depth of the minima decreases. If the interfacial layer is sufficiently thick, it may prevent the droplets from aggregating altogether, because the depth of the minima is relatively small compared to the thermal energy. This accounts for the effectiveness of uncharged polysaccharides (which form thick interfacial layers) in preventing droplet flocculation and coalescence, whereas uncharged protein molecules (which form thin interfacial layers) can prevent coalescence but not flocculation. As mentioned in Section 3.3.6, the adsorbed layer may have an additional influence on the overall interaction potential due to its modification of the van der Waals interactions.

3.11.4. van der Waals, Electrostatic, and Hydrophobic

The droplet surfaces in many food emulsions acquire some hydrophobic character during their manufacture, storage, or consumption. A typical example is a whey-protein–stabilized emulsion which is subjected to heat (Monahan et al. 1996, Demetriades et al. 1997b). Heating

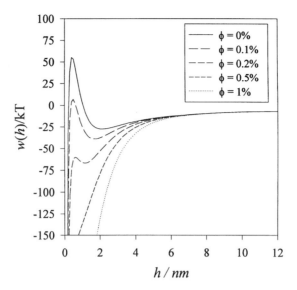

FIGURE 3.26 Influence of hydrophobic interactions on the interdroplet pair potential of electrostatically stabilized emulsions ($r = 1$ μm, $I = 0.1$ M).

the emulsion above 65°C causes the protein molecules adsorbed to the oil–water interface to partially unfold and thus expose some of the nonpolar amino acids to the aqueous phase. The overall interdroplet pair potential for this type of system is given by:

$$w(h) = w_{\text{VDW}}(h) + w_{\text{electrostatic}}(h) + w_{\text{hydrophobic}}(h) \qquad (3.32)$$

The dependence of the overall interdroplet pair potential on droplet separation for the same system as shown in Figure 3.22 but for droplets which have different degrees of surface hydrophobicity is shown in Figure 3.26. As the hydrophobicity of the droplet surface increases, the hydrophobic attraction increases, which causes a decrease in the height of the energy barrier. When the surface hydrophobicity is sufficiently large, the energy barrier becomes so small that the droplets aggregate into the primary minimum. This accounts for the experimental observation that whey-protein–stabilized emulsions become more susceptible to aggregation when they are heated above a temperature where the protein molecules unfold (Demetriades et al. 1997b).

3.11.5. van der Waals, Electrostatic, and Depletion

Depletion interactions are important when the continuous phase of an emulsion contains a significant concentration of small colloidal particles, such as surfactant micelles or nonadsorbing biopolymers (Dickinson and McClements 1995, Jenkins and Snowden 1996). The interdroplet pair potential for a system in which depletion interactions are important is given by:

$$w(h) = w_{\text{VDW}}(h) + w_{\text{electrostatic}}(h) + w_{\text{depletion}}(h) \qquad (3.33)$$

The variation of the interdroplet pair potential with droplet separation for this type of system is shown in Figure 3.27. At low concentrations of colloidal particles, the energy barrier is sufficiently large to prevent the droplets from falling into the primary minimum. As

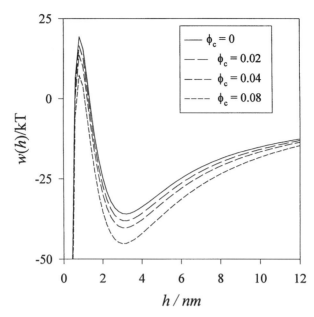

FIGURE 3.27   Influence of depletion interactions on the interdroplet pair potential of an electrostatically stabilized emulsion. As the volume fraction of colloidal particles increases, the depth of the secondary minima increases and the height of the energy barrier decreases.

the concentration of colloidal particles is increased, the attraction between the droplets increases, and eventually the energy barrier is reduced to such a low level that the droplets can fall into the primary minima and become coalesced or strongly flocculated. In addition, the depth of the secondary minimum increases, which will promote weak flocculation. A number of workers have shown that depletion interactions promote droplet flocculation in emulsions when the concentration of surfactant or biopolymer exceeds some critical concentration (Sperry 1982, Aronson 1992, McClements 1994, Dickinson et al. 1995, Jenkins and Snowden 1996).

3.12. PREDICTING COLLOIDAL INTERACTIONS IN FOOD EMULSIONS

In this chapter, we have examined the origin, magnitude, and range of the most important types of attractive and repulsive interactions which can arise between emulsion droplets. In principle, it is possible to predict the likelihood that the droplets in an emulsion will be in an aggregated or an unaggregated state using the theories given above. In practice, it is extremely difficult to make quantitative predictions about the properties of food emulsions for a number of reasons. First, food emulsions contain a huge number of different emulsion droplets (rather than just two) which interact with each other and with other components within the system, and it is difficult to quantify the overall nature of these interactions (Dickinson 1992). Second, there is often a lack of information about the relevant physical parameters needed to carry out the calculations (Hunter 1986). Third, certain simplifying assumptions often have to be made in the theories in order to derive tractable expressions for the interaction energies, and these are not always justified (Ninham and Yaminsky 1997). Fourth, food systems are not usually at thermodynamic equilibrium and so the above equations do not strictly apply (Israelachvili and Berman 1995). Fifth, covalent interactions are important in some systems, and these are not taken into account in the above analysis (Mohanan et al. 1996). Finally, food emulsions may be subjected to external forces which affect the interactions between the droplets (e.g., gravity, centrifugation, or mechanical agitation). Despite the limitations described above, an understanding of the various types of interactions which act between droplets gives food scientists a powerful tool for interpreting data and for systematically accounting for the effects of ingredient formulations and processing conditions on the properties of many food products. It is often possible to predict the major factors which determine the stability of emulsions (*albeit* in a fairly qualitative fashion). Rather than trying to theoretically predict the overall interaction between emulsion droplets, it is often easier to directly measure the forces which act between them. Over the past 30 years or so, a number of experimental techniques have been developed which allow scientists to accurately measure the forces between macroscopic surfaces down to separations of a fraction of a nanometer (Israelachvili 1992, Luckham and Costello 1993, Claesson et al. 1996). The application of these techniques is greatly advancing our understanding of this important area and leading to exciting new insights into the relative contributions of the various types of interaction mentioned above.

4 Emulsion Ingredients

4.1. INTRODUCTION

Food emulsions are compositionally complex materials which contain a wide variety of different chemical constituents (Friberg and Larsson 1997, Dickinson and Stainsby 1982, Dickinson 1992) (Table 4.1). The composition of an emulsion can be defined in a number of different ways: concentrations of specific atoms (e.g., H, C, O, N, Na, Mg, Cl), concentrations of specific molecules (e.g., water, sucrose, amylose, β-lactoglobulin), concentrations of groups of molecules (e.g., proteins, lipids, carbohydrates, minerals), and concentrations of specific ingredients (e.g., flour, milk, salt, egg). Food manufacturers are usually concerned with the concentration of specific ingredients, because food components are usually purchased and utilized in this form. On the other hand, research scientists may be more interested in the concentrations of specific atoms, molecules, or groups of molecules, depending on the purpose of their investigations.

The constituents in a food emulsion interact with each other, either physically or chemically, to determine the overall physicochemical and organoleptic properties of the final product. The efficient production of high-quality food emulsions therefore depends on a knowledge of the contribution that each individual constituent makes to the overall properties and how this contribution is influenced by the presence of the other constituents. Selection of the most appropriate constituents is one of the most important decisions that a food manufacturer must make during the formulation and production of a food product. Ingredients must exhibit the desired functional properties while being of a reliably high standard, low cost, and ready availability.

The formulation of food products has traditionally been more of a craft than a science. Many of the foods which are familiar to us today are the result of a long and complex history of development. Consequently, there has often been a rather poor understanding of the role (or multiple roles) that each chemical constituent plays in determining the overall quality. This century has seen the development of large-scale industrial manufacturing operations where foods are mass produced. Mass production has led to the availability of a wide variety of low-cost foods, which are quick and easy to prepare and are therefore appealing to the modern consumer. Nevertheless, increasing reliance on mass production has meant that food manufacturers have had to develop a more thorough understanding of the behavior of food ingredients before, during, and after processing (Hollingsworth 1995). This knowledge is required for a number of reasons:

1. The properties of the ingredients entering a food factory often vary from batch to batch. Food manufacturers utilize their knowledge of the behavior of food ingredients under different conditions to adjust the food-processing operations so that the final product has consistent properties.
2. There is a growing trend toward improving the quality, variety, and convenience of processed foods (Sloan 1994, 1996). Knowledge of ingredient properties en-

TABLE 4.1
Ingredients Typically Found in Food Emulsions

Macrocomponents	Microcomponents
Water	Emulsifiers
Lipids	Minerals
Proteins	Gums
Carbohydrates	Flavors
	Colors
	Preservatives
	Vitamins

ables food scientists to develop these foods in a more systematic and informed manner.

3. There is an increasing tendency toward removing or reducing the amounts of food constituents which have been associated with human health concerns, such as fat, salt, and cholesterol (Sloan 1994, 1996). The removal of certain ingredients causes significant changes in food properties which are perceived as being undesirable to consumers (e.g., many no-fat or low-fat products do not exhibit the desirable taste or textural characteristics of the full-fat products they are designed to replace) (O'Donnell 1995). Consequently, it is important to understand the role that each ingredient plays in determining the overall physicochemical and organoleptic properties of foods, so that this role can be mimicked by a healthier alternative ingredient.

4.2. FATS AND OILS

Fats and oils are part of a group of compounds known as *lipids* (Gunstone and Norris 1983, Weiss 1983, Nawar 1996). By definition, a lipid is a compound which is soluble in organic solvents but insoluble or only sparingly soluble in water (Nawar 1996). This group of compounds contains a large number of different types of molecules, including acylglycerols, fatty acids, and phospholipids (Nawar 1996). Triacylglycerols are by far the most common lipid in foods, and it is this type of molecule that is usually referred to as a fat or oil. Edible fats and oils come from a variety of different sources including plants, seeds, nuts, animals, and fish (Sonntag 1979c, Weiss 1983, Nawar 1996). By convention a *fat* is solid-like at room temperature, whereas an *oil* is liquid, although these terms are often used interchangeably (Walstra 1987). Because of their high natural abundance and their great importance in foods, we will be concerned almost exclusively with the properties of triacylglycerols in this section.

Fats and oils influence the nutritional, organoleptic, and physicochemical properties of food emulsions in a variety of ways. They are a major source of energy and essential nutrients; however, overconsumption of certain types of these constituents leads to obesity and human health concerns (Chow 1992, Smolin and Grosvenor 1994). Consequently, there has been a trend in the food industry to reduce the fat content of many traditional foods (O'Donnell 1995). The challenge for the food scientist is to create a product which has the same desirable quality attributes as the original, but with a reduced fat content, which is often extremely difficult. The perceived flavor of a food emulsion is determined by the type and concentration of lipids present (McNulty 1987, Overbosch et al. 1991, Landy et al. 1996). Lipids undergo a variety of chemical changes during the processing, storage, and handling of foods that generate products which can be either desirable or deleterious to their flavor profile (Nawar 1996). Controlling these reactions requires a knowledge of both lipid chemistry and emulsion science (Frankel 1991, Frankel et al. 1994, Coupland and McClements 1996). The taste and aroma of food emulsions are also governed by the partitioning of flavor compounds between the oil, water, interfacial, and gaseous phases (Chapter 9). The oil phase may also act as a solvent for various other important food components, including oil-soluble vitamins, antioxidants, preservatives, and essential oils. Reducing the fat content of an emulsion can therefore have a profound influence on its flavor profile, stability, and nutritional content.

Food emulsions usually appear either cloudy or opaque because the light passing through them is scattered by the droplets (Hernandez et al. 1991). Emulsion appearance is therefore the direct result of the immiscibility of oil and water. The characteristic texture of many food emulsions is due to the ability of the oil phase to crystallize (Mulder and Walstra 1974, Walstra 1987, Moran 1994). The "spreadability" of margarines and butters is determined by the formation of a three-dimensional network of aggregated fat crystals in the continuous phase which provides the product with mechanical rigidity (Moran 1994). The tendency of a cream to thicken or "clot" when it is cooled below a certain temperature is due to the formation of fat crystals in the oil droplets, which causes them to aggregate (Boode 1992). The melting of fat crystals in the mouth causes a cooling sensation, which is an important sensory attribute of many fatty foods (Walstra 1987). The ability of food scientists to improve the quality of food emulsions therefore depends on an improved understanding of the multiple roles that fats and oils play in determining their properties.

4.2.1. Molecular Structure and Organization

Chemically, triacylglycerols are esters of a glycerol molecule and three fatty acid molecules (Figure 4.1). Each of the fatty acids may contain different numbers of carbon atoms and may have different degrees of unsaturation and branching (Lawson 1995, Nawar 1996). Nevertheless, most naturally occurring fatty acids have an even number of carbon atoms (usually less than 24) and are straight chained. The fact that there are many different types of fatty acid molecules, and that these fatty acids can be located at different positions on the glycerol molecule, means that there are a huge number of possible triacylglycerol molecules present in foods. Indeed, edible fats and oils always contain a great many different types of triacylglycerol molecules, with the precise type and concentration depending on their origin (Weiss 1983).

Triacylglycerol molecules have a "tuning-fork" structure, with the two fatty acids at the ends of the glycerol molecule pointing in one direction and the fatty acid in the middle pointing in the opposite direction (Figure 4.1). Triacylglycerols are predominantly nonpolar molecules, and so the most important types of molecular interactions with their neighbors are

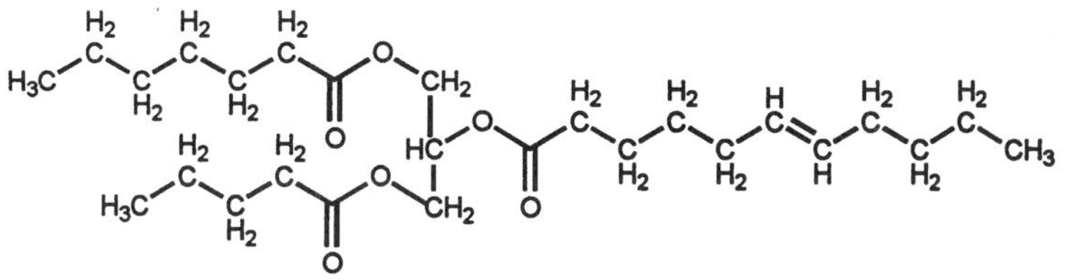

FIGURE 4.1 Chemical structure of a triacylglycerol molecule, which is assembled from three fatty acids and a glycerol molecule.

van der Waals attraction and steric overlap repulsion (Chapter 2). At a certain molecular separation, there is a minimum in the intermolecular pair potential whose depth is a measure of the strength of the attractive interactions which bind the molecules together in the solid and liquid states (Section 2.4). Whether a triacylglycerol exists as a gas, liquid, or solid at a particular temperature depends on a balance between these attractive interactions and the disorganizing influence of the thermal energy (Section 4.2.3).

4.2.2. Bulk Physicochemical Properties

The bulk physicochemical properties of edible fats and oils depend on the molecular structure and interactions of the triacylglycerol molecules they contain (Formo 1979; Gunstone and Norris 1983; Birker and Padley 1987; Timms 1991, 1995). The strength of the attractive interactions between molecules and the effectiveness of their packing in a condensed phase determine their melting point, boiling point, density, and compressibility (Israelachvili 1992). Triacylglycerols which contain branched or unsaturated fatty acids are not able to pack as closely together as those which contain linear saturated fatty acids, and so they have lower melting points, lower densities, and higher compressibilities than saturated triacylglycerols (Walstra 1987). The melting points of triacylglycerols increases with increasing chain length, is higher for saturated than for unsaturated fatty acids, is higher for straight-chain than branched fatty acids, and is higher for triacylglycerols with a more symmetrical distribution of fatty acids on the glycerol molecule (Table 4.2). The relatively low dielectric constant of triacylglycerol molecules compared to water is due to their small polarity (Table 4.3).

4.2.3. Fat Crystallization

One of the most important characteristics of fats and oils is their ability to undergo solid–liquid phase transitions at temperatures which occur during the processing, storage, and handling of food emulsions (Birker and Padley 1987; Walstra 1987; Timms 1991, 1995). The texture, mouthfeel, stability, and appearance of many food emulsions depend on the physical state of the lipid phase (Moran 1994). The conversion of milk into butter relies on the controlled destabilization of an oil-in-water emulsion (milk) into a water-in-oil emulsion (butter), which is initiated by the formation of crystals in the milk fat globules (Mulder and Walstra 1974, Boode 1992). The spreadability of the butter produced by this process is governed by the final concentration of fat crystals (Moran and Rajah 1994). If the percentage

TABLE 4.2
Melting Points of the Most Stable Polymorph
of Selected Triacylglycerol Molecules

Triglyceride	Melting point (°C)
LLL	46
MMM	54
PPP	64
SSS	73
OOO	5
SOS	42
SSO	38
SOO	24

TABLE 4.3

**Comparison of the Bulk Physicochemical Properties
of an Oil (Triolein) and Water at 20°C**

	Oil	Water
Molecular weight	885	18
Melting point (°C)	5	0
Density (kg m^{-3})	910	998
Compressibility (ms^2 kg^{-1})	5.03×10^{-10}	4.55×10^{-10}
Viscosity (mPa s)	≈ 50	1.002
Thermal conductivity (W m^{-1} K^{-1})	0.170	0.598
Specific heat capacity (J kg^{-1} K^{-1})	1980	4182
Thermal expansion coefficient (°C^{-1})	7.1×10^{-4}	2.1×10^{-4}
Dielectric constant	3	80.2
Surface tension (mN m^{-1})	≈ 35	72.8
Refractive index	1.46	1.333

of fat crystals is too high, the product is firm and difficult to spread, and if it is too low, the product is soft and tends to collapse under its own weight. The creation of food emulsions with desirable properties therefore depends on an understanding of the major factors which influence the crystallization and melting of lipids in foods (Birker and Padley 1987).

The arrangement of triacylglycerol molecules in the solid and liquid state is shown schematically in Figure 4.2. The physical state of a triacylglycerol at a particular temperature depends on its free energy, which is made up of contributions from enthalpic and entropic terms: $\Delta G_{S \to L} = \Delta H_{S \to L} - T\Delta S_{S \to L}$ (Atkins 1994). The enthalpy term ($\Delta H_{S \to L}$) represents the change in the overall strength of the molecular interactions between the triacylglycerols when they are converted from a solid to a liquid, whereas the entropy term ($\Delta S_{S \to L}$) represents the change in the organization of the molecules that is brought about by the melting process. The strength of the bonds between the molecules is greater in the solid than in the liquid state because the molecules are able to pack more efficiently, and so $\Delta H_{S \to L}$ is positive, which favors the solid state. On the other hand, the entropy of the molecules in the liquid state is greater than that in the solid state, and therefore $\Delta S_{S \to L}$ is positive, which favors the liquid state. At low temperatures, the enthalpy term dominates the entropy term ($\Delta H_{S \to L} > T\Delta S_{S \to L}$), and therefore the solid state has the lowest free energy (Atkins 1994). As the temperature increases, the entropic contribution becomes increasingly important. Above a certain temperature, known as the *melting point*, the entropy term dominates the enthalpy term ($T\Delta S_{S \to L} > \Delta H_{S \to L}$), and so the liquid state has the lowest free energy. A material therefore changes from a solid to a liquid when its temperature is raised above the melting point. This process

Solid fat   Liquid oil

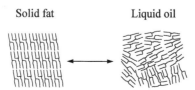

FIGURE 4.2   The arrangement of triacylglycerols in the solid and liquid states depends on a balance between the organizing influence of the attractive interactions between the molecules and the disorganizing influence of the thermal energy.

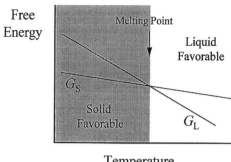

FIGURE 4.3 Temperature dependence of the free energies of the solid and liquid states. At low temperatures, the solid state is thermodynamically favorable, but above the melting point, the liquid state is more favorable.

leads to the release of heat because of the reduction in the amount of energy stored in the intermolecular interactions when the material changes from a solid to a liquid (i.e., it is an exothermic process).

The temperature dependence of the free energies of the solid and liquid states clearly shows that below the melting point, the solid state has the lowest free energy, but above it, the liquid state has the lowest (Figure 4.3). Thermodynamics informs us whether or not a phase transition can occur, but it tells us nothing about the rate at which this process occurs or about the physical mechanism by which it is accomplished (Atkins 1994). As will be seen below, an understanding of lipid phase transitions requires a knowledge of both the thermodynamics and kinetics of the process. The crystallization of fats can be conveniently divided into three stages: supercooling, nucleation, and crystal formation (Boistelle 1988, Mullin 1993).

4.2.3.1. Supercooling

Crystallization can only take place once a liquid phase is cooled below its melting point (Garside 1987, Walstra 1987). Even so, a material may persist as a liquid below its melting point for a considerable time before any crystallization is observed (Skoda and van den Tempel 1963, Phipps 1964, Mulder and Walstra 1974). This is because of an activation energy which must be overcome before the liquid–solid phase transition can occur (Figure 4.4). If the magnitude of this activation energy is sufficiently high compared to the thermal energy of the system, crystallization will not occur, even though the transition is thermody-

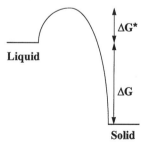

FIGURE 4.4 When there is a sufficiently high activation energy between the solid and liquid states, a liquid oil can persist in a metastable state below the melting point of a fat.

namically favorable (Turnbull and Cormia 1961). The system is then said to exist in a *metastable* state. The height of the activation energy depends on the ability of crystal nuclei which are stable enough to grow into crystals to be formed in the liquid oil (see Section 4.2.3.2.). The degree of supercooling of a liquid is defined as $\Delta T = T - T_{mp}$, where T is the temperature and T_{mp} is the melting point. The value of ΔT at which crystallization is observed depends on the chemical structure of the oil, the presence of any contaminating materials, the cooling rate, and the application of external forces (Dickinson and McClements 1995). Pure oils containing no impurities can often be supercooled by more than 10°C before any crystallization is observed (Turnbull and Cormia 1961; Dickinson et al. 1990, 1991c; McClements et al. 1993e).

4.2.3.2. Nucleation

Crystal growth can only occur after stable nuclei have been formed in a liquid. These nuclei are clusters of oil molecules that form small-ordered crystallites and are formed when a number of oil molecules collide and become associated with each other (Hernqvist 1984). There is a free energy change associated with the formation of one of these nuclei (Garside 1987). Below the melting point, the crystalline state is thermodynamically favorable, and so there is a decrease in free energy when some of the oil molecules in the liquid cluster together to form a nucleus. This negative free energy (ΔG_V) change is proportional to the volume of the nucleus formed. On the other hand, the formation of a nucleus leads to the creation of a new interface between the solid and liquid phases which requires an input of energy to overcome the interfacial tension (Chapter 5). This positive free energy (ΔG_S) change is proportional to the surface area of the nucleus formed. The total free energy change associated with the formation of a nucleus is therefore a combination of a volume and a surface term (Garside 1987):

$$\Delta G = \Delta G_V + \Delta G_S = \frac{4}{3}\pi r^3 \frac{\Delta H_{fus}\Delta T}{T_{mp}} + 4\pi r^2 \gamma_i \tag{4.1}$$

where r is the radius of the nuclei, ΔH_{fus} is the enthalpy change per unit volume associated with the liquid–solid transition (which is negative), and γ_i is the solid–liquid interfacial tension. The volume contribution becomes increasingly negative as the size of the nuclei increases, whereas the surface contribution becomes increasingly positive (Figure 4.5). The surface contribution dominates for small nuclei, while the volume term dominates for large

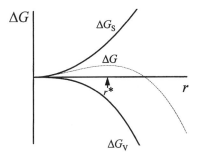

FIGURE 4.5 The critical size of a nucleus required for crystal growth depends on a balance between the volume and surface contributions to the free energy of nuclei formation.

nuclei. At a certain critical nucleus radius (r^*), the overall free energy has a maximum value given by:

$$\frac{d\Delta G}{dr} = 4\pi r^2 \frac{\Delta H_{\text{fus}}\Delta T}{T_{\text{mp}}} + 8\pi r\gamma_i = 0 \tag{4.2}$$

This equation can be rearranged to give an expression for the critical radius of the nucleus which must be achieved for crystallization to occur:

$$r^* = \frac{2\gamma_i T_{\text{mp}}}{\Delta H_{\text{fus}}\Delta T} \tag{4.3}$$

If a nucleus that has a radius below this critical size is formed, it will tend to dissociate so as to reduce the free energy of the system. On the other hand, if a nucleus that has a radius above this critical value is formed, it will tend to grow into a crystal. Equation 4.3 indicates that the critical size of nuclei required for crystal growth decreases as the degree of supercooling increases, which accounts for the increase in nucleation rate with decreasing temperature.

The rate at which nucleation occurs can be related to the activation energy (ΔG^*), which must be overcome before a stable nucleus is formed (Boistelle 1988):

$$J = A \exp(-\Delta G^*/kT) \tag{4.4}$$

where J is the nucleation rate, which is equal to the number of stable nuclei formed per second per unit volume of material; A is a preexponential factor; k is Boltzmann's constant; and T is the absolute temperature. The value of ΔG^* is calculated by replacing r in Equation 4.1 with the critical radius given in Equation 4.3. The variation of the nucleation rate predicted by Equation 4.4 with the degree of supercooling (ΔT) is shown in Figure 4.6. The formation of stable nuclei is negligibly slow at temperatures just below the melting point, but increases dramatically when the liquid is cooled below a certain temperature (T^*). In reality, the nucleation rate increases with cooling up to a certain temperature, but then decreases on further cooling. This is because the increase in viscosity of the oil which occurs as the

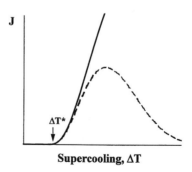

Supercooling, ΔT

FIGURE 4.6 Theoretically, the rate of the formation of stable nuclei increases with supercooling (solid line), but in practice, the nucleation rate decreases below a particular temperature because the diffusion of oil molecules is retarded by the increase in oil viscosity (broken line).

temperature is decreased slows down the diffusion of oil molecules toward the liquid–nucleus interface (Boistelle 1988). Consequently, there is a maximum in the nucleation rate at a particular temperature (Figure 4.6).

The type of nucleation described above occurs when there are no impurities present in the oil and is usually referred to as *homogeneous nucleation* (Boistelle 1988). If impurities such as dust particles are present in the oil, the internal surface of the vessel containing the oil, the interface of an emulsion droplet, the surface of an air bubble, or monoglyceride reverse micelles, then nucleation can be induced at a higher temperature than expected for a pure system (Walstra 1987, McClements et al. 1993e, Dickinson and McClements 1995). Nucleation in the presence of impurities is referred to as *heterogeneous nucleation* and can be divided into two types: primary and secondary (Boistelle 1988). Primary heterogeneous nucleation occurs when the impurities have a different chemical structure to that of the oil, whereas secondary heterogeneous nucleation occurs when crystals of the same chemical structure are present in the oil. Heterogeneous nucleation occurs when the impurities provide a surface where the formation of stable nuclei is more energetically favorable than in the pure oil (Boistelle 1988). As a result, the degree of supercooling required to initiate droplet crystallization is reduced. On the other hand, certain types of impurities are capable of decreasing the nucleation rate of oils because they are incorporated into the surface of the growing nuclei and therefore prevent any further oil molecules from being incorporated. Whether an impurity acts as a catalyst or an inhibitor of crystal growth depends on its molecular structure and interactions with the nuclei (Boistelle 1988).

4.2.3.3. *Crystal Growth*

Once stable nuclei have formed, they grow into crystals by incorporating molecules from the liquid oil at the solid–liquid interface (Garside 1987, Boistelle 1988). The crystal growth rate depends on the diffusion of a molecule from the liquid to the solid–liquid interface, as well as its incorporation into the crystal lattice. Either of these processes may be rate limiting, depending on the temperature and the molecular properties of the system. Molecules must encounter the interface at a location where crystal growth is favorable and also have the appropriate shape, size, and orientation to be incorporated (Timms 1991, 1995). At high temperatures, the incorporation of a molecule at the crystal surface is usually rate limiting, whereas at low temperatures, the diffusion step is rate limiting. This is because the viscosity of the liquid oil increases as the temperature is lowered, and so the diffusion of a molecule is retarded. The crystal growth rate therefore increases initially with supercooling, has a maximum rate at a certain temperature, and then decreases on further supercooling (i.e., it shows a similar trend to the nucleation rate) (Figure 4.6). Nevertheless, the maximum rate of nuclei formation may occur at a different temperature than the maximum rate of crystal growth. Experimentally, it has been observed that the rate of crystal growth is proportional to the degree of supercooling and inversely proportional to the viscosity of the melt (Timms 1991).

The crystal growth rate of fats can often be described by the following expression (Ozilgen et al. 1993):

$$\frac{1}{C(t)} = \frac{1}{C_0} + k_G t \tag{4.5}$$

where C_0 is the initial fraction of unsolidified oil, $C(t)$ is the fraction at time t, and k_G is the crystal growth rate. Thus, a plot of $1/C$ versus time should give a straight line with a slope equal to k_G. One of the major problems in determining k_G is that the nucleation rate and the crystal growth rate often have similar magnitudes, and thus it is difficult to unambiguously establish the contribution of each process (Ozilgen et al. 1993).

4.2.3.4. Crystal Morphology

The morphology of the crystals formed depends on a number of internal (e.g., packing of molecules and intermolecular forces) and external (e.g., temperature, cooling rate, mechanical agitation, and impurities) factors. When an oil is cooled rapidly to a temperature well below its melting point, a large number of small crystals are formed, but when it is cooled slowly to a temperature just below its melting point, a smaller number of larger crystals are formed (Moran 1994, Timms 1995). This is because the nucleation rate increases more rapidly with decreasing temperature than the crystallization rate (Timms 1991). Thus, rapid cooling produces many nuclei simultaneously which subsequently grow into small crystals, whereas slow cooling produces a smaller number of nuclei which have time to grow into larger crystals before further nuclei are formed. Crystal size has important implications for the rheology and organoleptic properties of foods. When crystals are too large, they are perceived as being "grainy" or "sandy" in the mouth (Walstra 1987). The efficiency of molecular packing in crystals also depends on the cooling rate. If a fat is cooled slowly, or the degree of supercooling is small, then the molecules have sufficient time to be efficiently incorporated into a crystal (Walstra 1987). At faster cooling rates, or high degrees of supercooling, the molecules do not have sufficient time to pack efficiently before another molecule is incorporated. Thus rapid cooling tends to produce crystals which contain more dislocations and in which the molecules are less densely packed (Timms 1991).

4.2.3.5. Polymorphism

Triacylglycerols exhibit a phenomenon known as *polymorphism*, which is the ability of a material to exist in a number of crystalline structures with different molecular packing (Hauser 1975, Garti and Sato 1988, Sato 1988, Hernqvist 1990). The three most commonly occurring types of packing in triacylglycerols are hexagonal, orthorhombic, and triclinic, which are usually designated as α, β', and β polymorphic forms (Nawar 1996). The thermodynamic stability of the three forms decreases in the order $\beta > \beta' > \alpha$. Even though the β form is the most thermodynamically stable, triacylglycerols often crystallize in one of the metastable states because they have a lower activation energy of nuclei formation (Figure 4.7). With time, the crystals transform to the most stable state at a rate which depends on environmental conditions such as temperature, pressure, and the presence of impurities (Timms 1991).

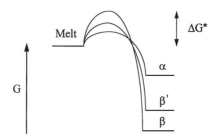

FIGURE 4.7   The polymorphic state that is initially formed when an oil crystallizes depends on the relative magnitude of the activation energies associated with nuclei formation.

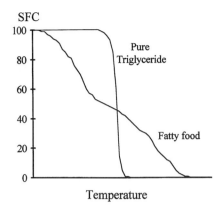

FIGURE 4.8 Comparison of the melting profile of a pure triacylglycerol and a typical edible fat. The edible fat melts over a much wider range of temperatures because it consists of a mixture of many different pure triacylglycerol molecules, each with different melting points.

4.2.3.6. Crystallization of Edible Fats and Oils

The melting point of a triacylglycerol depends on the chain length, branching, and degree of unsaturation of its constituent fatty acids, as well as their relative positions along the glycerol molecule (Table 4.3). Edible fats and oils contain a complex mixture of many different types of triacylglycerol molecules, each with a different melting point, and so they melt over a wide range of temperatures rather than at a distinct temperature, as would be the case for a pure triacylglycerol (Figure 4.8).

The melting profile of a fat is not simply the weighted sum of the melting profiles of its constituent triacylglycerols, because high-melting-point triacylglycerols are soluble in lower melting point ones (Timms 1995). For example, in a 50:50 mixture of tristearin and triolein, it is possible to dissolve 10% of solid tristearin in liquid triolein at 60°C (Walstra 1987, Timms 1995). The solubility of a solid component in a liquid component can be predicted, assuming they have widely differing melting points (>20°C):

$$\ln x = \frac{\Delta H_{\text{fus}}}{R} \left(\frac{1}{T_{\text{mp}}} - \frac{1}{T} \right) \tag{4.6}$$

Here, x is the solubility, expressed as a mole fraction, of the higher melting point component in the lower melting point component, and ΔH_{fus} is the molar heat of fusion (Walstra 1987). The structure and physical properties of crystals produced by cooling a complex mixture of triacylglycerols are strongly influenced by the cooling rate and temperature (Moran 1994). If an oil is cooled rapidly, all the triacylglycerols crystallize at approximately the same time and a *solid solution* is formed, which consists of homogeneous crystals in which the triacylglycerols are intimately mixed with each other (Walstra 1987). On the other hand, if the oil is cooled slowly, the higher melting point triacylglycerols crystallize first, while the low-melting-point triacylglycerols crystallize later, and so *mixed crystals* are formed. These crystals are heterogeneous and consist of some regions which are rich in high-melting-point triacylglycerols and other regions that are depleted in these triacylglycerols. Whether a crystalline fat forms mixed crystals or a solid solution influences many of its physicochemical properties, such as density, compressibility, and melting profile (Walstra 1987).

Once a fat has crystallized, the individual crystals may aggregate to form a three-dimensional network which traps liquid oil through capillary forces (Moran 1994). The interactions responsible for crystal aggregation in pure fats are primarily van der Waals interactions between the solid fat crystals, although "water bridges" between the crystals have also been suggested to play an important role (Dickinson and McClements 1995). Once aggregation has occurred, the fat crystals may partially fuse together, which strengthens the crystal network (Timms 1995). The system may also change over time due to the growth of larger crystals at the expense of smaller ones (i.e., Ostwald ripening) (Chapter 7).

4.2.3.7. *Crystallization in Oil-in-Water and Water-in-Oil Emulsions*

The influence of fat crystallization on the bulk physicochemical properties of food emulsions depends on whether the fat forms the continuous phase or the dispersed phase. The characteristic stability and rheological properties of water-in-oil emulsions, such as butter and margarine, are determined by the presence of a fat crystal network in the continuous phase (Moran 1994, Chyrsam 1996). The fat crystal network is responsible for preventing the water droplets from sedimenting under the influence of gravity, as well as determining the spreadability of the product. If there are too many fat crystals present, the product is firm and difficult to spread, but when there are too few crystals present, the product is soft and collapses under its own weight. Selection of a fat with the appropriate melting characteristics is therefore one of the most important aspects of margarine and spread production (Gunstone and Norris 1983). The melting profile of natural fats can be optimized for specific applications by various physical or chemical methods, including blending, interesterification, fractionation, and hydrogenation (Birker and Padley 1987).

Fat crystallization also has a pronounced influence on the physicochemical properties of many oil-in-water emulsions, such as milk or salad cream (Mulder and Walstra 1974, Boode 1992). When the fat droplets are partially crystalline, a crystal from one droplet can penetrate another droplet during a collision, which causes the two droplets to stick together (Walstra 1987, Boode 1992, Dickinson and McClements 1995). This phenomenon is known as *partial coalescence* and leads to a dramatic increase in the viscosity of an emulsion, as well as a decrease in the stability to creaming (Chapter 7). Extensive partial coalescence can eventually lead to phase inversion (i.e., conversion of an oil-in-water emulsion to a water-in-oil emulsion) (Mulder and Walstra 1974). This process is one of the most important steps in the production of butters, margarines, and spreads (Moran and Rajah 1994).

4.2.4. Chemical Changes

The two most important chemical changes which occur naturally in edible fats and oils are lipolysis and oxidation (Sonntag 1979b, Nawar 1996). Lipolysis is the process where ester bonds of fats and oils are hydrolyzed by certain enzymes or by a combination of heat and moisture. The result of lipolysis is the liberation of free fatty acids, which can be either detrimental or desirable to food quality. Lipolysis has deleterious effects on the quality of some food products because it leads to the generation of rancid off-flavors and off-odors ("hydrolytic rancidity"). In addition, free fatty acids are more surface active than triacylglycerols and therefore accumulate preferentially at an oil–water or air–water interface, which increases their susceptibility to oxidation and may increase the tendency for emulsion droplets to coalesce (Coupland and McClements 1996). On the other hand, a limited amount of lipolysis is beneficial to the quality of some foods because it leads to the formation of desirable flavors and aromas (e.g., cheese and yogurt) (Nawar 1996).

Lipid oxidation is one of the most serious causes of quality deterioration in many foods because it leads to the generation of undesirable off-flavors and off-odors ("oxidative rancidity"), as well as potentially toxic reaction products (Schultz et al. 1962, Simic et al. 1992, Nawar 1996). In other foods, a limited amount of lipid oxidation is beneficial because it leads to the generation of a desirable flavor profile (e.g., cheese). The term *lipid oxidation* describes an extremely complex series of chemical reactions which involves unsaturated lipids and oxygen (Halliwell and Gutterridge 1991, Nawar 1996). It has proved convenient to divide these reactions into three different types: initiation, propagation, and termination (Halliwell and Gutterridge 1991). Initiation occurs when a hydrogen atom is extracted from the methylene group (–CH=CH–) of a polyunsaturated fatty acid, leading to the formation of a free radical (–CH=C–). This process can be started by a variety of different initiators that are present in foods, including naturally occurring lipid peroxides, transition metal ions, UV light, and enzymes (Nawar 1996). It is worthwhile noting that many of these initiators are predominantly water soluble, which has important implications for the oxidation of emulsified oils, because the initiator must either travel through or interact across the interfacial membrane in order to come into contact with the oil (Coupland and McClements 1996). Once a free radical has formed, it reacts with oxygen to form a peroxy radical (–CH–COO–). These radicals are highly reactive and can extract hydrogen atoms from other unsaturated lipids and therefore propagate the oxidation reaction. Termination occurs when two radicals interact with each other to form a nonradical and thus end their role as propagators of the reaction. During lipid oxidation, a number of decomposition reactions occur simultaneously which leads to the formation of a complex mixture of reaction products, including aldehydes, ketones, alcohols, and hydrocarbons (Nawar 1996). Many of these products are volatile and therefore contribute to the characteristic odor associated with lipid oxidation. Some of the products are surface active and accumulate at oil–water or air–water interfaces, whereas others are water soluble and would therefore leach into the aqueous phase of an emulsion (Coupland and McClements 1996).

A considerable amount of fundamental research has been carried out by lipid chemists to establish the mechanism of lipid oxidation in bulk fats and oils (Nawar 1996). On the other hand, the understanding of lipid oxidation in emulsions is still in its infancy, and a great deal of research needs to be carried out to determine the relationship between emulsion structure and lipid oxidation (Frankel 1991, Frankel et al. 1994, Coupland and McClements 1996). Some of the work carried out in this area is discussed in the chapter on emulsion stability (Chapter 7).

4.3. WATER

Water plays an extremely important role in determining the bulk physicochemical and organoleptic properties of food emulsions. Its unique molecular and structural properties largely determine the solubility, conformation, and interactions of the other ingredients present in aqueous solutions. It is therefore important for food scientists to understand the contribution that water makes to the overall properties of food emulsions.

4.3.1. Molecular Structure and Organization

A water molecule is comprised of two hydrogen atoms covalently bonded to an oxygen atom (Figure 4.9). The oxygen atom is highly electronegative and pulls the electrons associated with the hydrogen atoms toward it (Eisenberg and Kauzmann 1969, Kern and Karplus 1972, Fennema 1996b). This leaves a partial positive charge ($\delta+$) on each of the hydrogen atoms

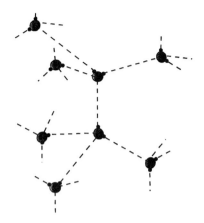

FIGURE 4.9   Molecular structure and tetrahedral organization of water molecules.

and a partial negative charge (δ–) on each of the lone pairs of electrons on the oxygen atom. The tetrahedral arrangement of the partial charges on a water molecule means that it can form hydrogen bonds with four of its neighbors (Figure 4.9). A hydrogen bond is formed between a lone pair of electrons on the oxygen atom of one water molecule and a hydrogen atom on a neighboring water molecule (i.e., O–H$^{\delta+}$... O$^{\delta-}$). A hydrogen bond is actually a composite of more fundamental interactions (i.e., dipole–dipole, van der Waals, steric, and partial charge transfer) (Baker and Hubbard 1984, Dill 1990). The magnitude of the hydrogen bonds in water is typically between 13 and 25 kJ mol^{-1} (5 to 10 kT), which is sufficiently strong to cause the water molecules to overcome the disorganizing influence of the thermal energy and become aligned with each other (Israelachvili 1992). In order to maximize the number of hydrogen bonds formed, water molecules organize themselves into a three-dimensional tetrahedral structure because this allows each water molecule to form hydrogen bonds with four of its nearest neighbors (Franks 1972–82, Fennema 1996b). In the solid state, the number of hydrogen bonds formed per molecule is four. In the liquid state, the disorganizing influence of the thermal energy means that the number of hydrogen bonds per molecule is between about 3 and 3.5 at room temperature and decreases with increasing temperature (Israelachvili 1992). The three-dimensional tetrahedral structure of water in the liquid state is highly dynamic, with hydrogen bonds continually being broken and reformed as the water molecules move about (Reichardt 1988, Franks 1991, Cramer and Truhlar 1994, Robinson et al. 1996). Water molecules which dissociate to form ions, such as H$_3$O$^+$ and OH$^-$, do not fit into the normal tetrahedral structure of water; nevertheless, they have little effect on the overall structure and properties of water because their concentration is so low (Fennema 1996b).

   In addition to forming hydrogen bonds with each other, water molecules are also capable of forming them with other polar molecules, such as organic acids, bases, proteins, and carbohydrates (Franks 1972–82). The strength of these interactions varies from about 2 to 40 kJ mol^{-1} (1 to 16 kT) depending on the electronegativity and orientation of the donor or acceptor groups (Baker and Hubbard 1984). Many ions form relatively strong ion–dipole interactions with water molecules, which has a pronounced influence on the structure and physicochemical properties of water (Franks 1973, 1975a,b; Israelachvili 1992; Fennema 1996b). It is the ability of water molecules to form relatively strong bonds with each other and with other types of polar or ionic molecules which determines many of the characteristic properties of food emulsions.

4.3.2. Bulk Physicochemical Properties

The bulk physicochemical properties of pure water are determined by the mass, dimensions, bond angles, charge distribution, and interactions of the water molecule (Eisenberg and Kauzmann 1969, Duckworth 1975, Simatos and Multon 1985, Fito et al. 1994). Water has a high dielectric constant because the uneven distribution of partial charges on the molecule means that it is easily polarized by an electric field (Hasted 1972). It has a high melting point, boiling point, enthalpy of vaporization and surface tension compared to other molecules of a similar size that also contain hydrogen (e.g., CH_4, NH_3, HF, and H_2S), because a greater amount of energy must be supplied to disrupt the strong hydrogen bonds holding the water molecules together in the condensed state (Israelachvili 1992, Fennema 1996b). Ice and liquid water have a relatively low density because the water molecules adopt a structure in which they are in direct contact with only four of their nearest neighbors, rather than forming a more closely packed structure (Franks 1975b, Fennema 1996b). The relatively low viscosity of water is due to the highly dynamic nature of hydrogen bonds compared to the time scale of a rheology experiment. Even though energy is required to break the hydrogen bonds between water molecules as they move past each other, most of this energy is regained when they form new hydrogen bonds with their new neighbors.

The crystallization of water has a pronounced effect on the bulk physiochemical properties of food emulsions. The presence of ice crystals in the aqueous phase of an oil-in-water emulsion, such as ice cream, contributes to the characteristic mouthfeel and texture of the product (Berger 1976). When these ice crystals grow too large, a product is perceived as being "grainy" or "sandy," which is commonly experienced when ice cream is melted and then refrozen (Berger 1976). Many foods are designed to be freeze–thaw stable (i.e., the quality should not be adversely affected once the product is frozen and then thawed) (Partmann 1975). Considerable care must be taken in the choice of ingredients and freezing/thawing conditions to create a food emulsion which is freeze–thaw stable. The basic principles of ice crystallization are similar to those described for fats and oils (Section 4.2.3.). Nevertheless, water does exhibit some anomalous behavior because of its unique molecular properties (e.g., it expands when it crystallizes, whereas most other substances contract) (Fennema 1996b). This is because the increased mobility of the water molecules in the liquid state means that they can get closer together, and so the density of the liquid state is actually greater than that of the solid state. Some of the most important bulk physicochemical properties of liquid water are compared with those of a liquid oil in Table 4.3. A more detailed discussion of the molecular basis of the physicochemical properties of water in relation to food quality is given by Fennema (1996b).

4.4. AQUEOUS SOLUTIONS

The aqueous phase of most food emulsions contains a variety of water-soluble constituents, including minerals, acids, bases, flavors, preservatives, vitamins, sugars, surfactants, proteins, and polysaccharides (Dickinson 1992). Interactions between these constituents and water molecules determine the solubility, partitioning, volatility, conformation, and chemical reactivity of many food ingredients. It is therefore important for food scientists to understand the nature of solute–water interactions and their influence on the bulk physicochemical and organoleptic properties of food emulsions.

When a solute molecule is introduced into pure water, the normal structural organization and interactions of the water molecules are altered, which results in changes in water properties such as density, compressibility, melting point, and boiling point (Franks 1973, Reichardt 1988, Israelachvili 1992, Murrell and Jenkins 1994, Fennema 1996b). The extent of these changes depends on the molecular characteristics of the solute (i.e., its size, shape, and polarity). The water molecules in the immediate vicinity of the solute experience the largest modification of their properties and are often referred to as being "bound" to the solute (Reichardt 1988, Murrell and Jenkins 1994). In reality, these water molecules are not permanently bound to the solute but rapidly exchange with the bulk water molecules, albeit with a reduced mobility (Franks 1991, Fennema 1996b). The mobility of "bound" water increases as the strength of the attractive interactions between it and the solute decreases (i.e., nonpolar–water > dipole–water > ion–water) (Israelachvili 1992). The amount of water "bound" to a solute can be defined as the number of water molecules whose properties are significantly altered by its presence. In practice, it is difficult to unambiguously define or stipulate the amount of "bound" water (Franks 1991), first because the water molecules "bound" to a solute do not all have the same properties (the water molecules closest to the solute are more strongly influenced by its presence than those farthest away) and, second, because each of the physicochemical properties that are measured in order to determine the amount of "bound" water is influenced to a different extent (e.g., density, compressibility, mobility, melting point). As a consequence, different analytical techniques often measure different amounts of "bound" water, depending on the physical principles on which they operate.

4.4.1. Interaction of Water with Ionic Solutes

Many of the solutes present in food emulsions are either ionic or capable of being ionized, including salts, acids, bases, proteins, and polysaccharides (Fennema 1996b). The degree of ionization of many of these solutes is governed by the pH of the aqueous solution, and so their interactions are particularly sensitive to pH. The ion–dipole interactions which occur between an ionic solute and a water molecule are usually stronger than the dipole–dipole interactions which occur between a pair of water molecules (Table 4.4). As a consequence, the water molecules in the immediate vicinity of an ion tend to orientate themselves so that their oppositely charged dipole faces the ion. Thus, a positively charged ion causes the water molecules to align themselves so that a δ^- group faces the ion, whereas the opposite is true for a negatively charged ion (Figure 4.10). The relatively strong nature of ion–dipole interactions means that the mobility of the water molecules near the surface of an ion is significantly less than that of bulk water (Reichardt 1988, Fennema 1996b, Robinson et al. 1996). The residence time of a water molecule in the vicinity of an ionic group is $\approx 10^{-8}$ s, whereas it is $\approx 10^{-11}$ s in bulk water (Fennema 1996b). The influence of an ion on the mobility and alignment of the water molecules is greatest at its surface because the electric field is strongest there. As one moves away from the ion surface, the strength of the electric field decreases, so that the ion–water interactions become progressively weaker. Thus, the water molecules become more mobile and are less likely to be aligned

TABLE 4.4
Typical Water–Solute Interactions Found in Food Emulsions

Interaction type	Typical example	Strength of interaction compared to water–H bond
Water–ion	Free ion (e.g., Na^+, Cl^-)	Greater
	Ionic group (e.g., $-CO_2^-$, $-NH_3^+$)	
Water–dipole	$-C=O$, $-NH$, $-OH$	Similar
Water–nonpolar	Alkyl group	Much smaller

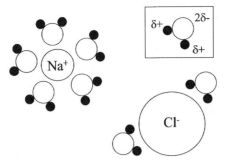

FIGURE 4.10   Organization of water molecules around ions in aqueous solutions.

toward the ion. At a sufficiently large distance from the ion surface, the water molecules are uninfluenced by its presence and have properties similar to those of bulk water (Franks 1973, Reichardt 1988). Alterations in the structural organization and interactions of water molecules in the vicinity of an ion cause significant changes in the physicochemical properties of water (Fennema 1996b). The water that is "bound" to an ionic solute is less mobile, less compressible, more dense, has a lower freezing point, and has a higher boiling point than bulk water. Most ionic solutes have a high water solubility because the formation of many ion–dipole bonds in an aqueous solution helps to compensate for the loss of the strong ion–ion bonds in the crystals, which is coupled with the favorable entropy of mixing contribution (Chapter 2).

The number of water molecules whose mobility and structural organization are altered by the presence of an ion increases as the strength of its electric field increases. The strength of the electrical field generated by an ion is determined by its charge divided by its radius (Israelachvili 1992). Thus, ions which are small and/or multivalent generate strong electric fields that influence the properties of the water molecules up to relatively large distances from their surface (e.g., Li^+, Na^+, H_3O^+, Ca^{2+}, Ba^{2+}, Mg^{2+}, Al^{3+} and OH^-). On the other hand, ions which are large and/or monovalent generate relatively weak electrical fields, and therefore their influence extends a much shorter distance into the surrounding water (e.g., K^+, Rb^+, Cs^+, NH_4^+, Cl^-, Br^-, and I^-). The number of water molecules "bound" to an ion is usually referred to as the *hydration number*. Thus the hydration number of small multivalent ions is usually larger than that of large monovalent ions.

When an ionic solute is added to pure water, it disrupts the existing tetrahedral arrangement of the water molecules but imposes a new order on the water molecules in its immediate vicinity (Franks 1973, Reichardt 1988). The overall structural organization of the water molecules in an aqueous solution can therefore either increase or decrease after a solute is added, depending on the amount of structure imposed on the water by the ion compared to that lost by disruption of the tetrahedral structure of bulk water (Israelachvili 1992). If the structure imposed by the ion is greater than that lost by the bulk water, the overall structural organization of the water molecules is increased, and the solute is referred to as a *structure maker* (Robinson et al. 1996). Ionic solutes that generate strong electric fields are structure makers, and the magnitude of their effect increases as the size of the ions decreases and/or their charge increases. If the structure imposed by an ion is not sufficiently large to compensate for that lost by disruption of the tetrahedral structure of bulk water, then the overall structural organization of the water molecules in the solution is decreased, and the solute is referred to as a *structure breaker* (Robinson et al. 1996). Ionic solutes that generate weak electric fields are structure breakers, and the magnitude of their effect increases as their size increases or they become less charged (Israelachvili 1992).

The influence of ionic solutes on the overall properties of water depends on their concentration (Israelachvili 1992, Robinson et al. 1996). At low solute concentrations, the majority of water is not influenced by the presence of the ions and therefore has properties similar to that of bulk water. At intermediate solute concentrations, some of the water molecules have properties similar to those of bulk water, whereas the rest have properties which are dominated by the presence of the ions. At high solute concentrations, all the water molecules may be influenced by the presence of the solute molecules and therefore have properties which are appreciably different from those of bulk water. The solubility of biopolymer molecules in aqueous solutions decreases as the concentration of ionic solutes increases above a certain level, which is known as "salting out," because the solutes compete with the biopolymers for the limited amount of water that is available to hydrate them.

The presence of electrolytes in the aqueous phase of an emulsion can influence the interactions between the droplets in a variety of ways. First, ions can screen electrostatic interactions (Section 3.4.3) or the zero-frequency contribution of the van der Waals interaction (Section 3.3.4). Second, multivalent ions may be able to form salt bridges between charged droplets (Section 3.4.4). Third, the ability of ions to alter the structural organization of water influences the strength of hydrophobic interactions (Section 3.7). Fourth, ions can cause changes in the size of biopolymer molecules in solution, which alters the strength of steric and depletion interactions (Sections 3.5 and 3.6). Finally, the binding of hydrated ions to the surface of emulsion droplets increases the hydration repulsion between them (Section 3.8). The fact that ions influence the interactions between emulsion droplets in many different ways means that it is difficult to accurately quantify their effect on emulsion properties.

4.4.2. Interaction of Water with Dipolar Solutes

Many food ingredients contain molecules which are either dipolar or contain dipolar groups, including alcohols, sugars, proteins, polysaccharides, and surfactants (Fennema 1996a). Water is capable of participating in dipole–dipole interactions with dipolar solutes (Franks 1973, 1991). By far the most important type of dipole–dipole interaction is between water and those solutes which have hydrogen bond donors (e.g., $-O-H^{\delta+}$) or acceptors (e.g., $^{\delta-}O-$). The strength of hydrogen bonds between water and this type of dipolar solute is similar to that between two water molecules (Table 4.4). The addition of a dipolar solute to water therefore has much less influence on the mobility and organization of the water molecules in its immediate vicinity than does a similarly sized ionic solute (Fennema 1996b). The influence of dipolar solutes on the properties of water is largely governed by the ease with which they can be accommodated into the existing tetrahedral structure of the water molecules (Figure 4.11). When a dipolar solute is of an appropriate size and shape, and has hydrogen bond acceptors and donors at positions where they can easily form bonds with the neighboring water molecules, it can fit into the tetrahedral structure (Brady and Ha 1975, Galema and Hoiland 1991, Franks 1991, Schmidt et al. 1994). For this type of solute, there need be little change in the number of hydrogen bonds formed per water molecule or the overall structural organization of the water molecules. This type of solute therefore tends to be highly water soluble because of the entropy of mixing (Chapter 2). If the solute molecule is not of an appropriate size and shape, or if its hydrogen bond donors and acceptors are not capable of aligning with those of neighboring water molecules, then it cannot easily fit into the tetrahedral structure of water (Figure 4.11). This causes a dislocation of the normal water structure surrounding the solute molecules, which is energetically unfavorable. In addition, there may be a significant alteration in the physicochemical properties of the water molecules in the vicinity of the solute. For this reason, dipolar solutes which are less compatible with the tetrahedral structure of water tend to be less soluble than those which are compatible.

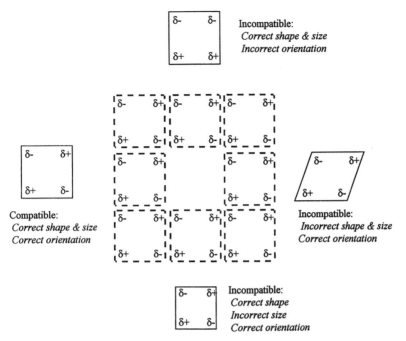

FIGURE 4.11   The effect of dipolar solutes on the structural organization of water molecules depends on their dimensions and the number and position of their hydrogen bonding groups.

Just as with ionic solutes, the effect of dipolar solutes depends on their concentration. At low solute concentrations, most of the water has the same properties as bulk water, but at high concentrations, a significant proportion of the water has properties which are altered by the presence of the solute. Nevertheless, it takes a greater concentration of a dipolar solute to cause the same effect as an ionic solute because of the greater strength of ion–water interactions compared to dipole–water interactions.

Interactions between dipolar groups and water determine a number of important properties of food components in emulsions. The hydration of the dipolar head groups of surfactant molecules is believed to be partly responsible for their stability to aggregation (Evans and Wennerstrom 1994). When surfactants are heated, the head groups become progressively dehydrated, which eventually causes the molecules to aggregate (Section 4.5). These hydration forces also play an important role in preventing the aggregation of emulsion droplets stabilized by nonionic surfactants (Section 3.8). The three-dimensional conformation and interactions of proteins and polysaccharides are influenced by their ability to form intramolecular and intermolecular hydrogen bonds (Section 4.6). The solubility, partitioning, and volatility of dipolar solutes depend on their molecular compatibility with the surrounding solvent: the stronger the molecular interactions between a solute and its neighbors in a liquid, the greater its solubility and the lower its volatility (Baker 1987).

4.4.3.   Interaction of Water with Nonpolar Solutes: The Hydrophobic Effect

The attraction between a water molecule and a nonpolar solute is much weaker than that between two water molecules, because nonpolar molecules are incapable of forming hydrogen bonds (Israelachvili 1992, Evans and Wennerstrom 1994). For this reason, when a

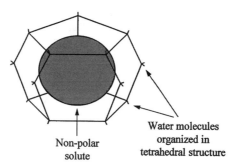

FIGURE 4.12 Schematic representation of the reorganization of water molecules near a nonpolar solute.

nonpolar molecule is introduced into pure liquid water, the water molecules surrounding it change their orientation so that they can maximize the number of hydrogen bonds formed with neighboring water molecules (Figure 4.12). The structural rearrangement and alteration in the physicochemical properties of water molecules in the immediate vicinity of a nonpolar solute is known as *hydrophobic hydration*. At relatively low temperatures, it is believed that a "cage-like" or clathrate structure of water molecules exists around a nonpolar solute, in which the water molecules involved have a coordination number of four, which is greater than that of the water molecules in the bulk phase (3 to 3.5) (Israelachvili 1992). Despite gaining some order, the water molecules in the cage-like structures are still highly dynamic, having residence times of the order of 10^{-11} s (Evans and Wennerstrom 1994). The alteration in the organization and interactions of water molecules surrounding a nonpolar solute has important implications for the solubility and interactions of nonpolar groups in water (Tanford 1980, Dill 1990, Israelachvili 1992, Cramer and Tuhlar 1994, Fennema 1996b).

The behavior of nonpolar solutes in water can be understood by considering the transfer of a nonpolar molecule from an environment where it is surrounded by similar molecules to one where it is surrounded by water molecules (Tanford 1980). When a nonpolar solute is transferred from a nonpolar solvent into water, there are changes in both the enthalpy ($\Delta H_{transfer}$) and entropy ($\Delta S_{transfer}$) of the system. The enthalpy change is related to the alteration in the overall strength of the molecular interactions, whereas the entropy change is related to the alteration in the structural organization of the solute and solvent molecules. The overall free energy change ($\Delta G_{transfer}$) depends on the relative magnitude of these two contributions (Evans and Wennerstrom 1994):

$$\Delta G_{transfer} = \Delta H_{transfer} - T\Delta S_{transfer} \qquad (4.7)$$

The relative contribution of the enthalpic and entropic contributions to the free energy depends on temperature (Table 4.5). An understanding of the temperature dependence of the free energy of transfer is important for food scientists because it governs the behavior of many food components during food processing, storage, and handling. At relatively low temperatures (<15°C), the number of hydrogen bonds formed by the water molecules in the cage-like structure surrounding the nonpolar solute is slightly higher than in bulk water, and so $\Delta H_{transfer}$ is negative (i.e., favors transfer). On the other hand, the water molecules in direct contact with the nonpolar solute are more ordered than those in bulk water, and so the entropy term is positive (i.e., opposes transfer). Overall, the entropy term dominates, and so the transfer of a nonpolar molecule into water is thermodynamically unfavorable (Tanford 1980, Israelachvili 1992).

TABLE 4.5
Temperature Dependence of the Enthalpic and Entropic Contributions to the Free Energy Change That Occurs When a Nonpolar Molecule (Benzene) Is Transferred from the Neat Liquid to Water

T (°C)	ΔH (kJ mol^{-1})	$-T\Delta S$ (J °C^{-1} mol^{-1})	ΔG (kJ mol^{-1})
15	−0.15	18.89	18.75
25	2.08	17.30	19.38
30	3.16	16.50	19.66
35	4.37	15.55	19.92

Adapted from Baldwin 1986.

As the temperature is raised, the water molecules become more thermally agitated, and so their organization within the cage-like structure is progressively lost, which has consequences for both the enthalpy and entropy contributions. First, some of the partial charges on the water molecules face toward the nonpolar group and are therefore unable to form hydrogen bonds with the surrounding water molecules. Thus, the number of hydrogen bonds formed by the water molecules in the cage-like structure decreases with increasing temperature. At a certain temperature, the number of hydrogen bonds formed by the water molecules in the cage-like structure becomes less than that of bulk water. Below this temperature, the enthalpy associated with transferring a nonpolar molecule into water is negative (favorable to transfer), but above it, it is positive (unfavorable to transfer). The enthalpy term therefore makes an increasing contribution to opposing the transfer of nonpolar molecules into water as the temperature rises. Second, the increasing disorganization of the water molecules surrounding a nonpolar molecule as the temperature is raised means that the entropy difference between the water molecules in the cage-like structure and those in the bulk water is lessened. Thus, as the temperature is increased, the contribution of the entropy term becomes progressively less important. In summary, at low temperatures, the major contribution to the unfavorable free energy change associated with transfer of a nonpolar molecule into water is the entropy term, but at higher temperatures it is the enthalpy term. Overall, the transfer of a nonpolar molecule from an organic solvent into water becomes increasingly thermodynamically unfavorable as the temperature is raised (Table 4.5).

The free energy associated with transferring a nonpolar molecule from an environment where it is surrounded by similar molecules to one in which it is surrounded by water molecules has been shown to be a product of its surface area and the interfacial tension between the bulk nonpolar liquid and water (i.e., $\Delta G = \gamma \Delta A$) (Tunon et al. 1992). An aqueous solution containing a nonpolar solute can decrease its free energy by reducing the unfavorable contact area between the nonpolar groups and water, which is known as the *hydrophobic effect*. The strong tendency for nonpolar molecules to associate with each other in aqueous solutions is a result of the attempt of the system to reduce the contact area between water and nonpolar regions and is known as the *hydrophobic interaction* (Section 3.7). The hydrophobic effect is responsible for many of the characteristic properties of food emulsions, including the aggregation of proteins, the formation of surfactant micelles, the adsorption of emulsifiers at oil–water and air–water interfaces, the aggregation of hydrophobic particles, and the immiscibility of oil and water.

The strength of the hydrophobic interaction between nonpolar substances in water is affected by the presence of ions in the aqueous phase separating them. Ions can either increase or decrease the structural organization of water molecules in an aqueous solution, depending on whether they are structure makers or structure breakers (Section 4.4.1). As one of the major driving forces for hydrophobic interactions is the difference in structural organization (entropy) between the water molecules in the immediate vicinity of the nonpolar solute and those in the bulk water, then changes in the organization of the water molecules in bulk water alter its strength (Israelachvili 1992). Structure makers decrease the magnitude of the hydrophobic interaction and therefore increase the water solubility of nonpolar solutes because the difference in structural organization is reduced, whereas structure breakers have the opposite effect. The strength of the hydrophobic interaction also depends on temperature, increasing as the temperature is raised.

4.5. SURFACTANTS

4.5.1. Molecular Characteristics

The principal role of surfactants in food emulsions is to enhance their formation and stability (Charalambous and Doxastakis 1989, Dickinson 1992, Hasenhuettl and Hartel 1997); however, they may also alter emulsion properties in a variety of other ways (e.g., by interacting with proteins or polysaccharides, by forming surfactant micelles, or by modifying the structure of fat crystals (Dickinson and McClements 1995, Bergenstahl 1997, Bos et al. 1997, Deffenbaugh 1997). By definition, a surfactant is an amphiphilic molecule that has a hydrophilic "head" group which has a high affinity for water and a lipophilic "tail" group which has a high affinity for oil (Myers 1988, Hasenhuettl 1997). Surfactants can therefore be represented by the formula RX, where X represents the hydrophilic head and R the lipophilic tail (Dickinson and McClements 1995). The characteristics of a particular surfactant depend on the nature of its head and tail groups (Table 4.6). The head group may be anionic, cationic, zwitterionic, or nonionic (Myers 1988, St. Angelo 1989, Zielinski 1997). Surfactants used in the food industry are mainly nonionic (e.g., monoacylglycerols, sucrose esters of fatty acids), anionic (e.g., fatty acids), or zwitterionic* (e.g., lecithin). The tail group usually consists of one or more hydrocarbon chains, with between 10 and 20 carbon atoms per chain (St. Angelo 1989, Bergenstahl 1997, Zielinski 1997). The chains may be saturated or unsaturated, linear or branched, aliphatic and/or aromatic. Most surfactants used in foods have either one or two linear aliphatic chains, which may be saturated or unsaturated. Each type of surfactant has functional properties that are determined by its unique chemical structure (Myers 1988). Thus, there is no single surfactant which is appropriate for every application. Instead, it is necessary to choose a surfactant which is the most appropriate for each particular application.

Surfactants aggregate spontaneously in solution to form a variety of thermodynamically stable structures known as *association colloids* (e.g., micelles, bilayers, vesicles, and reverse micelles) (Figure 4.13). These structures are adopted because they minimize the unfavorable contact area between the nonpolar tails of the surfactant molecules and water (Hiemenz 1986, Evans and Wennerstrom 1994). The type of association colloid formed by a surfactant depends principally on its polarity and molecular geometry (Section 4.5.3). Association colloids are held together by physical interactions that are relatively weak compared to the thermal energy, and therefore they have highly dynamic and flexible structures (Israelachvili 1992). In addition, their structures are particularly sensitive to changes in environmental conditions, such as temperature, pH, ionic strength, and ion type (MacKay 1987, Myers 1988, Israelachvili 1992). Because surfactant micelles are the most important type of association

---

* A zwitterionic surfactant is one which has both positively and negatively charged groups on the same molecule.

TABLE 4.6
Selected HLB Group Numbers

Hydrophilic group	Group number	Lipophilic group	Group number
$-SO_4Na^+$	38.7	$-CH-$	0.475
$-COO^-H^+$	21.2	$-CH_2-$	0.475
Tertiary amine	9.4	$-CH_3$	0.475
Sorbitan ring	6.8		
$-COOH$	2.1		
$-O-$	1.3		

Adapted from Davis 1994b.

colloid formed in many food emulsions, we will focus principally on their properties (Dickinson and McClements 1995).

4.5.2. Functional Properties

4.5.2.1. Critical Micelle Concentration

A surfactant forms micelles in an aqueous solution when its concentration exceeds some critical level, known as the *critical micelle concentration* or CMC (Myers 1988). Below the CMC, the surfactant molecules are dispersed predominantly as monomers, but once the CMC is exceeded, any additional surfactant molecules form micelles, and the monomer concentration remains fairly constant (Hiemenz 1986). Despite the highly dynamic nature of their structure, surfactant micelles have a fairly well-defined average size and shape under a given set of environmental conditions. Thus, when surfactant is added to a solution above the CMC, the *number* of micelles tends to increase, rather than the size or shape of the individual micelles (although this may not be true at high surfactant concentrations). There is an abrupt change in the physicochemical properties of a surfactant solution when the CMC is exceeded (e.g., surface tension, electrical conductivity, turbidity, and osmotic pressure) (Rosen 1978, Hiemenz 1986). This is because the properties of surfactant molecules dispersed as monomers are different from those in micelles. For example, surfactant monomers are amphiphilic

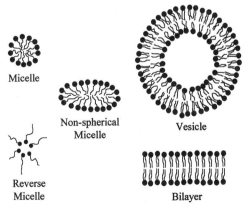

FIGURE 4.13 Some typical structures formed due to the self-association of surfactant molecules.

and have a high surface activity, whereas micelles have little surface activity because their surface is covered with hydrophilic head groups. Consequently, the surface tension of a solution decreases with increasing surfactant concentration below the CMC, but remains fairly constant above it (Chapter 5).

4.5.2.2. Cloud Point

When a surfactant solution is heated above a certain temperature, known as the *cloud point,* it becomes turbid (Myers 1988). This occurs because the hydrophilic head groups of the surfactant molecules become progressively dehydrated as the temperature is raised, which alters their molecular geometry (Section 4.5.3.3) and decreases the hydration repulsion between them (Israelachvili 1992, Evans and Wennerstrom 1994). Above a certain temperature, known as the cloud point, the micelles form aggregates which are large enough to scatter light and therefore make the solution appear turbid. As the temperature is increased further, the aggregates may grow so large that they sediment under the influence of gravity and form a separate phase which can be observed visually. The cloud point increases as the hydrophobicity of a surfactant molecule increases (i.e., the length of its hydrocarbon tail increases or the size of its hydrophilic head group decreases) (Myers 1988).

4.5.2.3. Solubilization

Nonpolar molecules, which are normally insoluble or only sparingly soluble in water, can be solubilized in an aqueous surfactant solution by incorporation into micelles or other types of association colloids (Elworthy et al. 1968, Mukerjee 1979, MacKay 1987, Christian and Scamehorn 1995). The resulting system is thermodynamically stable; however, it may take an appreciable time to reach equilibrium because of the time taken for molecules to diffuse through the system and because of the activation energy associated with transferring a nonpolar molecule from a bulk phase into a micelle (Dickinson and McClements 1995, Kabalnov and Weers 1996). Micelles containing solubilized materials are referred to as *swollen micelles* or *microemulsions,* whereas the material solubilized within the micelle is referred to as the *solubilizate.* The ability of micellar solutions to solubilize nonpolar molecules has a number of potentially important applications in the food industry, including selective extraction of nonpolar molecules from oils, controlled ingredient release, incorporation of nonpolar substances into aqueous solutions, transport of nonpolar molecules across aqueous membranes, and modification of chemical reactions (Dickinson and McClements 1995). There are three important factors which determine the functional properties of swollen micellar solutions: (1) the location of the solubilizate within the micelles, (2) the maximum amount of material that can be solubilized per unit mass of surfactant, and (3) the rate at which solubilization proceeds. Food manufacturers must therefore select a micellar system which is most appropriate for their particular application.

4.5.2.4. Surface Activity and Droplet Stabilization

Surfactants are used widely in the food industry to enhance the formation and stability of food emulsions (St. Angelo 1989, Dickinson 1992, Bergenstahl 1997). To do this, they must adsorb to the surface of emulsion droplets during homogenization and form a protective membrane which prevents the droplets from aggregating with each other during a collision (Walstra 1993a, 1996a,b). Surfactant molecules adsorb to oil–water interfaces because they can adopt an orientation in which the hydrophilic part of the molecule is located in the water while the hydrophobic part is located in the oil. This minimizes the contact area between hydrophilic and hydrophobic regions and therefore reduces the interfacial tension (Chapter

5). This reduction in interfacial tension is important during homogenization because it facilitates the further disruption of emulsion droplets (i.e., less energy is needed to break up a droplet when the interfacial tension is lowered) (Chapter 6). Once adsorbed to the surface of a droplet, the surfactant must provide a repulsive force which is strong enough to prevent the droplet from aggregating with its neighbors (Chapters 3 and 7). Ionic surfactants provide stability by causing all the emulsion droplets to have the same electric charge and therefore electrostatically repel each other. Nonionic surfactants provide stability by generating a number of short-range repulsive forces which prevent the droplets from getting too close together, such as steric, hydration, and thermal fluctuation interactions. Some surfactants form multilayers (rather than monolayers) at the surface of an emulsion droplet, which greatly enhances the stability of the droplets against aggregation (Friberg and El-Nokaly 1983). In summary, surfactants must have three characteristics to be effective at enhancing the formation and stability of emulsions (Chapter 6). First, they must rapidly adsorb to the surface of the freshly formed emulsion droplets during homogenization. Second, they must reduce the interfacial tension by a significant amount. Third, they must form an interfacial layer that prevents the droplets from aggregating. It should also be noted that the ability of surfactants to form micelles in the continuous phase of an emulsion can have a negative impact on emulsion stability, because they induce depletion flocculation or facilitate the transport of oil molecules between droplets (Dickinson and McClements 1995). The ability of surfactants to regulate the interactions between droplets can also have a pronounced influence on the rheological properties of emulsions (Chapter 8).

4.5.3. Surfactant Classification

A food manufacturer must consider a variety of factors when selecting a surfactant for a particular product, including its legal status as a food ingredient; its cost; the reliability of the supplier; the consistency of its quality from batch to batch; its ease of handling and dispersion; its shelf life; its compatibility with other ingredients; the processing, storage, and handling conditions it will experience; and the expected shelf life and physicochemical properties of the final product. How does a food manufacturer decide which surfactant is most suitable for a product? There have been various attempts to develop systems to classify surfactants according to their physicochemical properties. Classification schemes have been developed which are based on a surfactant's solubility in oil and/or water (Bancroft's rule), its ratio of hydrophilic to lipophilic groups (HLB number), and its molecular geometry (Davis 1994b, Dickinson and McClements 1995, Bergenstahl 1997). Ultimately, all of these properties depend on the chemical structure of the surfactant, and so the different classification schemes are closely related.

4.5.3.1. Bancroft's Rule

One of the first empirical rules developed to describe the type of emulsion that could be stabilized by a given surfactant was proposed by Bancroft (Davis 1996). Bancroft's rule states that the phase in which the surfactant is most soluble will form the continuous phase of an emulsion. Hence a water-soluble surfactant should stabilize oil-in-water emulsions, whereas an oil-soluble surfactant should stabilize water-in-oil emulsions. This rule is applicable to a wide range of surfactants, although there are many important exceptions. For example, many amphiphilic molecules are highly soluble in either one phase or the other, but do not form stable emulsions because they are not particularly surface active or do not protect droplets against aggregation. Bancroft's rule is a useful empirical method for classifying surfactants; however, it provides little insight into the relationship between the molecular structure of a surfactant and its ability to form or stabilize emulsions.

4.5.3.2. Hydrophile–Lipophile Balance

The hydrophile–lipophile balance (HLB) concept is a semiempirical method which is widely used for classifying surfactants. The HLB is described by a number which gives an indication of the relative affinity of a surfactant molecule for the oil and aqueous phases (Davis 1994b). Each surfactant is assigned an HLB number according to its chemical structure. A molecule with a high HLB number has a high ratio of hydrophilic groups to lipophilic groups and vice versa. The HLB number of a surfactant can be calculated from a knowledge of the number and type of hydrophilic and lipophilic groups it contains, or it can be estimated from experimental measurements of its cloud point (Shinoda and Friberg 1986). The HLB numbers of many surfactants have been tabulated in the literature (Shinoda and Kunieda 1983; Becher 1985, 1996). A widely used semiempirical method for calculating the HLB number of a surfactant is as follows (Davis 1994b):

$$\text{HLB} = 7 + \Sigma(\text{hydrophilic group numbers}) - \Sigma(\text{lipophilic group numbers}) \quad (4.8)$$

Group numbers have been assigned to many different types of hydrophilic and lipophilic groups (Table 4.6). The sums of the group numbers of all the lipophilic groups and of all the hydrophilic groups are substituted into Equation 4.8 and the HLB number is calculated. Despite originally being developed as a semiempirical equation, Equation 4.8 has been shown to have a thermodynamic basis, with the sums corresponding to the free energy changes in the hydrophilic and lipophilic parts of the molecule when micelles are formed (Becher 1985).

The HLB number of a surfactant gives a useful indication of its solubility in either the oil and/or water phase and can be used to predict the type of emulsion that will be formed (Davis 1994b). A surfactant with a low HLB number (3 to 6) is predominantly hydrophobic, dissolves preferentially in oil, stabilizes water-in-oil emulsions, and forms reverse micelles in oil. A surfactant with a high HLB number (8 to 18) is predominantly hydrophilic, dissolves preferentially in water, stabilizes oil-in-water emulsions, and forms micelles in water. A surfactant with an intermediate HLB number (6 to 8) has no particular preference for either oil or water. Molecules with HLB numbers below 3 and above 18 are not particularly surface active and tend to accumulate preferentially in bulk oil or aqueous phases, rather than at an oil–water interface. Emulsion droplets are particularly prone to coalescence when they are stabilized by surfactants that have extreme or intermediate HLB numbers. At very high or low HLB numbers, a surfactant has such a low surface activity that it does not accumulate appreciably at the droplet surface and therefore does not provide protection against coalescence. At intermediate HLB numbers (6 to 8), emulsions are unstable to coalescence because the interfacial tension is so low that very little energy is required to disrupt the membrane. Maximum stability of emulsions is obtained for oil-in-water emulsions using surfactants with an HLB number around 10 to 12 and for water-in-oil emulsions around 3 to 5. This is because the surfactants are surface active but do not lower the interfacial tension so much that the droplets are easily disrupted. It is possible to adjust the effective HLB number by using a combination of two or more surfactants with different HLB numbers (Becher 1957).

One of the major drawbacks of the HLB concept is that it does not take into account the fact that the functional properties of a surfactant molecule are altered significantly by changes in temperature or solution conditions (Davis 1994b). Thus a surfactant may be capable of stabilizing oil-in-water emulsions at one temperature but water-in-oil emulsions at another temperature, even though it has exactly the same chemical structure. The HLB concept could be extended to include temperature effects by determining the group numbers as a function of temperature, although this would be a rather tedious and time-consuming task.

4.5.3.3. *Molecular Geometry and the Phase Inversion Temperature*

The molecular geometry of a surfactant molecule can be described by a packing parameter, (*p*) (Israelachvili 1992, 1994; Kabalnov and Wennerstrom 1996):

$$p = \frac{v}{la_0} \tag{4.9}$$

where *v* and *l* are the volume and length of the hydrophobic tail and a_0 is the cross-sectional area of the hydrophilic head group (Figure 4.14). When surfactant molecules associate with each other, they tend to form monolayers that have a curvature which allows the most efficient packing of the molecules. At this *optimum curvature,* the monolayer has its lowest free energy, and any deviation from this curvature requires the expenditure of energy. The optimum curvature (H_0) of a monolayer depends on the packing parameter of the surfactant: for *p* = 1, monolayers with zero curvature ($H_0 = 0$) are preferred; for *p* < 1, the optimum curvature is convex ($H_0 < 0$); and for *p* > 1, the optimum curvature is concave ($H_0 > 0$) (Figure 4.14). Simple geometrical considerations indicate that spherical micelles are formed when *p* is less than *one-third,* nonspherical micelles when *p* is between *one-third* and *one-half,* and bilayers when *p* is between *one-half* and *one* (Israelachvili 1992, 1994). Above a certain concentration, bilayers join up to form vesicles because energetically unfavorable end effects can be eliminated. At values of *p* greater than 1, reverse micelles are formed, in which the hydrophilic head groups are located in the interior (away from the oil) and the hydrophobic tail groups are located at the exterior (in contact with the oil) (Figure 4.14). The packing parameter therefore gives a useful indication of the type of association colloid that a surfactant molecule forms in solution.

The packing parameter is also useful because it accounts for the temperature dependence of the physicochemical properties of surfactant solutions and of emulsions (Kabalnov and Wennerstrom 1996). The temperature at which a surfactant solution converts from a micellar to a reverse-micellar system or an oil-in-water emulsion changes to a water-in-oil emulsion

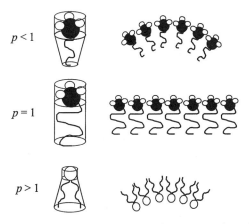

FIGURE 4.14   The physicochemical properties of surfactants can be related to their molecular geometry.

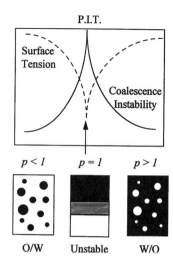

FIGURE 4.15   The phase inversion temperature occurs when the optimum curvature of a surfactant monolayer is zero.

is known as the phase inversion temperature or PIT (Shinoda and Kunieda 1983, Shinoda and Friberg 1986). Consider what happens when an emulsion that is stabilized by a surfactant is heated (Figure 4.15). At temperatures well below the PIT ($\approx 20°C$), the packing parameter is significantly less than unity, and so a system that consists of an oil-in-water emulsion in equilibrium with a swollen micellar solution is favored. As the temperature is raised, the hydrophilic head groups of the surfactant molecules become progressively dehydrated, which causes p to increase toward unity. Thus the emulsion droplets become more prone to coalescence and the swollen micelles grow in size. At the PIT, $p = 1$, and the emulsion breaks down because the droplets have an ultralow interfacial tension and therefore readily coalesce with each other (Aveyard et al. 1990, Kabalnov and Weers 1996). The resulting system consists of excess oil and excess water (containing some surfactant monomers), separated by a third phase that contains surfactant molecules aggregated into bilayer structures. At temperatures sufficiently greater than the PIT ($\approx 20°C$), the packing parameter is much larger than unity, and the formation of a system which consists of a water-in-oil emulsion in equilibrium with swollen reverse micelles is favored. A further increase in temperature leads to a decrease in the size of the reverse micelles and in the amount of water solubilized within them. The method of categorizing surfactant molecules according to their molecular geometry is now widely accepted as the most useful means of determining the type of emulsions they tend to stabilize (Kabalnov and Wennerstrom 1996).

4.5.3.4.   Other Factors

The classification schemes mentioned above provide information about the type of emulsion that a surfactant tends to stabilize (i.e., oil-in-water or water-in-oil), but they do not provide much insight into the size of the droplets which are formed during homogenization, the amount of surfactant required to form a stable emulsion, or the stability of the emulsion droplets once formed. In choosing a surfactant for a particular application, these factors must also be considered. The speed at which a surfactant adsorbs to the surface of the emulsion droplets produced during homogenization determines the minimum droplet size that can be produced: the faster the adsorption rate, the smaller the size (Chapters 5 and 6). The amount

Flexible Random-coil   Rigid Linear   Compact Globular
     Biopolymer   Biopolymer   Biopolymer

FIGURE 4.16   Biopolymers can adopt a number of different conformations in solution depending on their molecular structure. These can be conveniently categorized as random coil, rod-like, and globular.

of surfactant required to stabilize an emulsion depends on the total surface area of the droplets, as well as the surface area covered per unit mass of surfactant (Chapter 6). The magnitude and range of the repulsive interactions generated by an interfacial surfactant layer, as well as its viscoelasticity, determine the stability of emulsion droplets to aggregation (Chapters 3 and 7).

4.6.   BIOPOLYMERS

Proteins and polysaccharides are the two most important types of biopolymer used as ingredients in food emulsions (BeMiller and Whistler 1996, Damodaran 1996). Biopolymers provide an important source of energy and essential nutrients in the human diet (Smolin and Grosvenor 1994). In addition, they have the ability to modify the appearance, texture, stability, and taste of food emulsions because of their unique functional characteristics (i.e., their ability to stabilize emulsions and foams, to form gels, and to greatly enhance the viscosity of solutions) (Glicksman 1982a,b,c; Kinsella 1982; Dalgleish 1989; Damodaran 1994, 1996; Dickinson 1992; Dickinson and McClements 1995). A large number of ingredient companies manufacture and supply biopolymer ingredients to the food industry. A wide variety of different types of biopolymer ingredient are available, each with its own unique functional attributes. Selection of the most appropriate ingredient for a particular application is one of the most important decisions which a food manufacturer must make when creating a food emulsion.

4.6.1.   Molecular Characteristics

Proteins are polymers of amino acids, whereas polysaccharides are polymers of monosaccharides (MacGregor and Greenwood 1980, Lehninger et al. 1993, Damodaran 1996, BeMiller and Whistler 1996). The functional properties of biopolymers in foods are ultimately determined by molecular characteristics, such as their molecular weight, conformation, flexibility, polarity, and interactions (Dea 1982, Kinsella 1982, Damodaran 1996, BeMiller and Whitler 1996). These molecular characteristics are determined by the type, number, and sequence of the monomers that make up the polymer chain (van Holde 1977, MacGregor and Greenwood 1980). Monomers vary according to their polarity (ionic, dipolar, nonpolar, or amphiphilic), dimensions, interactions, and functional groups (Lehninger et al. 1993). If a biopolymer contains only one type of monomer, it is referred to as a *homopolymer* (e.g., starch or cellulose), but if it contains different types of monomer, it is referred to as a *heteropolymer* (e.g., xanthan, gum arabic, and any naturally occurring protein).

Both proteins and polysaccharides have covalent linkages between the monomers around which the polymer chain can rotate (Lehninger et al. 1993). Because of this and the fact that biopolymers contain large numbers of monomers (typically between 20 and 20,000), they can potentially take up a huge number of different configurations in solution. In practice, a biopolymer tends to adopt a conformation that minimizes its free energy under the prevailing environmental conditions.* A biopolymer does this by maximizing the number of favorable intermolecular and intramolecular interactions, minimizing the number of unfavorable ones, and maximizing its entropy, within the constraints set by its molecular structure and the prevailing environmental conditions (van Holde 1977, Dill 1990). Nevertheless, most foods are nonequilibrium systems, and so a biopolymer may exist in a metastable state, because there is a large activation energy preventing it from reaching the most stable state. The configurations that isolated biopolymers tend to adopt in aqueous solutions can be conveniently divided into three categories: globular, rod-like, or random coil (Figure 4.16). Globular biopolymers have fairly rigid compact structures, rod-like biopolymers have fairly rigid extended structures (often helical), and random-coil biopolymers have highly dynamic and flexible structures. Biopolymers can also be classified according to the degree of branching of the chain (Lehninger et al. 1993). Most proteins have linear chains, whereas polysaccharides can have either linear (e.g., amylose) or branched (e.g., amylopectin) chains. In practice, many biopolymers do not have exclusively one type of conformation, but have some regions which are random coil, some which are rod-like, and some which are globular. In addition, biopolymer molecules may change from one type of conformation to another if their environment is altered (e.g., pH, ionic strength, solvent composition, or temperature): helix ⇔ random coil or globular ⇔ random coil (Whitaker 1977, Sand 1982, Cesero 1994). In solution, a biopolymer may exist as an isolated molecule, or it may be associated with other types of molecules. The molecular conformation and association of biopolymers are governed by a delicate balance of interaction energies and entropy effects (Dickinson and McClements 1995).

4.6.2. Molecular Basis of Conformation and Aggregation

In this section, the most important types of molecular interaction and entropy effects which influence the conformation and aggregation of biopolymer molecules in aqueous solution are discussed.

4.6.2.1. Hydrophobic Interactions

Hydrophobic interactions play a major role in determining the molecular structure and aggregation of biopolymers which contain appreciable numbers of nonpolar groups (Alber 1989, Dill 1990). Protein molecules often have significant amounts of nonpolar amino acids along their polypeptide backbone, such as valine, leucine, isoleucine, and phenylalanine (Lehninger et al. 1993). Most polysaccharides are predominantly hydrophilic, but some do have nonpolar groups attached to their backbone (e.g., gum arabic and modified starch) (BeMiller and Whistler 1996). Biopolymers that have significant proportions of nonpolar groups tend to adopt an arrangement that minimizes the contact area between the nonpolar groups and water because of the hydrophobic effect (Section 4.4.3). This causes proteins that have a large proportion of nonpolar amino acids to fold into compact globular structures in which the nonpolar amino acids are located in the interior of the protein (away from the water), whereas the polar residues are located at the exterior (in contact with the water) (Damodaran 1996). Hydrophobic interactions are also believed to play an important role in stabilizing the rod-like helical structures formed in some biopolymers, because nonpolar groups can be located in the core of the helix, away from the water.** Hydrophobic inter-

---

* Helix formation is also favored because it maximizes the number of hydrogen bonds formed.
** It is possible that some biopolymers are "trapped" in a metastable state because there is a large energy barrier between it and the lowest free energy state.

actions are also largely responsible for the aggregation and surfactivity of amphiphilic biopolymers (Damodaran 1996). By aggregating to other nonpolar molecules, or by adsorbing to a nonpolar surface, a biopolymer can reduce the contact area between its hydrophobic groups and water.

4.6.2.2. Electrostatic Interactions

Electrostatic interactions play a major role in determining the molecular structure and aggregation of many biopolymers (MacGregor and Greenwood 1980, Dickinson and McClements 1995, Damodaran 1996). Proteins contain a number of amino acids that can ionize to form either positively charged ions (e.g., arginine, lysine, proline, histidine, and the terminal amino group) or negatively charged ions (e.g., glutamic and aspartic acids and the terminal carboxyl group) (Lehninger et al. 1993, Damodaran 1996). Some polysaccharides also have ionizable groups on their backbone (e.g., sulfate, carboxylate, or phosphate groups) (MacGregor and Greenwood 1980, BeMiller and Whistler 1996). The net charge on a biopolymer molecule depends on the pK values of its ionizable groups and the pH of the aqueous environment (Matthew 1985). Electrostatic interactions involving biopolymers are therefore particularly sensitive to the pH of the aqueous phase (Launay et al. 1986, Rha and Pradipasena 1986). In addition, they also depend on the concentration and type of counterions in the aqueous phase because of electrostatic screening effects (Section 3.4). If a biopolymer contains many similarly charged groups, it is more likely to adopt an extended configuration because this increases the average distance between the charges and therefore minimizes the unfavorable electrostatic repulsions (Launay et al. 1986, Rha and Pradipasena 1986). If, on the other hand, the biopolymer contains many oppositely charged groups, it is more likely to fold up into a compact structure that maximizes the favorable electrostatic attractions. For this reason, proteins are often extremely compact at their isoelectric point and unfold as the pH is either increased or decreased. Electrostatic interactions also play an important role in determining the aggregation of biopolymer molecules in solution. Similarly charged biopolymers repel each other and therefore tend to exist as individual molecules, whereas oppositely charged biopolymers attract each other and therefore tend to aggregate (depending on the strength of the various other types of interactions involved). The binding of low-molecular-weight ions, such as Na^+ and Ca^{2+}, to biopolymer molecules is also governed by electrostatic interactions and may influence the strength of the hydration repulsion between biopolymers in solution (Section 3.8).

4.6.2.3. Hydrogen Bonding

Both proteins and polysaccharides contain monomers that are capable of forming hydrogen bonds (Lehninger et al. 1993). Hydrogen bonds are a relatively strong type of molecular interaction, and therefore a system attempts to maximize the number and strength of the hydrogen bonds formed. A biopolymer molecule may adopt an arrangement that enables it to maximize the number of hydrogen bonds which are formed between the monomers within it, which leads to the formation of ordered regions such as helices, sheets, and turns (Lehninger et al. 1993). Alternatively, a biopolymer may adopt a less ordered structure where the monomers form hydrogen bonds with the surrounding water molecules. Thus, part or all of a biopolymer may exist in either a highly ordered conformation (which is entropically unfavorable) with extensive intramolecular hydrogen bonding or may exist in a more random-coil conformation (which is entropically more favorable) with extensive intermolecular hydrogen bonding. The type of structure formed by a biopolymer under a certain set of environmental conditions is governed by the relative magnitude of the hydrogen bonds compared to the various other types of interactions, most notably hydrophobic, electrostatic, and configurational entropy.

Hydrogen bond formation is also responsible for the association of many types of food biopolymer. Segments of one biopolymer may be capable of forming strong hydrogen bonds with segments on another molecule, which causes the molecules to aggregate (Lehninger et al. 1993). These junction zones usually involve hydrogen-bonded helical or sheet-like structures. Hydrogen-bonded junction zones tend to be stable at low temperatures but dissociate as the temperature is raised above a certain value because the configurational entropy term dominates (Section 4.6.2.7).

4.6.2.4. Steric Overlap and van der Waals Interactions

There is an extremely strong repulsive interaction between atoms or molecules at close separations because of the overlap of their electron clouds (Section 2.3.4). This steric overlap interaction determines the minimum distance that two atoms or molecules can come to each other. Steric overlap interactions greatly restrict the spatial distribution of the monomers within a polymer because two groups cannot occupy the same space. At the molecular level, van der Waals interactions operate between all types of molecules and tend to have fairly similar magnitudes (Israelachvili 1992). Consequently, they only play a minor role in determining the conformation of biopolymers in solution, since there is little change in the van der Waals interactions in the folded and unfolded states. Nevertheless, if a biopolymer molecule is large enough to act as a colloidal particle, then there may be a significantly strong van der Waals attraction, which favors its aggregation with other biopolymer molecules (Chapter 3).

4.6.2.5. Disulfide Bonds

The conformation and aggregation of many protein molecules are influenced by their ability to form disulfide bonds (MacGregor and Greenwood 1980, Dickinson and McClements 1995, Damodaran 1996). Cystine is an amino acid that has a thiol group (–SH) which is capable of forming disulfide bonds (–S–S–) with other thiol groups by an oxidation reaction (Friedman 1973). Free thiol groups may also participate in thiol–disulfide interchanges with disulfide bonds. Proteins are therefore capable of forming both intramolecular and intermolecular disulfide bonds under the appropriate conditions. Intramolecular bonds form when a pair of cysteine residues are brought into close proximity by the folding of a protein molecule and are important for stabilizing the structure of globular proteins against unfolding. Intermolecular disulfide bonds form when two different protein molecules come close to each other and are important in determining the aggregation of proteins in gels and at interfaces (Dickinson and Matsumura 1991, McClements et al. 1993d).

The chemistry of the sulfhydryl group (–SH) plays an important role in the molecular structure and interactions of proteins (Friedman 1973). The pK value of the –SH group of free L-cysteine is about 8.3. In proteins, the pK value is often higher and depends on the local environment of the –SH group and the degree of unfolding of the protein molecule (Friedman 1973). The –S⁻ ion, which predominates at high pH, is strongly nucleophilic and is capable of interacting with disulfide bonds by nucleophilic displacement.

$$P_1\text{–S–S–}P_2 + R\text{–S}^- \rightarrow P_1\text{–S}^- + R\text{–S–S–}P_2 \tag{4.10}$$

where P_1 and P_2 refer to polypeptide chains that may be on the same protein molecule or on different protein molecules. The R–S⁻ group may also be on the same or another protein molecule, or it may even be some other chemical group (e.g., β-mercaptoethanol). If there is an excess of the reactive sulfhydryl group present, the disulfide bond between two polypeptide chains may be broken:

$$P_1\text{--S--S--}P_2 + R\text{--S--S--}R \rightarrow P_1\text{--S--S--}R + R\text{--S--S--}P_2 \qquad (4.11)$$

Disulfide interchange does not occur spontaneously in acidic conditions (pH <6) because the thiol group is below its pK value and is therefore not ionized (Friedman 1973). The sulfyhydryl and disulfide groups in proteins are usually located in the interior of the folded molecule and are therefore unavailable for interaction, even under conditions that would normally favor thiol/disulfide interchange. Unfolding of protein molecules exposes reactive sulfyhydryl groups and allows thiol–disulfide interchanges to take place under the appropriate conditions (Dickinson and Matsumura 1991, McClements et al. 1993d). This explains the increase in sulfhydryl–disulfide interchange which occurs on heating proteins (McClements et al. 1994), when proteins are adsorbed to an interface (Dickinson and Matsumura 1991), or after proteins are treated with proteolytic enzymes such as trypsin and chymotrypsin (Friedman 1973).

4.6.2.6. *Configurational Entropy*

One of the most important factors determining the conformation and aggregation of biopolymer molecules in solution is their configurational entropy (Alber 1989, Dill 1990). Biopolymers have both local and nonlocal contributions to their configurational entropy. The local entropy refers to the number of conformations which sequences of monomers connected to one another along the chain can adopt. The formation of ordered structure, such as helices or sheets, is entropically unfavorable. The nonlocal entropy is determined by the possible configurations that the whole of the biopolymer chain can adopt. Thus, highly flexible random-coil biopolymers, which can occupy a large number of different conformations, have a much higher configurational entropy than compact globular biopolymers, which can occupy far fewer different conformations. Steric overlap interactions play an important role in determining the magnitude of the nonlocal entropy. Two segments of a biopolymer chain cannot occupy the same volume, and therefore the number of configurations that the molecule can adopt is restricted, which is entropically unfavorable.

The free energy associated with the configurational entropy ($-T\Delta S$) increases as the temperature is raised, which explains why globular proteins unfold at high temperatures, even though the hydrophobic attraction is stronger and therefore favors the folded state (Dickinson and McClements 1995).

An important consequence of the nonlocal entropy contribution to the conformation of a biopolymer is the effect of covalent cross-links between different segments of the molecule. Introducing a covalent bond, such as a disulfide bond in a protein, severely restricts the number of conformations that the unfolded molecule can adopt and therefore reduces the entropic driving force that favors the random-coil conformation (Kinsella 1982).

4.6.2.7. *Molecular Conformation and Aggregation*

The conformation and aggregation of biopolymer molecules depend on the relative magnitude of the various attractive and repulsive interactions which occur within and between molecules, as well as their configurational entropy (Dickinson and McClements 1995). Consider a biopolymer that can exist in two different states:

$$\text{State (1)} \leftrightarrow \text{State (2)} \qquad (4.12)$$

These states may be a folded and an unfolded state or an aggregated and an unaggregated state. The biopolymer will tend to exist in the state which has the lowest free energy under the prevailing environmental conditions (e.g., pH, ionic strength, ion type, temperature, and solvent type). The free energy of each state is governed by the various types of molecular interactions and entropy contributions mentioned above. Thus, the overall free energy change which occurs when a biopolymer molecule undergoes a transition from one state to another can be represented by the following equation:

$$\Delta G_{transition} = \Delta G_{VDW} + \Delta G_{H} + \Delta G_{HB} + \Delta G_{S} + \Delta G_{E} - T\Delta S_{CE} \qquad (4.13)$$

where ΔG_{VDW}, ΔG_{H}, ΔG_{HB}, ΔG_{S}, and ΔG_{E} are the differences in free energy between the two states associated with van der Waals, hydrophobic, hydrogen bonding, steric, and electrostatic interactions, and ΔS_{CE} is the change in configurational entropy. If $\Delta G_{transition}$ is negative, the biopolymer will tend to undergo the transition; otherwise, it will not. It is convenient to group together the various contributions that favor the transition and those that oppose it:

$$\Delta G_{transition} = \Delta G_{oppose} + \Delta G_{favor} \qquad (4.14)$$

If the contributions that favor the transition dominate, then the biopolymer will move from state 1 to state 2; otherwise, it will remain in state 1 (Equation 4.12). For many systems, the factors favoring the transition have a similar magnitude, but a different sign, than those opposing the transition (i.e., $\Delta G_{oppose} \approx -\Delta G_{favor}$) (Dickinson and McClements 1995). Thus the magnitude of $\Delta G_{transition}$ is much smaller than either ΔG_{oppose} or ΔG_{favor}, because it is the difference between these two relatively large numbers. This accounts for the fact that small changes in environmental conditions can cause a biopolymer to undergo a transition from one state to another. For example, a globular protein may unfold when it is heated above a certain temperature, or a random-coil protein may form a gel when it is cooled below a certain temperature.

It must be stressed that most foods are nonequilibrium systems, and so the biopolymer molecules are not in their lowest free energy state. Often a biopolymer is "trapped" in a metastable state because there is a large activation energy which prevents it from reaching its equilibrium state (Section 1.2.1). For example, if a globular biopolymer is heated above a temperature where it unfolds, it may aggregate with its neighbors. When the solution is cooled, the molecules are not able to disentangle themselves from their neighbors and so they remain aggregated, even though a solution of individual molecules may have a lower free energy.

The conformation of a biopolymer in solution often determines its functional characteristics. For example, alterations in the conformation of β-lactoglobulin during the preparation of powdered whey-protein concentrates or isolates is one of the major reasons for the inconsistency of its quality from batch to batch. Thus, two ingredients that have exactly the same composition may have very different functional properties because of alterations in the structure of the proteins during processing.

4.6.3. Functional Properties

4.6.3.1. Protein Hydration and Water Solubility

Many of the functional properties of biopolymers in food emulsions are governed by their interactions with water (e.g., solubility, dispersibility, swelling, thickening, emulsification, foaming, and gelling) (Suggett 1975a,b; Damodaran 1996; Fennema 1996b). Biopolymer ingredients are usually added to the aqueous phase of food emulsions in a powdered form. The functional properties of many of these biopolymers are only exhibited when they are fully dissolved and evenly distributed throughout the aqueous phase. Consequently, the effective dissolution of the powdered biopolymer ingredient in an aqueous solution is an important part of emulsion preparation. This process involves a number of stages, including dispersion, wetting, swelling, and dissolution. The effectiveness and rate of dissolution depend on many factors, including the pH, ionic strength, temperature, and composition of the aqueous phase, as well as the application of shearing forces.

It is useful to examine the way in which successive water molecules bind to dried biopolymer molecules (Fennema 1996b). First, the charged groups are hydrated, then the polar groups, and finally the nonpolar groups. Charged groups bind more water molecules per group (3 to 7) than polar groups (2 to 3), which in turn bind more than nonpolar groups (≈ 1). The water in a food containing biopolymers may exist in a number of different environments, including physically bound water, hydrodynamic water, capillary water, and free water (Fennema 1996b).

The *water solubility* of a biopolymer molecule depends on its compatibility with the aqueous solvent in which it is dispersed, which is governed by the relative magnitude of biopolymer–biopolymer, water–biopolymer, and water–water interactions. A biopolymer molecule has a low water solubility when the strength of the water–biopolymer interactions is significantly weaker than the average strength of the water–water and biopolymer–biopolymer interactions, because the molecules tend to associate with each other rather than with the solvent molecules (Section 2.6). An appreciation of the factors which determine the water solubility of a protein depends on an understanding of the various types of molecular interaction and entropy effects discussed in Section 4.6.2.

In general, the water solubility of biopolymer molecules is determined by a combination of van der Waals, electrostatic, hydrogen bonding, steric overlap, and hydrophobic interactions, as well as by the configurational entropy. In practice, the water solubility is usually dominated by only one or two of these interactions, the most common being the hydrophobic and electrostatic interactions (Damodaran 1996). It is therefore important for food scientists to identify the interactions which are most important for each type of biopolymer and to establish the influence of solvent conditions on these interactions.

Hydrophobic interactions promote aggregation and therefore lead to poor water solubility. They are particularly important when biopolymer molecules have high proportions of nonpolar groups on their surface. This accounts for the decreasing solubility of proteins with increasing surface hydrophobicity (Damodaran 1996) and for the fact that many globular proteins become insoluble in water when they are heated above a temperature where the protein molecule unfolds and exposes nonpolar groups (Damodaran 1996).

Electrostatic interactions play a major role in determining the water solubility of biopolymers that have charged groups and are important for all proteins and many polysaccharides (BeMiller and Whistler 1996, Damodaran 1996). Electrostatic interactions may be either attractive or repulsive, depending on the signs of the charges involved, and they may therefore either increase or decrease water solubility. A protein molecule is positively charged at pH values below its isoelectric point and negatively charged at pH values above it. As a consequence, there is an electrostatic repulsion between similarly charged molecules which prevents them from coming close enough together to aggregate and therefore enhances their water solubility (de Wit and van Kessel 1996). In addition, the molecules are also prevented from aggregating because the ionized groups are highly hydrated, which leads to a short-range hydration repulsion (Section 3.8). The water solubility of most proteins decreases dramatically at their isoelectric point (Franks 1991, Damodaran 1996). A protein molecule has no *net* charge at its isoelectric point, but it does have some groups that are positively charged and some that are negatively charged. Proteins are therefore particularly susceptible to aggregation at their isoelectric point because there is no electrostatic repulsion to prevent them from coming close together. In fact, there may even be an electrostatic attraction between the positive groups on one protein molecule and the negative groups on another, which promotes protein aggregation and insolubility. A number of proteins are highly soluble across the whole pH range, because the attractive forces between them are not sufficiently strong to overcome the thermal energy of the system and so the molecules remain dispersed (Damodaran 1996).

The water solubility of biopolymer molecules is highly dependent on the type and concentration of electrolyte ions in solution (Franks 1991). These ions can influence the water solubility of biopolymers in a number of different ways. At relatively low ionic strengths, electrolyte ions screen electrostatic interactions between molecules (Section 3.4.2). At the isoelectric point, one would expect screening to decrease the strength of the electrostatic attraction between protein molecules and therefore increase solubility. On the other hand, at pH values away from the isoelectric point, one would expect screening to decrease the electrostatic repulsion between charged biopolymers and therefore decrease their solubility. At intermediate ionic strengths, electrolyte ions may bind to the surface of biopolymers, thus increasing the short-range hydration repulsion interactions between them and thereby increasing their solubility (Israelachvili 1992). Alternatively, the electrolyte may alter the structural organization of the water molecules, which can either increase or decrease the magnitude of the hydrophobic attraction, depending on the nature of the ions involved (Israelachvili 1992, Damodaran 1996). At high ionic strengths, biopolymer molecules are often precipitated out of solution above a critical salt concentration, which is known as "salting out," because the majority of water molecules are strongly "bound" to the electrolyte ions and are therefore not available to hydrate the biopolymers (Damodaran 1996).

The temperature of an aqueous solution also plays an important role in determining the water solubility of biopolymers (Damodaran 1996). Altering the temperature of a solution changes the relative magnitude of the various kinds of molecular interaction and entropy effects mentioned in Section 4.6.2. This alters the balance between the forces favoring solubility and those favoring insolubility. An increase in temperature usually causes an increase in the strength of the hydrophobic attraction, a decrease in the strength of hydrogen bonds, a decrease in the magnitude of the hydration repulsion, an increase in the strength of any electrostatic interactions, and an increase in the configurational entropy. The effect of temperature on the solubility of a biopolymer therefore depends on the relative importance of the various types of interactions under a given set of experimental conditions.

The water solubility of a biopolymer can be characterized by a solubility index (SI), which is a measure of the percentage of biopolymer which is soluble in an aqueous solution under a specified set of solvent and environmental conditions (e.g., pH, ionic strength, temperature, and centrifugation speed) (Damodaran 1996). A known concentration of biopolymer is dispersed in solution, and then the solution is ultracentrifuged at a specific speed and time. The concentration of biopolymer remaining in solution is measured using an appropriate technique:

$$\text{SI} = 100 \times \frac{M_{remaining}}{M_{initial}} \tag{4.15}$$

where $M_{remaining}$ and $M_{initial}$ are the concentrations of protein in solution after and before centrifugation.

The solubility index is often a good indication of the degree of denaturation of a protein ingredient: the lower the solubility index, the greater the degree of denaturation.

4.6.3.2. Emulsification

Biopolymers that have significant amounts of both polar and nonpolar groups tend to be surface active (i.e., they are able to accumulate at oil–water or air–water interfaces (Dickinson 1992, Dalgleish 1996a, Damodaran 1996). The major driving force for adsorption is the hydrophobic effect. When the biopolymer is dispersed in an aqueous phase, some of the nonpolar groups are in contact with water, which is thermodynamically unfavorable, because of hydrophobic interactions (Section 4.4.3). When a biopolymer adsorbs to an interface, it can adopt a conformation where the nonpolar groups are located in the oil phase (away from the water) and the hydrophilic groups are located in the aqueous phase (in contact with the water). Adsorption also reduces the contact area between the oil and water molecules at the oil–water interface, which lowers the interfacial tension (Chapter 5). Both of these factors favor the adsorption of an amphiphilic molecule to an oil–water interface. The conformation that a biopolymer adopts at an interface and the physicochemical properties of the membrane formed depend on its molecular structure and interactions (Das and Kinsella 1990, Dickinson 1992, Dalgleish 1996a, Damodaran 1996). Flexible random-coil biopolymers adopt an arrangement where the predominantly nonpolar segments protrude into the oil phase, the predominantly polar segments protrude into the aqueous phase, and the neutral regions lie flat against the interface (Figure 4.17). The membranes formed by these types of molecules tend to be relatively open, thick, and of low viscoelasticity. Globular biopolymers (usually proteins) adsorb to an interface so that the predominantly nonpolar regions on the surface of the molecule face the oil phase, while the predominantly polar regions face the aqueous phase, and so they tend to have a definite orientation at an interface (Figure 4.18). Once they have adsorbed to an interface, biopolymers often undergo structural rearrangements so that they can maximize the number of contacts between nonpolar groups and oil. Random-coil biopolymers are relatively flexible molecules and can therefore rearrange their structures fairly rapidly, whereas globular biopolymers are more rigid molecules and therefore rearrange more slowly. The unfolding of a globular protein at an interface often exposes amino acids that were originally located in the hydrophobic interior of the molecule, which can lead to enhanced interactions with neighboring protein molecules through hydrophobic attraction or disulfide bond formation (Dickinson and Matsumura 1991, McClements et al. 1993d). Consequently, globular proteins tend to form relatively thin and compact membranes that have high viscoelasticities (Dickinson 1992).

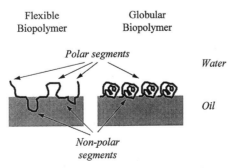

FIGURE 4.17   The structure of the interfacial membrane depends on the molecular structure and interactions of the surface-active molecules.

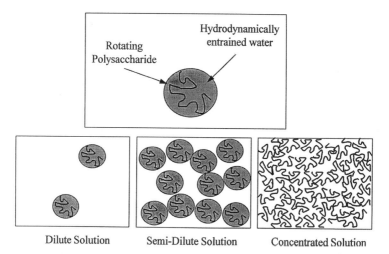

FIGURE 4.18 Extended biopolymers sweep out a large volume of water as they rotate in solution, and so they have a large effective volume fraction.

Membranes formed by globular proteins therefore tend to be more resistant to rupture than those formed from random-coil proteins.

To be effective emulsifiers, biopolymers must rapidly adsorb to the surface of the emulsion droplets created during homogenization and then form a membrane that prevents the droplets from aggregating with one another (Chapter 6). Biopolymer membranes stabilize emulsion droplets against aggregation by a number of different mechanisms. All biopolymers provide short-range steric repulsive forces that are usually sufficiently strong to prevent droplets from getting close enough together to coalesce (Section 3.5). If the membrane is sufficiently thick, the steric repulsive forces can also prevent droplets from flocculating. Otherwise, it must be electrically charged so that it can prevent flocculation by electrostatic repulsion (Section 3.4). The properties of emulsions stabilized by charged biopolymers are particularly sensitive to pH and ionic strength. At pH values near the isoelectric point of proteins, or at high ionic strengths, the electrostatic repulsion between droplets may not be sufficiently large to prevent the droplets from aggregating.

A wide variety of proteins are used as emulsifiers in foods because they naturally have a high proportion of nonpolar groups and are therefore surface active (Charalambous and Doxstakis 1989, Dickinson 1992, Damodaran 1996). Most polysaccharides are predominantly hydrophilic and are therefore not particularly surface active (BeMiller and Whistler 1996). However, a small number of naturally occurring polysaccharides do have some hydrophobic character (e.g., gum arabic) or have been chemically modified to introduce nonpolar groups (e.g., some modified starches), and these biopolymers can be used as emulsifiers (BeMiller and Whistler 1996).

4.6.3.3. Thickening and Stabilization

The second major role of biopolymers in food emulsions is to increase the viscosity of the aqueous phase (Mitchell and Ledward 1986). Viscosity enhancement modifies the texture and mouthfeel of food products ("thickening"), as well as reducing the rate at which particles sediment or cream ("stabilization"). Biopolymers used to increase the viscosity of aqueous solutions are usually highly hydrated and extended molecules or molecular aggregates. Their ability to increase the viscosity of a solution depends principally on their molecular weight, degree of branching, conformation, and flexibility (Launay et al. 1986, Rha and Pradipasena 1986).

The viscosity of a dilute solution of particles increases linearly as the concentration of the particles increases (Section 8.4):

$$\eta = \eta_0(1 + 2.5\phi) \tag{4.16}$$

where η is the viscosity of the particulate solution, η_0 is the viscosity of the pure solvent, and ϕ is the volume fraction of particles in solution. The ability of some biopolymers to greatly enhance the viscosity of aqueous solutions when used at very low concentrations is because their *effective* volume fraction is much greater than their *actual* volume fraction. A biopolymer rapidly rotates in solution because of its thermal energy, and so it sweeps out a spherical volume of water that has a diameter approximately equal to its end-to-end length (Figure 4.18). The volume of the biopolymer molecule is only a small fraction of the total volume of the sphere that is swept out, and so the effective volume fraction of a biopolymer is much greater than its actual volume fraction. The effectiveness of a biopolymer at increasing the viscosity increases as the volume fraction that it occupies within this sphere decreases. Thus, large highly extended linear biopolymers increase the viscosity more effectively than small compact branched biopolymers (Launay et al. 1986, Rha and Pradipasena 1986).

In a dilute biopolymer solution, the individual molecules (or molecular aggregates) do not interact with each other (Dickinson 1992, Lapasin and Pricl 1995). As the concentration of biopolymer increases above some critical value (c^*), the viscosity of the solution increases rapidly because the spheres swept out by the biopolymers begin to interact with each other. This type of solution is known as a *semidilute* solution, because even though the molecules are interacting with one another, each individual biopolymer is still largely surrounded by solvent molecules. At still higher polymer concentrations, the molecules pack so closely together that they become entangled with each other and the system has more gel-like characteristics. Biopolymers which are used to thicken the aqueous phase of emulsions are often used in the semidilute concentration range (Dickinson 1992).

Solutions containing extended biopolymers often exhibit strong shear-thinning behavior (pseudoplasticity); that is, their apparent viscosity decreases with increasing shear stress (Lapasin and Pricl 1995). Shear thinning occurs because the biopolymer molecules become aligned with the shear field or because the weak physical interactions responsible for biopolymer–biopolymer interactions are disrupted. Some biopolymer solutions have a characteristic yield stress. If such a biopolymer solution experiences an applied stress which is below its yield stress, it acts like an elastic solid, but when it experiences a stress that exceeds the yield stress, it acts like a liquid (Section 8.2.3). The characteristic rheological behavior of biopolymer solutions plays an important role in determining their functional properties in food emulsions. For example, a salad dressing must be able to flow when it is poured from a container, but must maintain its shape under its own weight after it has been poured onto a salad. The amount and type of biopolymer used must therefore be carefully selected so that it provides a low viscosity when the salad dressing is poured (high applied stress) but a high viscosity when the salad dressing is allowed to sit under its own weight (low applied stress). The viscosity of biopolymer solutions is also related to the mouthfeel of a food product. Liquids that do not exhibit extensive shear-thinning behavior at the shear stresses experienced within the mouth are perceived as being "slimy." On the other hand, a certain amount of viscosity is needed to contribute to the "creaminess" of a product. The shear-thinning behavior of biopolymer solutions is also important for determining the stability of food emulsions to creaming. As an oil droplet moves through an aqueous phase, it only exerts a very small shear stress on the surrounding liquid. As a result of the shear-thinning behavior of the solution, it experiences a very high viscosity, which greatly slows down the rate at which it creams.*

Many biopolymer solutions also exhibit thixotropic behavior (their viscosity decreases with time when they are sheared at a constant rate), because the weak physical interactions that cause biopolymer molecules to aggregate are disrupted. Once the shearing stress is removed from the sample, the biopolymer molecules may be able to reform the weak physical bonds with their neighbors, and so the system regains its original structure and rheological properties. This type of system is said to be reversible, and the speed at which the structure is regained may be important for the practical application of a biopolymer in a food. If the bonds are unable to reform once they are disrupted or if they are only partially reformed, then the system is said to be irreversible or partially reversible, respectively.

A food manufacturer must therefore select an appropriate biopolymer or combination of biopolymers to produce a final product that has a desirable mouthfeel, stability, and texture. Both proteins and polysaccharides can be used as thickening agents, but polysaccharides are usually preferred because they tend to have higher molecular weights and be more extended so that they can be used at much lower concentrations.

4.6.3.4. Gelation

Biopolymers are used as functional ingredients in many food emulsions because of their ability to cause the aqueous phase to gel (e.g., yogurts, cheeses, desserts, egg and meat products) (Morris 1986, Ledward 1986, Clark and Lee-Tuffnell 1986, Ziegler and Foegedding 1990). Gel formation imparts desirable textural and sensory attributes, as well as preventing the droplets from creaming. A biopolymer gel consists of a three-dimensional network of aggregated or entangled biopolymers that entraps a large volume of water, giving the whole structure some "solid-like" characteristics (Mitchell and Ledward 1986, Lapasin and Pricl 1995).

The properties of biopolymer gels depend on the type, structure, and interactions of the molecules they contain (Dea 1982, Ziegler and Foegedding 1990). Gels may be transparent or opaque, hard or soft, brittle or rubbery, homogeneous or heterogeneous, and exhibit syneresis or have good water-holding capacity. Gelation may be induced by a variety of different methods, including altering the temperature, pH, ionic strength, or solvent quality or by adding denaturing or cross-linking agents (Ziegler and Foegedding 1990). Biopolymers may be cross-linked to one another either by covalent and/or noncovalent bonds. The type of cross-links formed depends on the nature of the molecules involved, as well as the prevailing environmental conditions. It is often convenient to distinguish between two different types of gel: particulate and filamentous (Figure 4.19). Particulate gels consist of biopolymer aggregates (particles or clumps) which are assembled together to form a three-dimensional network (Doi 1993). This type of gel tends to be formed when the biopolymer molecules are able to interact with their neighbors at any point on their surface. Particulate gels are optically opaque because the particles are large enough to strongly scatter light and are prone to syneresis because the relatively large pore sizes between the particles means that the water is not held tightly within the gel network by capillary forces. Filamentous gels consist of thin filaments of individual or aggregated biopolymer molecules (Doi 1993). Filamentous gels tend to be optically transparent because the filaments are so thin that they

---

* It should be noted that biopolymers can actually promote creaming at certain concentrations because they cause depletion flocculation (Section 3.6).

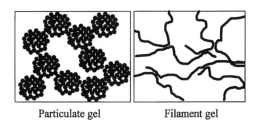

Particulate gel   Filament gel

FIGURE 4.19   It is often convenient to categorize gels as being either particulate or filamentous, depending on the structural organization of the molecules.

do not scatter light significantly. They also tend to have good water-holding capacity because the small pore size of the gel network means that the water molecules are held tightly by capillary forces.

In some foods, a gel is formed on heating (*heat-setting* gels), while in others it is formed on cooling (*cold-setting* gels) (Ziegler and Foegedding 1990). Gels may also be either *thermo-reversible* or *thermo-irreversible,* depending on whether or not gelation is reversible. Gelatin is an example of a cold-setting thermo-reversible gel: when a solution of gelatin molecules is cooled below a certain temperature, a gel is formed, but when it is reheated, the gel melts (Ledward 1986). Egg white is an example of a heat-setting thermo-irreversible gel: when egg white is heated above a certain temperature, a characteristic white gel is formed, but when it is cooled back to room temperature, it remains as a white gel rather than reverting back into the liquid from which it was formed (Ziegler and Foegedding 1990, Doi 1993). Whether a gel is reversible or irreversible depends on the type of bonds holding the biopolymer molecules together, as well as any changes in the molecular structure and organization of the molecules during gelation. Biopolymer gels that are stabilized by noncovalent interactions, and which do not involve large changes in the structure of the individual molecules prior to gelation, tend to be reversible. On the other hand, gels that are held together by covalent bonds, or which involve large changes in the structure of the individual molecules prior to gelation, tend to form irreversible gels.

The type of interaction holding the molecules together in gels varies from biopolymer to biopolymer. Some proteins and polysaccharides form helical junction zones through extensive hydrogen bond formation (e.g., gelatin and gums) (Dea 1982, Ledward 1986, Morris 1986). This type of junction zone tends to form when a gel is cooled and be disrupted when it is heated and is thus responsible for cold-setting gels. Below the gelation temperature, the attractive hydrogen bonds favor junction-zone formation, but above this temperature, the configurational entropy favors a random-coil-type structure. Biopolymers with extensive nonpolar groups tend to associate via hydrophobic interactions (e.g., caseins or denatured whey proteins) (Ziegler and Foegedding 1990). Electrostatic interactions play an important role in determining the gelation behavior of many biopolymers, and so gelation is particularly sensitive to the pH and ionic strength of solutions containing these biopolymers (McClements and Keogh 1995). For example, at pH values sufficiently far away from their isoelectric point, proteins may be prevented from gelling because of the strong electrostatic repulsion between the molecules; however, if the pH is adjusted near to the isoelectric point, or if salt is added, the proteins tend to aggregate and form a gel (Mulvihill and Kinsella 1987, 1988). The addition of multivalent ions, such as Ca^{2+}, can promote gelation of charged biopolymer molecules by forming salt bridges between the molecules (Morris 1986, Mulvihill and Kinsella 1988). Proteins with thiol groups are capable of forming covalent linkages through thiol–disulfide interchanges, which help to strengthen and enhance the stability of gels (Mulvihill and Kinsella 1987). The tendency for a biopolymer to form a gel under certain conditions and the physical properties of the gel formed depend on a delicate balance of various kinds of biopolymer–biopolymer, biopolymer–solvent, and solvent–solvent interactions.

The properties of food emulsions that have a gelled aqueous phase are dependent on the nature of the interactions between the emulsifier adsorbed to the surface of the droplets and the biopolymer molecules in the gel network (McClements et al. 1993c, Dickinson et al. 1996). If there is a strong attractive interaction between the droplet membrane and the gel network, then the network is reinforced and a strong gel is formed. On the other hand, if the emulsifier membrane does not interact favorably with the gel network, then the droplets may disrupt the network and weaken the gel strength. The magnitude of this effect depends on the size of the emulsion droplets (McClements et al. 1993c). The larger the droplets compared to the pore size of the gel network, the greater the disruptive effect. Specific interactions between the proteins and surfactants may also influence the properties of the gels formed (Dickinson and Yamamoto 1996, Dickinson ct al. 1996).

4.6.4. Modification of Biopolymers

A wide variety of natural biopolymers could be used as functional ingredients in foods; however, many of these are too difficult to extract economically or only naturally occur in small quantities. For this reason, only those biopolymers which occur in natural abundance, are easy to extract, and are relatively inexpensive tend to be used as food ingredients. Even so, these ingredients often do not have the functional properties required for a particular product, and so there has been a considerable effort to identify methods of creating new ingredients from existing biopolymers.

Food biopolymers can be modified in a variety of ways to improve their functional properties (e.g., enzymatically, genetically, chemically, or physically) (Das and Kinsella 1990, Dickinson and McClements 1995, Magdassi 1996, Damodaran 1996, BeMiller and Whistler 1996). The molecular structure of a biopolymer can be altered by breaking it down into smaller subunits, polymerizing it, substituting certain segments of its backbone, covalently attaching side groups, or physically altering its conformation. Consequently, it is possible to alter the molecular weight, conformation, charge, hydrophobicity, amphiphilicity, and branching of biopolymer molecules. By carrying out these modifications in a systematic fashion, it is possible to create a range of biopolymers with specific functional properties. The main challenge for the food scientist is to develop methods of modifying biopolymers that are both economically viable and legally acceptable.

In addition to their utilization as food ingredients, modified biopolymers have also been used to study the relationship between the molecular characteristics of biopolymers and their functional properties (Dickinson and McClements 1995). For example, the role of conformation, size, charge, or hydrophobicity can be investigated by altering specific monomers along the polymer chain. These studies have provided extremely valuable insights into the molecular basis of the functional properties of biopolymers.

4.6.5. Ingredient Selection

A wide variety of natural and modified biopolymers are available as ingredients in foods, each with its own unique functional properties and optimum range of applications. Food manufacturers must decide which biopolymer is the most suitable for each type of food product. The selection of the most appropriate ingredient is often the key to the success of a particular product. The factors that a manufacturer must consider include the desired properties of the final product (appearance, rheology, mouthfeel, stability); the composition of the product; the processing, storage, and handling conditions the food will experience during its lifetime; as well as the legal status, cost, availability, consistency from batch to batch, ease of handling, dispersibility, and functional properties of the biopolymer ingredient.

5   Interfacial Properties and Their Characterization

5.1. INTRODUCTION

The interfacial region which separates the oil from the aqueous phase constitutes only a small fraction of the total volume of an emulsion (see Table 1.3) Nevertheless, it has a direct influence on the bulk physicochemical and sensory properties of food emulsions, including their formation, stability, rheology, and flavor. A better understanding of this influence will allow food scientists to create emulsion-based foods with improved quality. In this chapter, we consider the nature of the interfacial region, experimental techniques for characterizing its properties, and the role that it plays in determining the bulk physicochemical properties of emulsions.

An interface is a narrow region that separates two phases, which could be a gas and a liquid, a gas and a solid, two immiscible liquids, a liquid and a solid, or two solids. The two phases may consist of different kinds of molecules (e.g., oil and water) or different physical states of the same kind of molecule (e.g., liquid water and solid ice). By convention, the region separating two condensed phases (solids or liquids) is referred to as an *interface*, while the region separating a condensed phase and a gas is called a *surface* (Everett 1988). A number of different types of surfaces and interfaces commonly occur in food emulsions, including oil and water (droplets), air and water (gas cells), ice crystals in water (ice cream), and fat crystals in oil (butter or margarine). In this chapter, the main focus is on the oil–water interface because it is present in all food emulsions. Nevertheless, various other types of surfaces and interfaces will be considered where appropriate, and it should be recognized that much of the discussion about oil–water interfaces is also applicable to these systems.

5.2. MOLECULAR BASIS OF INTERFACIAL PROPERTIES

5.2.1. Interfaces Between Two Pure Liquids

The interface that separates the oil and water phases is often assumed to be a planar surface of infinitesimal thickness (Figure 5.1a). This assumption is convenient for many purposes, but it ignores the highly dynamic nature of the interfacial region, as well as the structure and organization of the various types of molecules involved (Figure 5.1b). On the molecular level, the oil and water molecules intermingle with each other over distances of the order of a few molecular diameters (Everett 1988, Evans and Wennerstrom 1994). The composition of the system therefore varies smoothly across the interfacial region (Figure 5.1b), rather than changing abruptly (Figure 5.1a). The thickness and dynamics of the interfacial region depend on the relative magnitude of the interactions between the molecules involved (oil–oil, water–water, and water–oil), which can be characterized by the effective interaction parameter (w) discussed in Section 2.6.2 (Evans and Wennerstrom 1994). If this parameter is positive and much larger than the thermal energy ($w/kT \gg 1$), the interactions between the different types

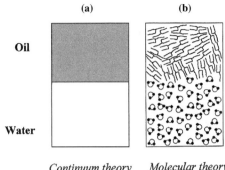

FIGURE 5.1   Interfaces are often assumed to be planar surfaces of infinitesimally small thickness (a), but in reality they are highly dynamic and have a certain thickness which depends on the dimensions and interactions of the molecules (b).

of molecules are highly unfavorable and the interfacial region is relatively thin because the protrusion of a molecule from one phase into another involves a large expenditure of energy. If the effective interaction parameter is approximately equal to the thermal energy ($w/kT \approx 1$), the liquids are partially miscible and the interfacial region increases in thickness. When the effective interaction parameter is negative and much smaller than the thermal energy ($w/kT \ll 1$), the two liquids are miscible and the interfacial region disappears altogether. The thickness and mobility of the interfacial region are therefore governed by a balance between the interaction energies and the thermal energy of the system (Everett 1988, Israelachvili 1992).

Oil molecules are incapable of forming hydrogen bonds with water molecules, and so the mixing of oil and water is strongly unfavorable because of the hydrophobic effect (Section 4.4.3). It is therefore necessary to supply energy to the system in order to increase the contact area between oil and water molecules. The amount of energy which must be supplied is proportional to the increase in contact area between the oil and water molecules (Hiemenz 1986):

$$\Delta G = \gamma_i \Delta A \qquad (5.1)$$

where ΔG is the free energy required to increase the contact area between the two immiscible liquids by ΔA (at constant temperature and pressure), and γ_i is a constant of proportionality called the *interfacial tension*. If one of the phases is a gas, the interfacial tension is replaced by the *surface tension* (γ_s). The interfacial tension is a physical characteristic of a system which is determined by the imbalance of molecular forces across an interface: the greater the interfacial tension, the greater the imbalance of forces (Israelachvili 1992, Evans and Wennerstrom 1994). Many of the most important macroscopic properties of food emulsions are governed by this imbalance of molecular forces at an interface, including the tendency for droplets to be spherical, the surface activity of emulsifiers, the nucleation and growth of ice and fat crystals, meniscus formation, and the rise of liquids in a capillary tube (Section 5.13).

5.2.2. Interfaces with Adsorbed Emulsifiers

So far, we have only considered the molecular characteristics of an interface that separates two pure liquids. In practice, food emulsions contain various types of surface-active molecules which can accumulate at the interface and therefore alter its properties (e.g., proteins, polysaccharides, alcohols, and surfactants) (Dickinson and Stainsby 1982, Dickinson 1992).

5.2.2.1. *Surface Activity and the Reduction of Interfacial Tension*

The *surface activity* of a molecule is a measure of its ability to accumulate at an interface. A molecule tends to accumulate at an interface when the free energy of the adsorbed state is significantly lower than that of the unadsorbed state (Hiemenz 1986). The difference in free energy between the adsorbed and unadsorbed states (ΔG_{ads}) is determined by changes in the interaction energies of the molecules involved, as well as by various entropy effects (Shaw 1980). The change in the interaction energies which occurs as a result of adsorption comes from two sources, one associated with the interface and the other with the emulsifier molecule itself. First, by adsorbing to an oil–water interface, an emulsifier molecule is able to "shield" the oil molecules from the water molecules. The direct contact between oil and water molecules is replaced by contacts between the nonpolar segments of the emulsifier and oil molecules and between the polar segments of the emulsifier and water molecules (Israelachvili 1992). These interactions are less energetically unfavorable than the direct interactions between oil and water molecules. Second, emulsifier molecules usually have both polar and nonpolar segments, and when they are dispersed in bulk water, some of the nonpolar segments come into contact with water, which is energetically unfavorable because of the hydrophobic effect. By adsorbing to an interface, they are able to maximize the number of energetically favorable interactions between the polar segments and water while minimizing the number of unfavorable interactions between the nonpolar segments and water (Figure 5.2). The major driving force favoring the adsorption of an amphiphilic molecule at an interface is therefore the hydrophobic effect. Nevertheless, various other types of interaction may also contribute to the surface activity, which may either favor or oppose adsorption (e.g., hydration repulsion, electrostatic interactions, steric interactions, and hydrogen bonding).

The entropy effects associated with adsorption are mainly due to the fact that when a molecule adsorbs to an interface, it is confined to a region which is considerably smaller than the volume it would occupy in a bulk liquid and that its molecular rotation is restricted. Both of these effects are entropically unfavorable, and so a molecule will only adsorb to an interface if the energy gained by optimizing the interaction energies is sufficiently large to offset the entropy lost. When the adsorption energy is large compared to the thermal energy (i.e., $-\Delta G_{ads} \gg kT$), a molecule "binds" strongly to the surface and has a high surface activity. When the adsorption energy is small compared to the thermal energy (i.e., $-\Delta G_{ads} \ll kT$), a molecule tends to be located mainly in the bulk liquid and has a low surface activity. When

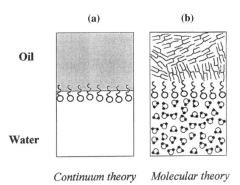

Continuum theory   *Molecular theory*

FIGURE 5.2   Surface-active molecules accumulate in the interfacial region because this minimizes the free energy of the system.

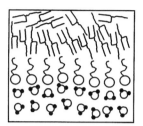 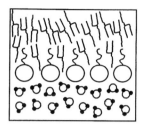

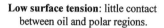

Low surface tension: little contact   **High surface tension**: Contact
between oil and polar regions.   between oil and polar regions.

FIGURE 5.3   The decrease in interfacial tension caused by surface-active molecules depends on how effectively they "shield" the oil molecules from the water molecules. The interfacial tension of the system shown on the left is lower than that on the right, because the contact between the polar and nonpolar molecules is shielded more effectively.

the free energy of adsorption is highly positive (i.e., $\Delta G_{ads} \gg kT$), there is a deficit of solute in the interfacial region, which is referred to as *negative adsorption* (Shaw 1980).

The decrease in the free energy of a system which occurs when a surface-active molecule adsorbs to an interface manifests itself as a decrease in the interfacial tension (i.e., less energy is required to increase the surface area between the oil and water phases). The extent of this decrease depends on the effectiveness of the molecule at "shielding" the direct interactions between the oil and water molecules, as well as on the strength of the interactions between the hydrophilic segments and water, and between the hydrophobic segments and oil (Israelachvili 1992). The more efficient an emulsifier is at shielding the interaction between oil and water, the lower the interfacial tension (Figure 5.3).

The ability of surfactant molecules to shield direct interactions between two immiscible liquids is governed by their optimum packing at an interface, which depends on their molecular geometry (Chapter 4). When the curvature of an interface is equal to the optimum curvature of a surfactant monolayer (i.e., optimum packing is possible), the interfacial tension is ultralow because direct interactions between the oil and water molecules are effectively eliminated (Section 4.5.3.3). On the other hand, when the curvature of an interface is not at its optimum, the interfacial tension increases because some of the oil molecules are exposed to the polar regions of the surfactant or some of the water molecules come into contact with the hydrophobic part of the surfactant. Surfactants can usually screen the interactions between the oil and water phases more efficiently than biopolymers, which means they are more effective at reducing the interfacial tension. This is because biopolymers cannot pack as efficiently at the interface and because they often have some nonpolar regions on their surface exposed to the water phase and some polar regions exposed to the oil phase (Damodaran 1996).

The reduction of the interfacial tension by the presence of an emulsifier is referred to as the *surface pressure*: $\pi = \gamma_{o/w} - \gamma_{emulsifier}$, where $\gamma_{o/w}$ is the interfacial tension of a pure oil–water interface and $\gamma_{emulsifier}$ is the interfacial tension in the presence of the emulsifier (Hiemenz 1986).

5.2.2.2. Adsorption Kinetics of Emulsifiers to Interfaces

The tendency of an emulsifier to exist in either the bulk or interfacial regions is governed by thermodynamics; however, the rate at which an emulsifier adsorbs to an interface is determined by various mass transport processes (e.g., diffusion and convection) and energy barriers associated with the adsorption process (e.g., availability of free sites, electrostatic repulsion, steric repulsion, hydrodynamic repulsion, and micelle dynamics). The major factors which influence the adsorption kinetics of an emulsifier at an interface are discussed in some detail in Section 5.7.

5.2.2.3.   *Conformation of Emulsifiers at Interfaces*

The conformation and orientation of molecules at an interface are governed by their attempt to reduce the free energy of the system (Evans and Wennerstrom 1994). Amphiphilic molecules arrange themselves so that the maximum number of nonpolar groups are in contact with the oil phase, while the maximum number of polar groups are in contact with the aqueous phase (Figure 5.4). For this reason, small-molecule surfactants tend to have their polar head groups protruding into the aqueous phase and their hydrocarbon tails protruding into the oil phase (Myers 1988). Similarly, biopolymer molecules adsorb so that predominantly nonpolar segments are located within the oil phase, whereas predominantly polar segments are located within the water phase (Dickinson 1992, Damodaran 1996, Dalgleish 1996b). Biopolymer molecules often undergo structural rearrangements after adsorption to an interface in order to maximize the number of favorable interactions. In aqueous solution, globular proteins adopt a three-dimensional conformation in which the nonpolar amino acids are predominantly located in the hydrophobic interior of the molecule so that they can be away from the water (Dill 1990). When they adsorb to an oil–water interface, they are no longer completely surrounded by water, and so the protein can reduce its free energy by altering its conformation so that more of the hydrophobic amino acids are located in the oil phase and more of the polar amino acids are located in the water phase (Dalgleish 1996b). The rate at which the conformation of a biopolymer changes at an oil–water interface depends on its molecular structure (Dickinson 1992). Flexible random-coil molecules can rapidly alter their conformation, whereas rigid globular molecules change more slowly because of various kinetic constraints. Immediately after adsorption to an interface, a globular protein has a conformation that is similar to that in the bulk aqueous phase. With time, it alters its conformation so that it can optimize the number of favorable interactions between the nonpolar amino acids and the oil molecules. An intermediate stage in this unfolding process is the exposure of some of the nonpolar amino acids to water, which is energetically unfavorable because of the hydrophobic effect, and so there is an energy barrier which must be overcome before unfolding can occur. In this case, the rate of any conformational changes will depend on the height of the energy barriers compared to the thermal energy.

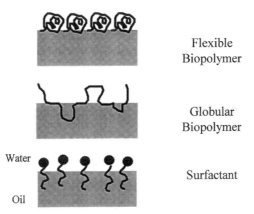

FIGURE 5.4   The orientation and conformation of molecules at an interface are determined by their tendency to reduce the free energy of the system.

The configuration of emulsifier molecules at an interface can have an important influence on the bulk physicochemical properties of food emulsions (Dalgleish 1996b). The coalescence stability of many oil-in-water emulsions depends on the unfolding and interaction of protein molecules at the droplet surface. When globular proteins unfold, they expose reactive amino acids that are capable of forming hydrophobic and disulfide bonds with their neighbors, thus generating a highly viscoelastic membrane that is resistant to coalescence (Dickinson and Matsumura 1991, Dickinson 1992). The susceptibility of certain proteins to enzymatic hydrolysis depends on which side of the adsorbed molecule faces toward the droplet surface (Dalgleish 1996a,b). The susceptibility of surfactants with unsaturated hydrocarbon tails to lipid oxidation depends on whether their tails are perpendicular or parallel to the droplet surface, the latter being more prone to oxidation by free radicals generated in the aqueous phase.

5.3.   THERMODYNAMICS OF INTERFACES

5.3.1.   Gas–Liquid Interface in the Absence of an Emulsifier

In the previous section, the relationship between the molecular structure of emulsifiers and their surface activity and interfacial conformation was highlighted. In this section, mathematical quantities and thermodynamic relationships which can be used to describe the properties of interfaces are introduced. As a whole, emulsions are thermodynamically unstable systems because of the unfavorable contact between oil and water molecules (Section 7.2). Nevertheless, their interfacial properties can often be described by thermodynamics because the adsorption–desorption of emulsifier molecules occurs at a rate which is much faster than the time scale of the kinetic destabilization processes (Hunter 1986). Thermodynamically, it is convenient to assume that an interface is a smooth, infinitesimally thin plane which separates two homogeneous liquids (Figure 5.1a). In order to model a real system, it is necessary to decide exactly where this imaginary plane should be located (Hiemenz 1986). For simplicity, consider a system which consists of liquid water in equilibrium with its vapor (Figure 5.5). The volume fraction of water molecules in the liquid water is approximately unity and decreases to approximately zero as one moves up through the interfacial region and into the vapor phase.

The imaginary plane interface could be located anywhere in the interfacial region indicated in Figure 5.5 (Hiemenz 1986, Hunter 1986). In practice, it is convenient to assume that the interface is located at a position where the excess concentration of the substance on one side of the interface is equal to the deficit concentration of the substance on the other side of the interface: $c_{excess} = c_{deficit}$. In our example, the *excess concentration* corresponds to the amount of water above the interface, which is greater than that which would have been present if the concentration of water was the same as that in the bulk vapor phase right up to the interface. Similarly, the *deficit concentration* corresponds to the amount of water which is below the interface, which is lower than that which would have been present if the concentration of the water was the same as that in the bulk liquid phase right up to the interface. This location of the interface is known as the *Gibbs dividing surface*, after the scientist who first proposed this convention.

5.3.2.   Gas–Liquid Interface in the Presence of an Emulsifier

The concept of the Gibbs dividing surface is particularly useful for defining the amount of emulsifier which accumulates at an interface (Hunter 1986). Consider a system which consists of a surfactant solution in contact with its vapor (Figure 5.6). The emulsifier is distributed among the bulk aqueous phase, the vapor, and the interfacial region. The excess

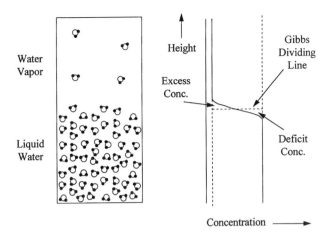

FIGURE 5.5 From a thermodynamic standpoint, it is convenient to locate the interface (Gibbs dividing surface) separating a liquid and its vapor where $c_{excess} = c_{deficit}$.

emulsifier concentration at the surface (n_i) corresponds to the total amount of emulsifier present in the system minus that which would be present if the emulsifier were not surface active and equals the shaded area shown in Figure 5.6. The accumulation of emulsifier molecules at an interface is characterized by a *surface excess concentration* (Γ), which is equal to the excess emulsifier concentration divided by the surface area: $\Gamma = n_i/A$. Food emulsifiers typically have Γ values of a few milligrams per meter squared (Dickinson 1992, Dalgleish 1996b). It is important to note that the emulsifier molecules are not actually concentrated at the Gibbs dividing surface (which is infinitely thin), because of their finite size and the possibility of multilayer formation. Nevertheless, this approach is extremely convenient for thermodynamic descriptions of the properties of surfaces and interfaces (Hunter 1986). The surface excess concentration is often identified with an experimentally measurable parameter called the *surface load,* which is the amount of emulsifier adsorbed to the surface of emulsion droplets per unit area of interface.

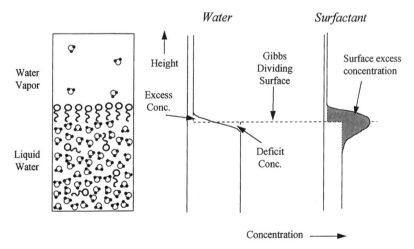

FIGURE 5.6 When an emulsifier is present, the Gibbs dividing surface is conveniently located at the position where $c_{excess} = c_{deficit}$ for the liquid in which the emulsifier is most soluble, and the surface excess concentration is equal to the shaded region.

5.3.3. Liquid–Liquid Interface

For an interface between pure oil and pure water, the Gibbs dividing surface could be positioned at the point where the excess and deficit concentrations of either the oil or the water were equal on either side of the interface, which will in general be different (Tadros and Vincent 1983). For convenience, it can be assumed that the phase in which the emulsifier is most soluble is the one used to decide the position of the Gibbs dividing surface. The surface excess concentration of the emulsifier is then equal to that which is present in the system minus that which would be present if there were no accumulation at the interface (Hunter 1986).

5.3.4. Measurement of the Surface Excess Concentration

The surface excess concentration of an emulsifier can be determined from measurements of the variation in the surface tension of an air–liquid interface as the emulsifier concentration in the bulk liquid is increased (Figure 5.7). There is an equilibrium between emulsifier molecules at the interface and those in the bulk liquid. As the concentration of emulsifier in the bulk liquid is increased, so does their concentration at the interface. The presence of the emulsifier molecules at the interface shields the unfavorable contact between the oil and water molecules and therefore reduces the surface tension. At a certain concentration, the surface tension reaches a constant value because the surface becomes saturated with emulsifier molecules. For small-molecule surfactants, the saturation of the surface occurs at approximately the same concentration as the surfactant molecules form micelles in the bulk liquid (i.e., the critical micelle concentration) (Section 4.5.2). A relationship between the decrease in surface tension with emulsifier concentration and the amount of emulsifier present at the surface can be derived from a mathematical treatment of the thermodynamics of the system (Hiemenz 1986, Hunter 1986). This relationship is known as the *Gibbs isotherm equation*, which is given by the following expression for an ideal solution:

$$\Gamma = -\frac{1}{RT}\left(\frac{d\gamma}{d\,\ln(x)}\right) \qquad \text{for nonionic emulsifiers} \qquad (5.2)$$

$$\Gamma = -\frac{1}{2RT}\left(\frac{d\gamma}{d\,\ln(x)}\right) \qquad \text{for ionic emulsifiers} \qquad (5.3)$$

where x is the concentration of emulsifier in the aqueous phase, R is the gas constant, and T is the absolute temperature. These equations are used to determine the surface excess concentration of emulsifiers from experimental measurements of the surface tension versus emulsifier concentration, with Γ as the slope of the initial part of the curve (Figure 5.7). The factor 2 appears in the denominator of Equation 5.3 because the counterions associated with the head groups of ionic surfactants also accumulate at the interface (Hunter 1986). Equation 5.3 is only applicable at low ionic strengths where the interaction between the head groups and counterions is strong. As the ionic strength increases, the electrostatic interactions between the head groups and counterions are screened, and so Equation 5.2 becomes more applicable. Knowledge of the surface excess concentration is important for formulating food emulsions because it determines the minimum amount of emulsifier which can be used to create an emulsion with a given size distribution. The smaller the value of Γ, the greater the area of oil–water interface which can be covered per gram of emulsifier, and therefore the smaller

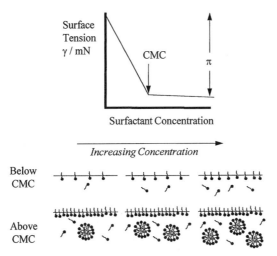

FIGURE 5.7 Dependence of the surface tension on the concentration of emulsifier. This type of plot can be used to measure the critical micelle concentration (CMC), the surface excess concentration, and maximum surface pressure of emulsifiers.

the size of droplets which can be effectively stabilized by the emulsifier. Plots of surface tension versus emulsifier concentration are also useful because they indicate the maximum surface pressure (π_{max}) which can be achieved when the surface is saturated by an emulsifier, which has important consequences for the formation and stability of food emulsions (Chapters 6 and 7).

5.4. PROPERTIES OF CURVED INTERFACES

The majority of surfaces or interfaces found in food emulsions are curved rather than planar. The curvature of an interface alters its characteristics in a number of ways. The interfacial tension tends to cause an emulsion droplet to shrink in size so as to reduce the unfavorable contact area between the oil and water phases (Hunter 1986, Everett 1988). As the droplet shrinks, there is an increase in its internal pressure because of the compression of the water molecules. Eventually, an equilibrium is reached where the inward stress due to the interfacial tension is balanced by the outward stress associated with compressing the bonds between the liquid molecules inside the droplet.* At equilibrium, the pressure within the droplet is larger than that outside and can be related to the interfacial tension and radius of the droplets using the *Laplace equation* (Atkins 1994):

$$\Delta p = \frac{2\gamma}{r} \tag{5.4}$$

This equation indicates that the pressure difference across the surface of an emulsion droplet increases as the interfacial tension increases or the size of the droplet decreases. The properties of a material depend on the pressure exerted on it, and so the properties of a

---

* The shrinkage of a droplet due to the interfacial tension is usually negligibly small, because liquids have a very low compressibility.

material within a droplet are different from those of the same material in bulk (Atkins 1994). This effect is usually negligible for liquids and solids which are contained within particles that have radii greater than a few micrometers, but it does become significant for smaller particles (Hunter 1986). For example, the water solubility of oil increases as the radius of an oil droplet decreases (Dickinson 1992, Atkins 1994):

$$\frac{S}{S^*} = \exp\left(\frac{2\gamma v}{rRT}\right) \tag{5.5}$$

where S is the water solubility of the oil in the droplet, S^* is the water solubility of bulk oil, and v is the molar volume of the oil. For a typical food oil ($v = 10^{-3}$ m^3 mol^{-1}, $\gamma = 10$ mJ m^{-2}), the value of S/S^* is 2.24, 1.08, 1.01, and 1.0 for oil droplets with radii of 0.01, 0.1, 1, and 10 µm, respectively. Equation 5.5 has important implications for the stability of emulsion droplets, fat crystals, and ice crystals to Ostwald ripening (Chapter 7).

So far, we have assumed that the interfacial tension of a droplet is independent of its radius. Experimental work has indicated that this assumption is valid for oil droplets, even down to sizes where they only contain a few molecules, but that it is invalid for water droplets below a few nanometers because of the disruption of long-range hydrogen bonds (Israelachvili 1992).

Emulsifier molecules have a major influence on the properties of curved surfaces and interfaces. Each type of surfactant has an optimum curvature which is governed by its molecular geometry and interactions with its neighbors (Section 4.5.3). When the curvature of an interface is equal to the optimum curvature of a surfactant monolayer, the interfacial tension is extremely low because the interactions between the emulsifier molecules are optimized and shielding between the oil and water molecules is extremely efficient. On the other hand, when the curvature of the interface is not close to the optimum curvature of the surfactant monolayer, the interfacial tension increases because the interactions between the surfactant molecules are not optimum and the shielding between oil and water molecules is less efficient. The interfacial tension of an interface therefore depends on the molecular geometry of the surfactant used.

5.5. CONTACT ANGLES AND WETTING

In food systems, we are often interested in the ability of a liquid to spread over or "wet" the surface of another material. In some situations, it is desirable for a liquid to spread over a surface (e.g., when coating a food with an edible film), while in other situations it is important that a liquid does not spread (e.g., when designing waterproof packaging). When a drop of liquid is placed on the surface of a material, it may behave in a number of ways, depending on the nature of the interactions between the various types of molecules present. The two extremes of behavior that are observed experimentally are outlined below (Figure 5.8):

1. *Poor wetting.* The liquid gathers up into a lens, rather than spreading across the surface of a material.
2. *Good wetting.* The liquid spreads over the surface of the material to form a thin film, which has a liquid–gas interface and a liquid–solid interface.

The situation that occurs in practice depends on the relative magnitude of the interactions between the various types of molecules involved (i.e., solid–liquid, solid–gas, and liquid–gas). A system tends to organize itself so that it can maximize the number of favorable

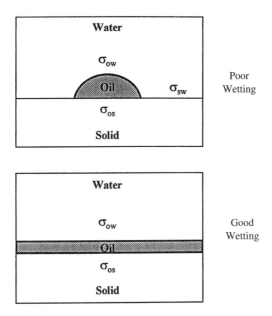

FIGURE 5.8 The wetting of a surface by a liquid depends on a delicate balance of molecular interactions among solid, liquid, and gas phases.

interactions and minimize the number of unfavorable interactions between the molecules. Consider what may happen when a drop of liquid is placed on a solid surface (Figure 5.8). If the liquid remained as a lens, there would be three different interfaces: solid–liquid, solid–gas, and liquid–gas, each with its own interfacial or surface tension. If the liquid spread over the surface, there would be a decrease in the area of the solid–gas interface but an increase in the areas of both the liquid–gas and solid–liquid interfaces. The tendency for a liquid to spread therefore depends on the magnitude of the solid–gas interactions (γ_{SG}) compared to the magnitude of the solid–liquid and liquid–gas interactions that replace it ($\gamma_{SL} + \gamma_{LG}$). This situation is conveniently described by a *spreading coefficient*, which is defined as (Hunter 1993):

$$S = \gamma_{SG} - (\gamma_{SL} + \gamma_{LG}) \tag{5.6}$$

If the energy associated with the solid–gas interface is greater than the sum of the energies associated with the solid–liquid and liquid–gas interfaces ($\gamma_{SG} > \gamma_{SL} + \gamma_{LG}$), then S is positive and the liquid tends to spread over the surface to reduce the energetically unfavorable contact area between the solid and the gas. On the other hand, if the energy associated with the solid–gas interface is less than that associated with forming the solid–liquid and liquid–gas interfaces ($\gamma_{SG} < \gamma_{SL} + \gamma_{LG}$), then S is negative and the liquid tends to form a lens.

The shape of a droplet can be predicted by carrying out a force balance at the point on the surface where the solid, liquid, and gas meet (Figure 5.9) using the Young equation (Hiemenz 1986):

$$\gamma_{SG} = \gamma_{SL} + \gamma_{LG} \cos \theta \tag{5.7}$$

so that:

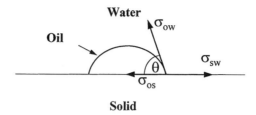

FIGURE 5.9   Force balance of a droplet at a solid–gas interface.

$$\cos \theta = \frac{\gamma_{SG} - \gamma_{SL}}{\gamma_{LG}} \tag{5.8}$$

Here, θ is known as the contact angle, which is the angle of a tangent drawn at the point where the liquid contacts the surface. By convention, this angle is measured from the side of the droplet (Figure 5.9). The shape of a droplet on a surface can therefore be predicted from a knowledge of the contact angle: the smaller θ, the greater the tendency for the liquid to spread over the surface.

So far, we have only considered the situation where a liquid spreads over a solid surface, but similar equations can be used to consider other three-component systems, such as a liquid spreading over the surface of another liquid (e.g., oil, water, and air) or a crystal at an interface between two other liquids (e.g., a fat crystal at an oil–water interface). The latter case is important when considering the nucleation and location of fat crystals in oil droplets and has a pronounced influence on the stability and rheology of many important food emulsions, including milk, cream, butter, and whipped cream (Walstra 1987, Boode 1992, Dickinson and McClements 1995).

The above equations assume that the materials involved are completely insoluble in each other, so that the values of γ_{SG}, γ_{SL}, and γ_{LG} (or the equivalent terms for other three-component systems) are the same as those for pure systems. If the materials are partially miscible, then the interfacial tensions will change over time until equilibrium is reached (Hunter 1993). The solubility of one component in another generally leads to a decrease in the interfacial tension. This means that the shape that a droplet adopts on a surface may change with time (e.g., a spread liquid may gather into a lens, or vice versa, depending on the magnitude of the changes in the various surface or interfacial tensions).

The contact angle of a liquid can conveniently be measured using a microscope, which is often attached to a computer with video image analysis software (Hunter 1986). A droplet of the liquid to be analyzed is placed on a surface and its shape is recorded via the microscope. The contact angle is determined by analyzing the shape of the droplet. The advantages and disadvantages of a variety of other techniques available for measuring contact angles have been considered by Hunter (1986).

The concepts of a contact angle and a spreading coefficient are useful for explaining a number of important phenomena which occur in food emulsions. The contact angle determines the distance that a fat crystal protrudes from the surface of a droplet into the surrounding water (Boode 1992) and whether nucleation occurs within the interior of a droplet or at the oil–water interface (Dickinson and McClements 1995). It also determines the amount of liquid which is drawn into a capillary tube and the shape of the meniscus at the top of the liquid (Hunter 1986, Hiemenz 1986). A knowledge of the contact angle is also often required in order to make an accurate measurement of the surface or interfacial tension of a liquid (Couper 1993).

5.6. CAPILLARY RISE AND MENISCUS FORMATION

The surface tension of a liquid governs the rise of liquids in capillary tubes and the formation of menisci (curved surfaces) at the top of liquids (Hiemenz 1986, Hunter 1986). When a glass capillary tube is dipped into a beaker of water, the liquid climbs up the tube and forms a curved surface (Figure 5.10). The origin of this phenomenon is the imbalance of intermolecular forces at the various surfaces and interfaces in the system (Evans and Wennerstrom 1994). When water climbs up the capillary tube, some of the air–glass contact area is replaced by water–glass contact area, while the air–water contact area remains fairly constant. This occurs because the imbalance of molecular interactions between glass and air is much greater than that between glass and water. Consequently, the system attempts to maximize the glass–water contacts and minimize the glass–air contacts by having the liquid climb up the inner surface of the capillary tube. This process is opposed by the downward gravitational pull of the liquid. When the liquid has climbed to a certain height, the surface energy it gains by optimizing the number of favorable water–glass interactions is exactly balanced by the potential energy that must be expended to raise the mass of water up the tube (Hiemenz 1986). A mathematical analysis of this equilibrium leads to the derivation of the following equation:

$$\gamma = \frac{\Delta\rho g h r}{2 \cos \theta} \tag{5.9}$$

where g is the gravitational constant, h is the height that the meniscus rises above the level of the water, r is the radius of the capillary tube, $\Delta\rho$ is the difference in density between the water and the air, and θ is the contact angle (Figure 5.10). This equation indicates that the surface tension can be estimated from a measurement of the height that a liquid rises up a capillary tube and the contact angle: the greater the surface tension, the higher the liquid rises up the tube. This is one of the oldest and simplest methods of determining the surface tension of pure liquids, but it has a number of problems which limit its application to emulsifier solutions (Couper 1993).

Capillary forces are responsible for the entrapment of water in biopolymer networks and oil in fat crystal networks. When the gaps between the network are small, the capillary force is strong enough to hold relatively large volumes of liquid, but when the gaps exceed a certain size, the capillary forces are no longer strong enough and syneresis or "oiling off"

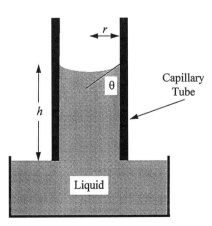

FIGURE 5.10 The rise of a liquid up a capillary tube is a result of its surface tension.

occurs. A knowledge of the origin of capillary forces is therefore important for an under-standing of the relationship between the microstructure of foods and many of their quality attributes.

5.7. ADSORPTION KINETICS

The rate at which an emulsifier adsorbs to an interface is one of the most important factors determining its efficacy as a food ingredient (Magdassi and Kamyshny 1996, Walstra 1996b). The adsorption rate depends on the molecular characteristics of the emulsifier (e.g., size, conformation, and interactions), the nature of the bulk liquid (e.g., viscosity), and the prevail-ing environmental conditions (e.g., temperature and mechanical agitation). It is often conve-nient to divide the adsorption process into two stages: (1) movement of the emulsifier molecules from the bulk liquid to the interface and (2) incorporation of the emulsifier molecules at the interface. In practice, emulsifier molecules are often in a dynamic equilib-rium between the adsorbed and unadsorbed states, and so the rate at which emulsifier molecules leave the interface must also be considered (Hunter 1993, Magdassi and Kamyshny 1996).

5.7.1. Movement of Molecules to an Interface

In this section, we assume that an emulsifier molecule is adsorbed to an interface as soon as it encounters it (i.e., there are no energy barriers that retard adsorption). In an isothermal quiescent liquid, emulsifier molecules move from a bulk liquid to an interface by molecular diffusion, with an initial adsorption rate given by (Tadros and Vincent 1983, Magdassi and Kamyshny 1996):

$$\frac{d\Gamma}{dt} = c\sqrt{\frac{D}{\pi t}} \tag{5.10}$$

where D is the translational diffusion coefficient, Γ is the surface excess concentration, t is the time, and c is the concentration of emulsifier initially in the bulk liquid. The variation of the surface excess concentration with time is obtained by integrating this equation with respect to time:

$$\Gamma(t) = 2c\sqrt{\frac{Dt}{\pi}} \tag{5.11}$$

Thus, a plot of the surface excess concentration *versus* $\sqrt{t}$ should be a straight line that passes through the origin. This equation indicates that the accumulation of an emulsifier at an interface occurs more rapidly as the concentration of emulsifier in the bulk liquid increases or as the diffusion coefficient of the emulsifier increases. The diffusion coefficient increases as the size of molecules decreases, and one would therefore expect smaller molecules to adsorb more rapidly than larger ones. Experiments with proteins have shown that Equation 5.11 gives a good description of the early stages of adsorption to clean interfaces (Damodaran 1989, Walstra 1996b). After the initial stages, the adsorption rate decreases because the interface becomes saturated with emulsifier molecules and therefore there are fewer sites available for the emulsifier to adsorb to (Figure 5.11). In practice, the rate may be faster than that given by Equation 5.10 because of convection currents caused by temperature gradients within a liquid. Consequently, considerable care must be taken to

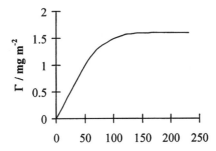

FIGURE 5.11   Adsorption kinetics for a diffusion-controlled system.

ensure that the temperature within a sample is uniform when measuring diffusion-controlled adsorption processes.

The above equations do not apply during the homogenization of emulsions, because homogenization is a highly dynamic process and mass transport is governed mainly by convection rather than diffusion (Dickinson 1992, Walstra 1996b). Under isotropic turbulent conditions, the initial increase of the surface excess concentration with time is given by (Dukhin et al. 1995):

$$\Gamma(t) = Cr_d c\left(1 + \frac{r_e}{r_d}\right)^3 t \tag{5.12}$$

where C is a constant which depends on the experimental conditions, and r_d and r_e are the radii of the droplet and emulsifier, respectively. This equation predicts that the adsorption rate increases as the concentration of emulsifier increases, the size of the emulsion droplets increases, or the size of the emulsifier molecules increases relative to the size of the droplets. This equation implies that when an emulsion is homogenized, the emulsifier molecules initially adsorb preferentially to the larger droplets and that larger emulsifier molecules adsorb more rapidly than smaller ones (which is the opposite of diffusion-controlled adsorption). This explains why large casein micelles adsorb faster than individual casein molecules during the homogenization of milk (Mulder and Walstra 1974).

5.7.2.   Incorporation of Emulsifier Molecules at an Interface

So far, we have assumed that as soon as an emulsifier molecule reaches an interface, it is immediately adsorbed. In practice, there may be one or more energy barriers that must be overcome before a molecule adsorbs, and so only a fraction of the encounters between an emulsifier molecule and an interface lead to adsorption (Damodaran 1996, Magdassi and Kamyshny 1996). In these systems, adsorption kinetics may be governed by the height of the energy barrier, rather than by the rate at which the molecules reach the interface (Magdassi and Kamyshny 1996).

There are a number of reasons why an energy barrier to adsorption may exist:

1. As an emulsifier molecule approaches an interface, there may be various types of repulsion interactions between it and the emulsifier molecules already adsorbed to the interface (e.g., electrostatic, steric, hydration, or thermal fluctuation) (Chapter 3).
2. Some surface-active molecules will only be adsorbed if they are in a specific orientation when they encounter the interface. For example, it has been suggested

that globular proteins which have hydrophobic patches on their surface must face toward an oil droplet during an encounter (Damodaran 1996).

3. The ability of surfactant molecules to form micelles plays a major role in determining their adsorption kinetics (Dukhin et al. 1995, Stang et al. 1994, Karbstein and Schubert 1995). Surfactant monomers are surface active because they have a polar head group and a nonpolar tail, but micelles are not surface active because their exterior is surrounded by hydrophilic head groups. The adsorption kinetics therefore depends on the concentration of monomers and micelles present, as well as the dynamics of micelle formation–disruption.

Micelle dynamics can be characterized by two relaxation times, one associated with the movement of individual monomers in and out of the micelles (τ_{fast}) and the other associated with the complete dissolution of a whole micelle into a number of individual molecules (τ_{slow}) (Land and Zana 1987). Consider the processes that occur when a fresh oil–water interface is brought into contact with an aqueous solution containing surfactant monomers and micelles (Kabalnov and Weers 1996). When a monomer adsorbs to the surface, the local monomer concentration is depleted, and so there is a concentration gradient between the region in the immediate vicinity of the interface and the bulk solution. This gradient can be equalized by the diffusion of surfactant monomers from the bulk liquid into the depleted zone. However, this disturbs the equilibrium between the monomers and micelles, leading to the partial dissociation of the micelles. Thus the micelles influence the adsorption kinetics indirectly rather than directly. The adsorption kinetics therefore depends on the diffusion coefficient of the monomers and micelles, as well as the dynamic properties of the micelles (i.e., the relaxation times). When the micelle relaxation mechanisms that release the monomers are much slower than the diffusion of the monomers, the adsorption kinetics is governed mainly by the critical micelle concentration, rather than the actual micelle concentration present. On the other hand, if the micelle relaxation mechanisms occur rapidly, then the micelles act as an additional source of monomers, and the adsorption rate is enhanced when the surfactant concentration is increased above the critical micelle concentration.

The adsorption of emulsifier molecules at a surface or interface can be measured using a variety of experimental methods (Couper 1993, Kallay et al. 1993, Dukhin et al. 1995). The most commonly used method is to measure the variation in surface or interfacial tension with time using a tensiometer (Section 5.9). A number of workers have also used radiolabeled emulsifier molecules to measure adsorption kinetics (Damodaran 1989). A radioactivity detector is placed immediately above a water–air surface. The radiolabeled emulsifier is injected into the water and the increase in the radioactivity at the surface is recorded over time by the detector. The radioactivity is highly attenuated by the water, and so only those molecules which are close to the air–water surface are detected. A variety of other experimental methods that can detect changes in interfacial properties due to the adsorption of emulsifier molecules have also been used to monitor adsorption kinetics, including ultraviolet-visible spectroscopy, infrared spectroscopy, fluorescence spectroscopy, ellipsometry, and interfacial rheology (Kallay et al. 1993).

5.8. COMMON TYPES OF INTERFACIAL MEMBRANES

A variety of different types of interfacial membranes are commonly found in food emulsions as a result of the different types of surface-active materials in the system. The four most common types of interface are discussed briefly below:

1. ***Solid particles.*** Many foods contain tiny solid particles that are capable of adsorbing to the surface of emulsion droplets (e.g., mustard powder and spices) (Dickinson

and Stainsby 1982, Dickinson 1992). The arrangement of these particles at an interface depends on whether they are preferentially wetted by the continuous or the dispersed phase. Those particles which are wetted best by the dispersed phase protrude into the droplet, whereas those which are wetted best by the continuous phase protrude into the surrounding liquid. Obviously, those particles which protrude into the continuous phase are most suitable for providing protection against droplet aggregation.

2. ***Flexible biopolymers.*** A number of surface-active proteins and polysaccharides have fairly flexible random-coil-type structures (Section 4.6). These biopolymers tend to form relatively thick interfacial membranes with fairly open structures. To be effective at preventing droplet aggregation, part of the molecules must have a high affinity for the continuous phase and extend a significant distance into it.

3. ***Globular biopolymers.*** Many surface-active proteins have a compact globular structure and form relatively thin but dense interfacial layers. These proteins usually provide protection against aggregation by a combination of electrostatic and steric interactions (Chapter 3).

4. ***Small-molecule surfactants.*** Surfactants form a fairly dense interfacial membrane which prevents molecular contact between the liquid in the droplets and that in the continuous phase. At low surfactant concentrations, this membrane is usually a monolayer, but at high concentrations, it may be a multilayer (Friberg and El-Nokaly 1983).

5.9. INTERFACIAL COMPOSITION AND COMPETITIVE ADSORPTION

Most food emulsions contain droplets which are coated by a mixture of different types of surface-active components (e.g., proteins, polysaccharides, phospholipids, and surfactants), rather than a single chemically pure type (Dickinson 1992; Dalgleish 1996a,b; Walstra 1996b). Bulk physicochemical properties of emulsions, such as their ease of formation, stability, and texture, are governed by the nature of the interface, and therefore it is important for food scientists to understand the factors which determine the composition of the interfacial region (Dalgleish 1996a). Interfacial composition is determined by the type and concentration of surface-active components present, their relative affinity for the interface, the method used to prepare the emulsion, solution conditions (e.g., temperature, pH, and ionic strength), and the history of the emulsion. In this section, some of the most important factors which influence the interfacial composition of food emulsions are reviewed.

Interfacial composition depends on the relative adsorption rates of the various types of surface-active components which are present during the homogenization process, as well as any changes that may occur after homogenization (Dickinson 1992). Immediately after homogenization, the droplets tend to be coated by those surface-active molecules which absorb to the interface most rapidly under turbulent conditions (Section 5.7). Nevertheless, the interfacial composition may change with time after homogenization because some of the surface-active molecules that were initially present at the droplet surface are displaced by molecules in the bulk liquid that have a greater affinity for the surface. Alternatively, additional surface-active components may be added to the continuous phase of an emulsion after homogenization, and these may displace some of the emulsifier molecules at the droplet surface. The displacement of emulsifier molecules from an interface may be retarded if they are capable of undergoing some form of conformational change that enables them to bind strongly to their neighbors. For example, when an emulsion is heated under alkaline conditions, β-lactoglobulin is capable of unfolding and forming extensive covalent (disulfide) bonds with its neighbors, which makes it difficult to displace from an interface, even by more surface-active molecules (McClements et al. 1993d). The factors that determine the rate at

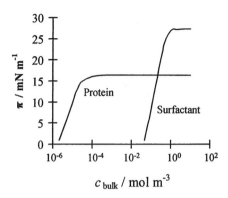

FIGURE 5.12 Comparison of the affinity of amphiphilic biopolymers and small-molecule surfactants for an oil–water interface.

which emulsifier molecules are adsorbed to an interface are discussed in Section 5.7. In this section, the factors which determine the relative affinity of surface-active molecules for an interface are considered.

The affinity of an emulsifier molecule for an interface can be described by its *adsorption efficiency* and its *surface activity* (Dickinson 1992). The adsorption efficiency is a measure of the minimum amount of emulsifier required to saturate an interface, whereas the surface activity is a measure of the maximum decrease in interfacial tension achievable when an interface is completely saturated (i.e., Π_{max}). Adsorption efficiencies and surface activities depend on the molecular structure of emulsifiers, as well as the prevailing environmental conditions. Amphiphilic biopolymers, such as proteins, tend to have higher adsorption efficiencies, but lower surface activities, than small-molecule surfactants (Figure 5.12).

Emulsifier molecules in solution partition themselves between the bulk solution and the interfacial region. The equilibrium constant between the adsorbed and unadsorbed states is proportional to $e^{E/kT}$, where E is a binding energy (Hiemenz 1986, Dickinson 1992). A small molecule tends to have one fairly strong binding site (its hydrophobic tail), whereas biopolymer molecules tend to have a large number of relatively weak binding sites (nonpolar amino acid side groups). The overall binding energy of a biopolymer molecule tends to be greater than that of a small-molecule surfactant, and therefore it binds more efficiently (i.e., less emulsifier must be added to the bulk aqueous phase before the interface becomes completely saturated). For the same reason, small-molecule surfactants tend to rapidly exchange between the adsorbed and unadsorbed states, whereas biopolymer molecules tend to remain at an interface for extended periods after adsorption. The relatively slow desorption of biopolymer molecules from an interface compared to small-molecule surfactants is best illustrated by an analogy. Consider a flock of birds that are all connected to each other by a piece of string. At any particular time, an individual bird may decide to fly, but it cannot get very far from the ground because it is held back by the other birds. It is only when a large number of birds simultaneously decide to take off together that they are all able to leave the ground, which is statistically very unlikely. Thus, at low emulsifier concentrations, biopolymers have a greater affinity for an interface than small-molecule surfactants.

On the other hand, small-molecule surfactants tend to decrease the interfacial tension by a greater amount than biopolymer molecules at concentrations where the interface is completely saturated, because they pack more efficiently and therefore screen the unfavorable interactions between the oil and water molecules more effectively. Thus, at high emulsifier concentrations, small-molecule surfactants have a greater affinity for an interface than biopolymers and will tend to displace them. This accounts for the ability of relatively high concentrations of surfactant molecules (e.g., 1% Tween 20) to displace proteins from the surface of oil droplets (Dickinson et al. 1993c; Courthaudon et al. 1991a,b,c,d; Dickinson and Tanai 1992; Dickinson 1992; Dickinson and Iveson 1993). Small-molecule surfactants are often added to ice cream premixes so as to displace the proteins from the surface of the milk fat globules prior to cooling and shearing (Goff et al. 1987). This causes the droplets to become more susceptible to partial coalescence, which leads to the formation of a network of aggregated droplets which stabilizes the air bubbles and gives the final product its characteristic shelf life (Berger 1976).

The ease with which proteins can be displaced from an oil–water interface often depends on the age of the interfacial membrane. Some globular proteins become surface denatured after adsorption to an interface because of the change in their environment (Dickinson and Matsumura 1991, McClements et al. 1993d). When the proteins unfold, they expose amino acids that are capable of forming disulfide bonds with their neighbors and thus form an interfacial membrane that is partly stabilized by covalent bonds. This accounts for the experimental observation that the ease with which β-lactoglobulin can be displaced from the surface of oil droplets decreases as the emulsion ages (Dalgleish 1996a).

The phase in which a surfactant is most soluble also determines its effectiveness at displacing proteins from an interface. For example, water-soluble surfactants have been shown to be more effective at displacing proteins from the surface of oil droplets than oil-soluble surfactants (Dickinson et al. 1993a,b).

Interfacial composition also depends on the relative concentration of the different types of emulsifiers present. A minor surface-active ingredient may make up a substantial fraction of the droplet membrane when the droplets are large (small total surface area), but have a negligible contribution when the droplets are small (large total surface area).

The thermal history of an emulsion often has an important influence on the composition of its interfacial layer. This is particularly true for globular proteins that are capable of undergoing conformational changes at elevated temperatures. The interfacial concentration of β-lactoglobulin in oil-in-water emulsions increases when they are heated to 70°C (Dickinson and Hong 1994). In addition, the protein becomes more difficult to displace from the interface using small-molecule surfactants. The most likely explanation for this behavior is the thermal denaturation of the β-lactoglobulin molecules at this temperature. Thermal denaturation causes the molecules to unfold and expose amino acids that were originally located in their interior, including amino acids with nonpolar and cysteine side groups. The increase in surface hydrophobicity of the protein enhances its affinity for the droplet surface, as well as for other protein molecules, which accounts for the increase in interfacial concentration. The ability of the protein to form disulfide bonds with its neighbors leads to a covalently bonded film, which accounts for the difficulty in displacing the protein with surfactants. Temperature has been found to influence the ability of small-molecule surfactants to displace β-casein from the surface of oil droplets (Dickinson and Tanai 1992). A minimum in the interfacial concentration of protein was observed at temperatures between 5 and 10°C.

The pH and ionic strength of the aqueous phase containing the emulsifier molecules also have an important influence on adsorption kinetics and interfacial composition (Hunt and Dalgleish 1994, 1996). Potassium chloride has been shown to affect the competition between proteins for an interface (Hunt and Dalgleish 1995). Increasing the concentration of KCl present in the aqueous phase prior to homogenization causes a decrease in β-lactoglobulin and an increase in α-lactalbumin at the interface of oil-in-water emulsions at pH 7, but has no influence on interfacial composition for emulsions at pH 3. Changes in the relative proportions of the different types of caseins were also observed for emulsions stabilized by caseinate when the KCl concentration was increased. Potassium chloride was also found to alter the competitive adsorption of proteins added after homogenization (Hunt and Dalgleish 1996).

The above discussion highlights the wide variety of factors which can influence interfacial composition, such as the emulsifier concentration, emulsifier type, solution conditions, temperature, and time. For this reason, a great deal of research is being carried out to establish the relative importance of each of these factors and to establish the relationship between interfacial composition and the bulk physicochemical properties of food emulsions.

5.10. MEASUREMENT OF SURFACE AND INTERFACIAL TENSIONS

By definition, surface tension is measured at a gas–fluid interface (e.g., air–water or oil–water), while interfacial tension is measured at a fluid–fluid interface (e.g., oil–water). These quantities are measured using instruments called surface or interfacial tensiometers, respectively (Couper 1993). Tensiometers can be used to provide valuable information about the characteristics of surfaces and interfaces and about the properties of emulsifiers in solution, such as the surface excess concentration, surface pressure, critical micelle concentration, and adsorption kinetics (Hiemenz 1986, Hunter 1986, Evans and Wennerstrom 1994). A wide variety of different types of tensiometer are available for providing information about the properties of surfaces and interfaces (Couper 1993). These instruments differ according to the physical principles on which they operate, their mechanical design, whether measurements are static or dynamic, and whether they are capable of measuring surface tension, interfacial tension, or both. Static measurements are carried out on surfaces or interfaces which are considered to be at equilibrium, whereas dynamic measurements are carried out on surfaces or interfaces which are not at equilibrium and can therefore be used to monitor adsorption kinetics.

5.10.1. Du Nouy Ring Method

The Du Nouy ring method is used to measure static surface and interfacial tensions of liquids (Hiemenz 1986, Couper 1993). The apparatus required to carry out this type of measurement consists of a vessel containing the liquid(s) to be analyzed and a ring which is attached to a sensitive force-measuring device (Figure 5.13). The vessel is capable of being moved

TABLE 5.1
Summary of the Instruments Used for Measuring Surface and Interfacial Tensions

Surface tension	Interfacial tension
Equilibrium	
Du Nouy ring	Du Nouy ring
Wilhelmy plate	Wilhelmy plate
Pendant drop	Pendant drop
Sessile drop	Sessile drop
Spinning drop	Spinning drop
Capillary rise	
Dynamic	
Maximum bubble pressure	Drop volume
Oscillating jet	Pendant drop
Drop volume	Spinning drop
Surface waves	

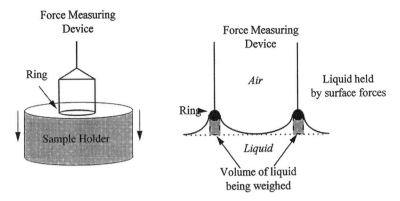

FIGURE 5.13   Du Nouy ring method of determining the interfacial and surface tension of liquids.

upward and downward in a controlled manner while the position of the ring is kept constant. Initially, the vessel is positioned so that the ring is submerged just below the surface of the liquid being analyzed. It is then slowly lowered and the force exerted on the ring is recorded. As the surface of the liquid moves downward, some of the liquid "clings" to the ring because of its surface tension (Figure 5.13). The weight of the liquid that clings to the ring is recorded by the force-measuring device and is related to the force that results from the surface tension.

The Du Nouy ring method is usually used in a "detachment" mode. The vessel is lowered until the liquid clinging to the ring ruptures and the ring becomes detached from the liquid. The force exerted on the ring at detachment is approximately equal to the surface tension multiplied by the length of the ring perimeter: $F = 4\pi R\gamma$, where R is the radius of the ring. In practice, this force has to be corrected because the surface tension does not completely act in the vertical direction and because some of the liquid remains clinging to the ring after it has become detached.

$$F = 4\pi R\gamma\beta \tag{5.13}$$

where β is a correction factor which depends on the dimensions of the ring and the density of the liquid(s) involved. Values of β have been tabulated in the literature or can be calculated using semiempirical equations (Couper 1993). One of the major problems associated with the Du Nouy ring method, as well as any other detachment method, is that serious errors may occur when measuring the surface tension of emulsifier solutions rather than pure liquids. When a ring detaches from a liquid, it leaves behind some fresh surface which is initially devoid of emulsifier. The measured surface tension therefore depends on the speed at which the emulsifier molecules diffuse from the bulk liquid to the fresh surface during the detachment process. If an emulsifier adsorbs rapidly compared to the detachment process, the surface tension measured by the detachment method will be the same as the equilibrium value, but if the emulsifier adsorbs relatively slowly, the surface tension will be greater than the equilibrium value because the surface has a lower emulsifier concentration than expected.

The Du Nouy ring method can also be used to determine the interfacial tension between two liquids (Couper 1993). In this case, the ring is initially placed below the surface of the most dense liquid (usually water). The force acting on the ring is then measured as it is pulled up through the interface and into the oil phase. A similar equation can be used to determine the interfacial tension from the force exerted on the ring, but one has to take into account the densities of the oil and water phases and use a different correction factor. For two liquids which are partially immiscible with each other, the interfacial tension may take an appreciable time to reach equilibrium because of the diffusion of water molecules into the oil phase and vice versa (Hunter 1993).

The Du Nouy ring method can also be used to determine surface or interfacial tensions by continuously monitoring the force acting on the ring as the vessel containing the liquid is lowered, rather than just measuring the detachment force. As the liquid is lowered, the force initially increases, but at a certain position it reaches a maximum (when the surface tension acts vertically), before decreasing slightly prior to detachment. In this case, the maximum force, rather than the detachment force, is used in the equations to calculate the surface or interfacial tension (Couper 1993). The advantage of this method is that it does not involve the rupture of the liquid, and therefore there are fewer problems associated with the kinetics of emulsifier adsorption during the detachment process.

For accurate measurements, it is important that the bottom edge of the ring be kept parallel to the surface of the fluid and that the contact angle between the liquid and the ring is close to zero. Rings are usually manufactured from platinum or platinum–iridium because these give contact angles that are approximately equal to zero. The Du Nouy ring method can be used to determine surface tensions to an accuracy of about 0.1 mN m^{-1} (Couper 1993).

5.10.2. Wilhelmy Plate Method

The Wilhelmy plate method is normally used to determine the static surface or interfacial tensions of liquids (Hiemenz 1986, Couper 1993). Nevertheless, it can also be used to monitor adsorption kinetics provided that the accumulation of the emulsifier at the surface is slow compared to the measurement time (Dickinson 1992). The apparatus consists of a vessel that contains the liquid being analyzed and a plate which is attached to a sensitive force-measuring device (Figure 5.14). The vessel is capable of being moved upward and downward, while the plate remains stationary. The vessel is positioned so that the liquid just comes into contact with the plate (i.e., the bottom of the plate is parallel to the surface of the bulk liquid). Some of the liquid "climbs" up the edges of the plate because this reduces the unfavorable contact area between the plate and the air. The amount of liquid which moves up the plate depends on its surface tension and density. If the force-measuring device is tarred

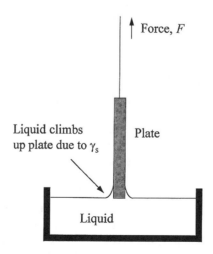

FIGURE 5.14   Wilhelmy plate method of determining the interfacial and surface tension of liquids.

prior to bringing the plate into contact with the liquid (to take into account the mass of the plate), then the force recorded by the device is equal to the weight of the liquid clinging to the plate. This weight is balanced by the vertical component of the surface tension multiplied by the length of the plate perimeter:

$$F = 2(l + L)\gamma \cos \theta \tag{5.14}$$

where l and L are the length and thickness of the plate and θ is the contact angle. Thus the surface tension of a liquid can be determined by measuring the force exerted on the plate. Plates are often constructed of materials which give a contact angle that is close to zero, such as platinum or platinum/iridium, as this facilitates the analysis. The Wilhelmy plate can also be used to determine the interfacial tension between two liquids (Hiemenz 1986). In this case, the plate is positioned so that its bottom edge is parallel to the interface between the two bulk liquids, and the force-measuring device is tarred when the plate is located in the less dense liquid (usually oil). A major advantage of the Wilhelmy plate method over the Du Nouy ring method is that it does not rely on the disruption of the liquid surface and therefore is less prone to errors associated with the adsorption kinetics of emulsifiers (Section 5.10.1).

The Wilhelmy plate method is widely used to study the adsorption kinetics of proteins (Dickinson 1992). It can be used for this purpose because their adsorption rate is much slower than the time required to carry out a surface tension measurement. The plate is positioned at the surface of the liquid at the beginning of the experiment, and then the force required to keep it at this position is recorded as a function of time. The Wilhelmy plate method is not suitable for studying the adsorption kinetics of small-molecule surfactants because they adsorb too rapidly to be followed using this technique. For accurate measurements, it is important that the bottom of the plate be kept parallel to the liquid surface and that the contact angle be either known or close to zero. Accuracies of about 0.05 mN m^{-1} have been reported using this technique (Couper 1993).

5.10.3. Sessile- and Pendant-Drop Methods

The sessile- and pendant-drop methods can be used to determine the static surface and interfacial tensions of liquids (Hunter 1986, Couper 1993). The shape of a liquid droplet depends on a balance between the gravitational and surface forces. Surface forces favor a spherical droplet because this shape minimizes the contact area between the liquid and its surroundings. On the other hand, gravitational forces tend to cause droplets to become flattened (if they are resting on a solid surface) or elongated (if they are hanging on a solid surface). A flattened drop is usually referred to as a sessile drop, whereas a hanging one is referred to as a pendent drop (Figure 5.15). The equilibrium shape that is adopted by a droplet is determined by its volume, density, and surface or interfacial tension.

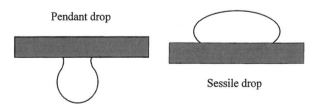

FIGURE 5.15 Sessile-drop and pendant-drop methods of determining the interfacial and surface tension of liquids.

The surface or interfacial tension is determined by measuring the shape of a drop using an optical microscope, often in conjunction with an image analysis program, and mathematical equations which describe the equilibrium shape (Couper 1993). This technique can be used to determine surface or interfacial tensions as low as 10^{-4} mN m^{-1} and can also be used to simultaneously determine the contact angle. The accuracy of the technique has been reported to be about 0.01% (Couper 1993).

5.10.4. Drop-Volume Method

The drop-volume method is used to measure surface and interfacial tensions of liquids by a detachment method. The liquid to be analyzed is pumped through the tip of a vertical capillary tube whose tip protrudes into air or an immiscible liquid (Figure 5.16). A droplet detaches itself from the tip of the capillary tube when the separation force (due to gravity) is balanced by the adhesion force (due to surface or interfacial tension). To measure surface tension, the tip should point downward into air, whereas to measure interfacial tension, the tip may point either upward or downward depending on the relative densities of the two liquids being analyzed. If the liquid in the tip has a higher density than the surrounding medium, the opening of the tip faces down and the drop moves downward once it becomes detached. On the other hand, if the liquid in the tip has a lower density, then the tip faces up and the drop moves upward when it becomes detached.

The surface or interfacial tension can be related to the volume of the detached droplet by analyzing the forces which act on it just prior to detachment (Couper 1993):

$$\gamma = \frac{V_{DROP}\Delta\rho g}{\pi d} \tag{5.15}$$

where d is the diameter of the tip, $\Delta\rho$ is the density difference between the liquid being analyzed and the surrounding liquid, V_{DROP} is the volume of the droplet, and g is the gravitational constant. The volume of the droplets can be determined using a graduated syringe or by weighing them using a sensitive balance (once the liquid density is known).

As with other detachment methods, the surface area of the droplet expands rapidly during the detachment process, and therefore the method is unsuitable for analyzing liquids that contain emulsifiers whose adsorption time is comparable to the detachment time. To obtain reliable results, it is normal practice to bring the droplet slowly to the detachment process. The accuracy of this technique has been reported to be about 0.1% (Couper 1993).

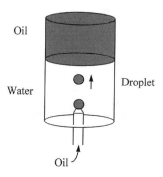

FIGURE 5.16 Drop-volume method of determining the interfacial and surface tension of liquids.

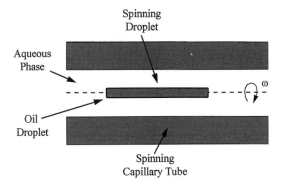

FIGURE 5.17   Spinning-drop method of determining the interfacial tension of liquids.

5.10.5.   Spinning-Drop Method

The spinning-drop method is used to measure static interfacial tensions of liquids (Couper 1993). It was designed to characterize liquids that have particularly low interfacial tensions (i.e., between about 10^{-6} to 5 mN m^{-1}). Most food emulsions have interfacial tensions that are above this range, and so this technique is not widely used in the food industry. Nevertheless, it may be useful for some fundamental studies, and therefore its operating principles are briefly outlined below.

A drop of oil is injected into a glass capillary tube which is filled with a more dense liquid (usually water). When the tube is made to spin at a particular angular frequency (ω), the droplet becomes elongated along its axis of rotation so as to reduce its rotational kinetic energy (Figure 5.17). The elongation of the droplet causes an increase in its surface area and is therefore opposed by the interfacial tension. Consequently, the equilibrium shape of the droplet at a particular angular frequency depends on the interfacial tension. The shape of the elongated droplet is measured using optical microscopy, and the interfacial tension is then calculated using the following equation:

$$\gamma = kr^3\omega^2\Delta\rho \qquad (5.16)$$

where k is an instrument constant which can be determined using liquids of known interfacial tension, r is the radius of the elongated droplet, and $\Delta\rho$ is the difference in density between the oil and the surrounding liquid. This equation is applicable when the droplet becomes so elongated that it adopts an almost cylindrical shape. More sophisticated equations are needed when the shape of the droplet does not change as dramatically (Couper 1993).

5.10.6.   Maximum Bubble Pressure Method

The maximum bubble pressure method is used to measure the static or dynamic surface tensions of liquids (Couper 1993). The apparatus required to carry out this type of measurement consists of a vertical capillary tube whose tip is immersed below the surface of the liquid being analyzed (Figure 5.18). Gas is pumped into the tube, which causes an increase in the pressure and results in the formation of a bubble at the end of the tip. The buildup of pressure in the tube is monitored using a pressure sensor inside the instrument. As the bubble grows, the pressure increases until it reaches a maximum when the bubble becomes hemispherical and the surface tension acts in a completely vertical direction. Any further bubble

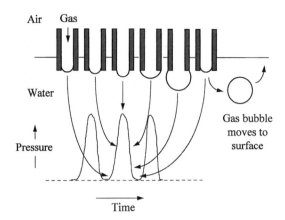

FIGURE 5.18 Maximum bubble pressure method of determining the surface tension of liquids.

growth beyond this point causes a decrease in the pressure. Eventually, the bubble formed breaks away from the tip and moves to the surface of the liquid; then another bubble begins to form and the whole process is repeated. The maximum bubble pressure can be related to the surface tension using suitable mathematical equations (Couper 1993).

The maximum bubble pressure method can be used to measure the static surface tension of pure liquids at any bubble frequency. It can also be used to monitor the dynamic surface tension of emulsifier solutions by varying the bubble frequency. The age of the gas–liquid interface is approximately half the time interval between the detachment of successive bubbles and can be varied by changing the flow rate of the gas. It is therefore possible to monitor adsorption kinetics of emulsifiers by monitoring the surface tension as a function of bubble frequency. The dynamic surface tension increases as the bubble frequency increases because there is less time for the emulsifier molecules to move to the surface of the droplets. The variation in the dynamic surface tension with bubble frequency therefore gives an indication of the speed at which emulsifier molecules are adsorbed to the surface. This information is important to food manufacturers because the adsorption rate of emulsifiers determines the size of the droplets produced during homogenization: the faster the rate, the smaller the droplet size (Chapter 6). Nevertheless, it should be pointed out that the maximum bubble pressure method can only be used to measure dynamic surface tensions down to about 50 ms, whereas many surfactants have faster adsorption rates than this. More rapid techniques are therefore needed to study these systems, such as the oscillating jet method (Section 5.10.7), the capillary wave method (Section 5.10.8), or the recently developed punctured membrane method (Stang et al. 1994).

One of the major advantages of the maximum bubble method is that it can be used to analyze optically opaque liquids because it is not necessary to visually observe the bubbles. In addition, it is not necessary to know the contact angle of the liquid because the maximum pressure occurs when the surface tension acts in a completely vertical direction.

5.10.7. Oscillating Jet Method

The oscillating jet method is used to determine the static and dynamic surface and interfacial tensions of liquids (Couper 1993). The apparatus used for this type of measurement consists of an elliptical orifice through which the liquid to be analyzed is pumped. The stream of liquid that emerges changes its shape from a vertical elliptical cross-section to a spherical cross-section as it moves away from the orifice because this enables it to reduce its surface area. Nevertheless, the liquid will continue to change its shape because of its momentum

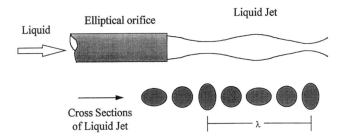

FIGURE 5.19   Oscillating jet method of determining the interfacial and surface tension of liquids.

until it gains a horizontal elliptical cross-section (Figure 5.19). As the liquid travels away from the orifice, its shape therefore oscillates between these different cross-sections. These oscillations have a characteristic wavelength which depends on the surface or interfacial tension of the liquid. The wavelength of the oscillating jet is measured using special optical techniques. The surface or interfacial tension can then be determined from the following equation:

$$\gamma = \frac{2\pi^2 r^3 \rho v^2}{3g\lambda^2}$$

(5.17)

where r is the radius of the orifice, v is the velocity of the liquid stream, g is the gravitational constant, and λ is the wavelength of oscillation. Surface tensions are measured by having the stream of liquid travel through air, whereas interfacial tensions are measured by having it travel through an immiscible liquid. The age of the surface can be varied by altering the flow rate of the liquid emerging from the elliptical orifice: the faster the flow rate, the shorter the age of the interface. Dynamic surface or interfacial tensions can be measured down to times of 1 ms using this method, and so it is particularly suitable for characterizing the adsorption kinetics of rapidly adsorbing emulsifiers. Nevertheless, there is some debate about whether the movement of the emulsifier molecules to the surface is due to convection or diffusion under the conditions prevailing within the instrument (Couper 1993). Pure liquids can be analyzed at any flow rate, because their surface or interfacial tensions should be independent of surface age. The major limitation of this technique is that it can only be used to analyze low-viscosity liquids (i.e., those with a viscosity less than about 5 mPa s).

5.10.8.   Capillary Wave Method

The capillary wave method is a fairly new technique which can be used to measure the static and dynamic surface tension of liquids (Couper 1993). A liquid is placed in a container that has a device which is capable of generating surface waves at a certain frequency. The amplitude and wavelength of the waves formed on the liquid surface are measured using a light-scattering technique and can be related to the surface tension and interfacial rheology of the liquid using an appropriate theory. The surface age can be varied by altering the frequency of the surface waves, and so it is possible to study adsorption kinetics. For pure liquids, the surface tension measured by this technique is independent of surface age and is in good agreement with that determined by static methods. For emulsifier solutions, the dynamic surface tension increases as the frequency of the surface waves is increased because the emulsifier molecules have less time to adsorb. The technique is particularly suitable for analyzing liquids that have extremely low surface tensions and can be used to study surface ages between about 5 and 100 ms.

TABLE 5.2
Rheological Characteristics of Interfaces

	Viscosity	Elasticity
Shear deformation	$\tau_i = \eta_i \dot{\sigma}_i$	$\tau_i = G_i \sigma_i$
Dilatational deformation	$d\gamma_i = \kappa_i A \dfrac{d\gamma}{dA}$	$d\gamma_i = \varepsilon_i \dfrac{dA}{A}$

5.11. INTERFACIAL RHEOLOGY

Interfacial rheology has been defined as the "study of the mechanical and flow properties of adsorbed layers at fluid interfaces" (Murray and Dickinson 1996). A knowledge of the factors which determine interfacial rheology is important to food scientists because the mechanical and flow properties of interfaces influence the formation, texture, and stability of food emulsions (Chapters 6 to 8). When an emulsion is subjected to mechanical agitation, the surfaces of the droplets experience a number of different types of deformation as a result of the stresses acting upon them. Stresses may cause different regions of the interface to move past each other, without altering the overall surface area, which is known as *interfacial shear deformation*. Alternatively, they may cause the surface area to expand or contract (like a balloon when it is inflated or deflated), which is known as *interfacial dilatational deformation*. An interface may have solid-like characteristics which are described by an *interfacial elastic constant* or fluid-like characteristics which are described by an *interfacial viscosity*. In practice, most interfaces have partly "solid-like" and partly "fluid-like" characteristics and therefore exhibit viscoelastic behavior.

It is convenient to assume that an interface is infinitesimally thin so that it can be treated as a two-dimensional plane, because its rheological characteristics can then be described using the two-dimensional analogs of the relationships used to characterize bulk materials (Chapter 8). A number of the most important rheological characteristics of interfaces are summarized in Table 5.2, where τ_i is the interfacial shear stress, σ_i is the interfacial shear strain, η_i is the interfacial shear viscosity, G_i is the interfacial shear modulus, ε_i is the interfacial dilatational elasticity, κ_i is the interfacial dilatational viscosity, A is the surface area, and γ is the surface or interfacial tension.

The stability of emulsion droplets to coalescence is largely governed by the resistance of the oil–water interface to deformation and rupture (Chapter 7). Emulsifiers that form highly viscoelastic membranes which are resistant to rupture often provide the greatest stability against coalescence. It is therefore important to have analytical techniques which can be used to characterize the interfacial rheology of different emulsifiers and to establish the factors that influence their interfacial properties.

5.11.1. Measurement of Interfacial Shear Rheology

A variety of experimental methods have been developed to measure the shear rheology of surfaces and interfaces (Murray and Dickinson 1996). One of the most commonly used methods is analogous to the concentric cylinder technique used to measure the shear properties of bulk materials (Chapter 8). The sample to be analyzed is placed in a thermostated vessel, and a thin disk is placed in the plane of the interface that separates the two phases (e.g., water-and-air or oil-and-water) (Figure 5.20). The vessel is then rotated and the torque on the disk is measured. The sample can be analyzed in a number of different ways depending on whether it is solid-like, liquid-like, or viscoelastic. For liquid-like interfaces, the torque on the

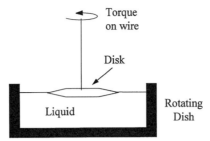

FIGURE 5.20   Experimental technique for measuring interfacial shear rheology.

disk is measured as the vessel is rotated continuously. For solid-like interfaces, the torque is measured after the vessel has been moved to a fixed angle. For viscoelastic interfaces, the torque is measured continuously as the vessel is made to oscillate backward and forward at a certain frequency and angle.

The interfacial shear viscosity or elasticity of surfactant membranes is usually several orders of magnitude less than that of biopolymer membranes because biopolymer molecules often become entangled or interact with each other through various covalent or physical forces. The rheology of bulk emulsions depends on the concentration, size, and interactions of the droplets. By analogy, the rheology of interfaces depends on the concentration, size, and interactions of the adsorbed emulsifier molecules. Interfacial shear rheology measurements are particularly useful for providing information about adsorption kinetics, competitive adsorption, and the structure and interactions of molecules at an interface, especially when they are used in conjunction with experimental techniques that provide information about the concentration of the emulsifier molecules at the interface (e.g., interfacial tension or radioactive labeling techniques). The concentration of emulsifier molecules at an interface often reaches a constant value after a particular time, while the shear modulus or viscosity continues to increase because of interactions between the adsorbed molecules (Dickinson 1992).

5.11.2. Measurement of Interfacial Dilatational Rheology

Most analytical techniques used to characterize interfacial dilatational rheology measure the change in interfacial tension as the surface area is increased or decreased in a controlled manner (e.g., trough and overflowing cylinder methods) (Murray and Dickinson 1996). Trough methods measure the surface or interfacial tension of a liquid using a Wilhelmy plate as the interfacial area is varied by changing the distance between two solid barriers which confine the liquid (Figure 5.21). The overflowing cylinder method can be used to measure the dynamic dilatational rheology of surfaces or interfaces. A Wilhelmy plate is used to measure the surface or interfacial tension of a liquid as it is continuously pumped into a cylinder so that it overflows at the edges. The dilatational viscosity is measured as a function of the surface age by altering the rate at which the liquid is pumped into the cylinder.

The capillary wave method described in Section 5.10.8 can also be used to measure the interfacial dilatational rheology of liquids. A laser beam reflected from the surface of a liquid is used to determine the amplitude and wavelength of the surface waves, which can then be related to the dilatational modulus or viscosity of the surface using an appropriate theory. These surface waves are believed to play an important role in the coalescence of emulsion droplets (Chapter 7), and therefore this technique may provide information that has direct practical importance for food scientists.

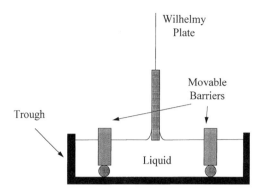

FIGURE 5.21 Experimental technique for measuring interfacial dilatational rheology.

It should be noted that the dilatational rheology of an interface can be influenced by the adsorption kinetics of emulsifiers (Murray and Dickinson 1996). When a surface undergoes a dilatational expansion, the concentration of emulsifier per unit area decreases and therefore its surface tension increases, which is energetically unfavorable. The dilatational elasticity or viscosity is a measure of this resistance of the surface to dilation (Table 5.2). When emulsifier molecules are present in the surrounding liquid, they may be adsorbed to the surface during dilation and thereby reduce the surface tension. The dilatational rheology therefore depends on the rate at which emulsifier molecules are adsorbed to a surface, which is determined by their concentration, molecular structure, and the prevailing environmental conditions (Section 5.4). The faster the molecules adsorb to the freshly formed interface, the lower is the resistance to dilation, and therefore the lower is the dilatational modulus or viscosity. When an interface undergoes dilatational compression, some of the emulsifier molecules may leave the interface to reduce the resulting strain, and therefore the desorption rate may also influence the rheological characteristics of an interface.

5.12. CHARACTERIZATION OF INTERFACIAL STRUCTURE

The structural organization of the emulsifier molecules at the surface of the droplets in an emulsion influences the properties of the whole emulsion. Consequently, it is important for food scientists to have analytical techniques that can be used to provide information about interfacial structure, such as the thickness of the adsorbed layer and the packing and orientation of the emulsifier molecules (Dalgleish 1996). A number of the most important experimental techniques that are available to provide information about interfacial structure are listed in Table 5.3. Each of these techniques works on a different physical principle and is sensitive to a different aspect of the properties of an interface (Kallay et al. 1993).

A variety of experimental techniques rely on the scattering of radiation or subatomic particles from a surface or interface (e.g., light, X-ray, or neutron scattering techniques). Some of these techniques can only be applied to planar interfaces, while others can also be applied to the study of interfaces on emulsion droplets. Analytical techniques based on the scattering of radiation or subatomic particles can be used to provide information about the thickness of the interfacial layer and sometimes also about the variation in the emulsifier concentration within the interface. They can therefore be used to compare the thickness of interfaces formed by different emulsifiers or to study changes in interfacial thickness in response to alterations in environmental conditions such as pH, ionic strength, or temperature. The application of these techniques has provided valuable insights into the structures formed

TABLE 5.3
Experimental Techniques to Characterize the Structural Properties of Interfaces

Techniques	Applications
Scattering techniques	
Light scattering	Thickness
Small-angle X-ray scattering	Thickness, concentration profile
Neutron scattering	Thickness, concentration profile
Reflection techniques	
Ellipsometry	Thickness, refractive index
Neutron reflection	Thickness, concentration profile
Techniques that utilize absorption	
of electromagnetic radiation	
Infrared	Concentration, conformational changes
Ultraviolet-visible	Concentration, conformational changes
Fluorescence	Concentration, conformational changes
Nuclear magnetic resonance	Concentration, conformational changes
Circular dichroism	Secondary structure
Biochemical techniques	
Enzyme hydrolysis	Orientation, conformation
Immunoassays	Orientation, conformation
Microscopic techniques	
Atomic force microscopy	Structural organization of molecules
Electron microscopy	Interfacial structure

by proteins at oil–water and air–water interfaces, which is useful for predicting the stability of emulsion droplets to flocculation or coalescence (Dalgleish 1996).

A number of spectroscopic techniques have been developed that rely on the absorption of electromagnetic radiation by the molecules at an interface (e.g., infrared, ultraviolet, visible, fluorescence, and X-ray techniques) (Kallay et al. 1993). These techniques can be used either in a transmission mode, where the electromagnetic wave is propagated through the sample, or in a reflection mode, where the electromagnetic wave is reflected from the interface. In both cases, the absorption of the radiation by the wave is measured and correlated to some property of the molecules at the interface. The different techniques are sensitive to different characteristics of the molecules. Infrared relies on characteristic molecular vibrations of different groups within a molecule, whereas ultraviolet-visible and fluorescence techniques rely on characteristic electronic transitions which occur in unsaturated groups. These techniques have been used to determine the concentration of emulsifier molecules at an interface, to study adsorption kinetics, and to provide information about changes in molecular conformation (Kallay et al. 1993). When a molecule undergoes a conformational change, the adsorption spectrum is altered because of the change in the environment of the chemical groups.

A number of techniques have been developed which rely on the interference pattern produced by the reflection of a beam of light or subatomic particles from a thin interface. Ellipsometry relies on the reflection of a polarized light beam from a highly reflective planar surface (Azzam and Bachara 1989, Malmsten et al. 1995). Initially, the phase and amplitude of the light beam reflected from the clean surface are measured. Emulsifier is then allowed to adsorb to the surface and the experiment is repeated. A mathematical analysis of the reflected wave provides information about the thickness and refractive index profile of the adsorbed layer. The refractive index profile can then be converted to a concentration profile.

Neutron reflection techniques work on a similar principle, but they use a beam of neutrons, rather than a beam of light, to probe the interface (Atkinson 1995, Dickinson and McClements 1995). The beam of neutrons is directed at an air–water or oil–water interface at an angle. The beam of neutrons reflected from the surface is analyzed and provides information about the thickness of the interface and the neutron refractive index profile, which can be related to the emulsifier concentration profile across the interface. These techniques have been widely used to study the characteristics of protein layers and the influence of pH and ionic strength on their properties.

A number of biochemical techniques have also been developed to provide information about the structure and organization of biopolymer molecules at the surface of emulsion droplets (Dalgleish 1996b). The susceptibility of proteins to hydrolysis by specific enzymes can be used to identify the location of particular peptide bonds. When a protein is dissolved in an aqueous solution, the whole of its surface is accessible to enzyme hydrolysis, but when it is adsorbed to an interface, it adopts a conformation where some of the amino acids are located close to the interface (and are therefore inaccessible to proteolysis), whereas others are exposed to the aqueous phase (and are therefore accessible to proteolysis). By comparing the bonds which are susceptible to enzyme hydrolysis and the rate at which hydrolysis of the adsorbed and unadsorbed proteins occurs, it is possible to obtain some idea about the orientation and conformation of the adsorbed proteins at the interface. This technique is more suitable for the study of flexible random-coil proteins than globular proteins, because the latter have compact structures which are not particularly accessible to enzyme hydrolysis in either the adsorbed or unadsorbed state. Immunological techniques can also be used to provide similar information. These techniques utilize the binding of antibodies to specific sites on a protein to provide information about its orientation at an interface or about any conformational changes which take place after adsorption.

The thickness of a layer of emulsifier molecules adsorbed to an interface can be determined by measuring the forces between two interfaces as they are brought into close contact (Claesson et al. 1995, 1996). There is usually a steep rise in the repulsive forces at close separations between the interfaces because of steric repulsion. The distance at which this rise is observed can be assumed to be the outer edge of the emulsifier layer.

Information about the conformation of proteins adsorbed to the surfaces of emulsion droplets has also been obtained using sensitive differential scanning calorimetry (DSC) techniques (Corredig and Dalgleish 1995; Dalgleish 1996a,b). The emulsion being analyzed is heated at a controlled rate and the amount of heat absorbed by the proteins when they undergo conformational changes is recorded by the DSC instrument. The temperature at which the conformation change occurs ($T_{transition}$) and the amount of energy adsorbed by the protein during the transition ($\Delta H_{transition}$) are determined. The values for the adsorbed protein are then compared with those for a solution containing a similar concentration of unadsorbed proteins. Experiments with oil-in-water emulsions containing adsorbed globular proteins, such as α-lactalbumin, lysozyme, and β-lactoglobulin, have shown that there is a decrease in $\Delta H_{transition}$ after adsorption, which indicates that they have undergone some degree of surface denaturation (Corredig and Dalgleish 1995). It was also observed that after the proteins had been desorbed from the surface, the denaturation of β-lactoglobulin was irreversible, but that of α-lactalbumin and lysozyme was at least partly reversible. Flexible biopolymers that have little structure, such as casein, cannot be studied using this technique because they do not undergo any structural transitions which adsorb or release significant amounts of heat. It should be noted that the data from DSC experiments must be interpreted carefully because a reduction in $\Delta H_{transition}$ may be because the protein has no secondary or tertiary structure or because the structure is so stable that it does not unfold at the same temperature as in water. In addition, the adsorption of a protein at an interface may also reduce $\Delta H_{transition}$ because its thermodynamic environment has been altered: in solution, it is completely surrounded by water, but at an interface, part of the molecule is in contact with a hydrophobic surface. Despite these limitations, the DSC technique is a powerful tool for studying structural changes caused by adsorption, especially when used in conjunction with other techniques.

5.13. PRACTICAL IMPLICATIONS OF INTERFACIAL PHENOMENA

To conclude this chapter, a brief outline of some of the most important practical implications of interfacial phenomena for food emulsions is presented. One of the most striking features of food emulsions when observed under a microscope is the sphericity of the droplets. Droplets tend to be spherical because this shape minimizes the energetically unfavorable contact area between oil and water molecules, which is described by the Laplace equation (Equation 5.4). Droplets become nonspherical when they experience an external force that is large enough to overcome the Laplace pressure (e.g., gravity, centrifugal forces or mechanical agitation). The magnitude of the force needed to deform a droplet decreases as the interfacial tension decreases or the radius increases. This accounts for the ease with which the relatively large droplets in highly concentrated emulsions, such as mayonnaise, are deformed into polygons (Dickinson and Stainsby 1982).

The thermodynamic driving force for coalescence and "oiling off" is the interfacial tension between the oil and water phases caused by the imbalance of molecular interactions across an oil–water interface (Section 7.2). The reason that surface-active molecules adsorb to an air–fluid or oil–water interface is because of their ability to reduce the surface or interfacial tension (Section 5.2.2.1). The tendency of a liquid to spread over the surface of another material or to remain as a lens depends on the relative magnitude of the interfacial and surface tensions between the different types of substance involved (Section 5.5).

The formation of stable nuclei in a liquid is governed by the interfacial tension between the crystal and the melt (Section 4.2). The larger the interfacial tension between the solid and liquid phases, the greater the degree of supercooling required to produce stable nuclei. The interfacial tension also determines whether an impurity is capable of promoting heterogenous nucleation.

The solubility of a substance increases as the size of the particle containing it decreases (Section 5.4). If a suspension contains particles (emulsion droplets, fat crystals, ice crystals, or air bubbles) of different sizes, there is a greater concentration of the substance dissolved in the region immediately surrounding the smaller particles than that surrounding the larger particles. Consequently, there is a concentration gradient which causes material to move from the smaller particles to the large particles. With time, this process manifests itself as a growth of the large particles at the expense of the smaller ones, which is referred to as Ostwald ripening (Chapter 7). This process is responsible for the growth in size of emulsion droplets, fat crystals, ice crystals, and air bubbles in food emulsions, which often has a detrimental effect on the quality of a food. For example, the growth of ice crystals in ice cream causes the product to be perceived as "gritty" (Berger 1976). It should be clear from this chapter that even though the interfacial region only comprises a small fraction of the total volume of an emulsion, it plays an extremely important role in determining its bulk physicochemical properties.

6 Emulsion Formation

6.1. INTRODUCTION

Fresh milk is an example of a naturally occurring emulsion that can be consumed directly by human beings (Swaisgood 1996). In practice, however, most milk is subjected to a number of processing operations prior to consumption in order to ensure its safety, to extend its shelf life, and to create new products. Processing operations, such as homogenization, pasteurization, whipping, and churning, are responsible for the wide range of properties exhibited by dairy products (e.g., homogenized milk, cream, ice cream, butter, and cheese) (Harper and Hall 1976, Desrosier 1977, Kessler 1981). Unlike dairy products, most other food emulsions are manufactured by combining raw materials which are not normally found together in nature (Friberg and Larsson 1997, Dickinson and Stainsby 1982). For example, a salad dressing may be prepared using water, proteins from milk, oil from soybeans, vinegar from apples, and polysaccharides from seaweed. The physicochemical and sensory properties of a particular food emulsion depend on the type and concentration of ingredients it contains, as well as the method used to create it. To improve the quality of existing products, develop new products, and reduce production costs, it is important for food manufacturers to have a thorough understanding of the physical processes which take place during emulsion formation. The physical principles of emulsion formation, the various techniques available for creating emulsions, and the factors which affect the efficiency of emulsion formation are discussed in this chapter.

6.2. OVERVIEW OF HOMOGENIZATION

The formation of an emulsion may involve a single step or a number of consecutive steps, depending on the nature of the starting material and the method used to create it. Prior to converting separate oil and aqueous phases into an emulsion, it is usually necessary to disperse the various ingredients into the phase in which they are most soluble. Oil-soluble ingredients, such as vitamins, colors, antioxidants, and surfactants, are usually mixed with the oil, whereas water-soluble ingredients, such as proteins, polysaccharides, sugars, salts, vitamins, colors, antioxidants, and surfactants, are usually mixed with the water. The intensity and duration of the mixing process depend on the time required to solvate and uniformly distribute the ingredients. Adequate solvation is important for the functionality of a number of food components (e.g., the emulsifying properties of proteins are often improved by allowing them to hydrate in water for a few minutes or hours prior to homogenization) (Kinsella and Whitehead 1989). If the lipid phase contains any crystalline material, it is necessary to warm it to a temperature where all the fat melts prior to homogenization; otherwise it is difficult, if not impossible, to create a stable emulsion (Mulder and Walstra 1974, Phipps 1985).

The process of converting two immiscible liquids into an emulsion is known as *homogenization*, and a mechanical device designed to carry out this process is called a *homogenizer*

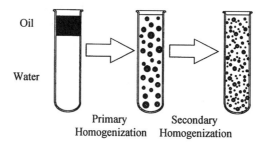

FIGURE 6.1 Homogenization can be conveniently divided into two categories: primary and secondary. Primary homogenization is the conversion of two bulk liquids into an emulsion, whereas secondary homogenization is the reduction in size of the droplets in an existing emulsion.

(Loncin and Merson 1979). To distinguish between the nature of the starting materials, it is convenient to divide homogenization into two categories. The creation of an emulsion directly from two separate liquids will be defined as *primary* homogenization, whereas the reduction in size of the droplets in an existing emulsion will be defined as *secondary* homogenization (Figure 6.1). The creation of a particular type of food emulsion may involve the use of either of these types of homogenization or a combination of both. For example, the preparation of a salad dressing in the kitchen is carried out by direct homogenization of the aqueous and oil phases and is therefore an example of primary homogenization, whereas homogenized milk is manufactured by reducing the size of the fat globules in raw milk and so is an example of secondary homogenization. In many food-processing operations and laboratory studies, it is more efficient to prepare an emulsion using two steps (Dickinson and Stainsby 1982). The separate oil and water phases are converted to a coarse emulsion which contains fairly large droplets using one type of homogenizer (e.g., a high-speed blender), and then the size of the droplets is reduced using another type of homogenizer (e.g., a high-pressure valve homogenizer). Many of the same physical processes occur during primary and secondary homogenization (e.g., mixing, droplet disruption, and droplet coalescence), and so there is no clear distinction between the two.

Emulsions which have undergone secondary homogenization usually contain smaller droplets than those which have undergone primary homogenization, although this is not always the case. Some homogenizers are capable of producing emulsions with small droplet sizes directly from the separate oil and water phases (e.g., ultrasound, microfluidizers, or membrane homogenizers) (see Section 6.4).

The physical processes which occur during homogenization can be highlighted by considering the formation of an emulsion from pure oil and pure water. When the two liquids are placed in a container, they tend to adopt their thermodynamically most stable state, which consists of a layer of oil on top of a layer of water (Figure 6.1). This arrangement is adopted because it minimizes the contact area between the two immiscible liquids and because oil has a lower density than water (Section 7.2). To create an emulsion, it is necessary to supply energy in order to disrupt and intermingle the oil and water phases, which is usually achieved by mechanical agitation (Walstra 1993b). The type of emulsion formed in the absence of an emulsifier depends primarily on the initial concentration of the two liquids: at high oil concentrations, a water-in-oil emulsion tends to be formed, but at low oil concentrations, an oil-in-water emulsion tends to be formed.* In this example, we assume that the oil concentration is so low that an oil-in-water emulsion is formed. Mechanical agitation can be applied

---

* In the presence of an emulsifier, the type of emulsion formed is governed mainly by the properties of the emulsifier (i.e., the HLB number and optimum curvature) (Chapter 4).

in a variety of different ways (Section 6.4), the simplest being to vigorously shake the oil and water together in a sealed container. Immediately after shaking, an emulsion is formed which appears optically opaque because light is scattered by the emulsion droplets (Farinato and Rowell 1983). The oil droplets formed during the application of the mechanical agitation are constantly moving around and frequently collide and coalesce with neighboring droplets (Walstra 1993b). As this process continues, the large droplets formed move to the top of the container due to gravity and merge together to form a separate layer. As a consequence, the system reverts back to its initial state — a layer of oil sitting on top of a layer of water (Figure 6.1). The thermodynamic driving forces for this process are the hydrophobic effect, which favors minimization of the contact area between the oil and water, and gravity, which favors the upward movement of the oil (Section 7.2).

To form an emulsion that is (kinetically) stable for a reasonable period of time, one must prevent the droplets from merging together after they have been formed (Walstra 1983, 1993b). This is achieved by having a sufficiently high concentration of *emulsifier* present during the homogenization process. The emulsifier adsorbs to the surface of the droplets during homogenization, forming a protective membrane which prevents them from coming close enough together to coalesce. The size of the droplets produced during homogenization depends on two processes: (1) the initial generation of droplets of small size and (2) the rapid stabilization of these droplets against coalescence once they are formed (Section 6.3).

Many of the bulk physicochemical and organoleptic properties of food emulsions depend on the size of the droplets they contain, including their stability, texture, appearance, and taste (Chapters 7 to 9). One of the major objectives of homogenization is therefore to create an emulsion in which the majority of droplets fall within an optimum range which has previously been established by the food manufacturer. It is therefore important for food scientists to appreciate the major factors which determine the size of the droplets produced during homogenization.

This brief introduction to homogenization has highlighted some of the most important aspects of emulsion formation, including the necessity to mechanically agitate the system, the competing processes of droplet formation and droplet coalescence, and the role of the emulsifier. These topics will be considered in more detail in the rest of the chapter.

6.3. PHYSICAL PRINCIPLES OF EMULSION FORMATION

As mentioned in the previous section, the size of the droplets produced by a homogenizer depends on a balance between two opposing physical processes: *droplet disruption* and *droplet coalescence* (Figure 6.2). A better understanding of the factors which influence these processes would enable food manufacturers to select the most appropriate ingredients and homogenization conditions required to produce a particular food product.

6.3.1. Droplet Disruption

The precise nature of the physical processes which occur during emulsion formation depends on the type of homogenizer used. Nevertheless, there are some common aspects of droplet disruption which apply to most types of homogenizer. The initial stages of primary homogenization involve the breakup and intermingling of the bulk oil and aqueous phases so that fairly large droplets of one of the liquids become dispersed throughout the other liquid (Walstra 1983, 1993b). The later stages of primary homogenization, and the whole of secondary homogenization, involve the disruption of larger droplets into smaller ones. It is therefore important to understand the nature of the forces which are responsible for the disruption of droplets during homogenization. Whether or not a droplet breaks up is determined by a balance between *interfacial forces,* which tend to hold the droplets together, and the *disrup-*

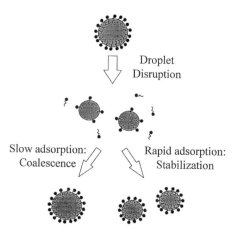

FIGURE 6.2 The size of the droplets produced during homogenization depends on the balance between the time for an emulsifier to adsorb to the surface of a droplet (τ_{adsorb}) and the time between droplet–droplet collisions ($\tau_{collision}$).

tive forces generated within the homogenizer, which tend to pull them apart (Walstra 1983, 1993b).

6.3.1.1. Interfacial Forces

An emulsion droplet tends to be spherical because this shape minimizes the energetically unfavorable contact area between the oil and aqueous phases (Chapter 5). Changing the shape of a droplet, or breaking it up into a number of smaller droplets, increases this contact area and therefore requires an input of energy. The interfacial force responsible for keeping a droplet in a spherical shape is characterized by the *Laplace pressure* (ΔP_L), which acts across the oil–water interface toward the center of the droplet so that there is a larger pressure inside the droplet than outside of it:

$$\Delta P_L \;=\; \frac{4\gamma}{d} \tag{6.1}$$

Here, γ is the interfacial tension between oil and water, and d is the droplet diameter. To deform and disrupt a droplet during homogenization, it is necessary to apply an external force which is significantly larger than the interfacial force (Walstra 1983, 1996b). Equation 6.1 indicates that the pressure required to disrupt a droplet grows as the interfacial tension increases or as the droplet size decreases. It also indicates that intense pressures must be generated within a homogenizer in order to overcome the interfacial forces holding the emulsion droplets together. For example, the Laplace pressure of a 1-μm droplet with an interfacial tension of 0.01 N m^{-1} is about 40 kPa, which corresponds to a pressure gradient of $\Delta P_L/d \approx 40 \times 10^9$ Pa m^{-1} across the droplet.

6.3.1.2. Disruptive Forces

The nature of the disruptive forces which act on a droplet during homogenization depends on the flow conditions it experiences (i.e., laminar, turbulent, or cavitational) and therefore on the type of homogenizer used (Phipps 1985, Walstra 1993b). For a droplet to be broken up during homogenization, the disruptive forces must exceed the interfacial forces, and their duration must be longer than the time required to deform and disrupt the droplet (τ_{deform})

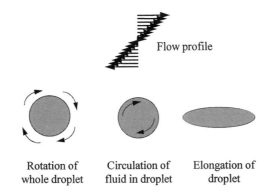

FIGURE 6.3   In the presence of a simple shear flow, droplets may rotate and become elongated. In addition, the fluid inside a droplet may circulate around the center of the droplet.

(Stone 1994, Karbstein and Schubert 1995). The susceptibility of emulsion droplets to disruption is conveniently characterized by the Weber number (We), which is the ratio of the disruptive forces to the interfacial forces (Walstra 1983). Droplets are disrupted when the Weber number exceeds some critical value (around unity), which depends on the physical characteristics of the oil and aqueous phases.

The flow profile of an emulsion within a homogenizer is usually extremely complex and is therefore difficult to model mathematically (Phipps 1985). For this reason, it is not easy to accurately calculate the disruptive forces that a droplet experiences during homogenization. Nevertheless, it is possible to gain much insight into the factors which affect the homogenization process by considering droplet disruption under simpler flow conditions which approximate those that occur in actual homogenizers (i.e., laminar, turbulent, or cavitational flow conditions) (Gopal 1968; Walstra 1983, 1993b; Williams et al. 1997).

Laminar Flow Conditions.   This type of flow profile is predominant at low flow rates, where the fluid moves in a regular and well-defined pattern (Curle and Davies 1968, Walstra 1993b). Different types of laminar flow profile are possible, depending on the direction and velocity at which different regions within the fluid move relative to one another (e.g., parallel, simple shear, rotational, and hyperbolic flow) (Gopal 1968). For convenience, only the disruption of droplets under simple shear flow conditions is considered here. In the presence of a simple shear field, a droplet experiences a combination of normal and tangential stresses (Loncin and Merson 1979). These stresses cause the droplet to rotate and become elongated, as well as causing the liquid within the droplet to circulate (Figure 6.3). At sufficiently high shear rates, the droplet becomes so elongated that it is broken up into a number of smaller droplets (Stone 1994, Williams et al. 1997). The manner in which the droplets break up depends on the ratio of the viscosities of the droplet and continuous phase (η_D/η_C). Experiments in which droplets were photographed under different flow conditions have shown that at low values of η_D/η_C the droplets break up at their edges, at intermediate values they break up near their middle, and at high values they may not break up at all, because there is insufficient time for the droplets to deform during the application of the disruptive forces (Gopal 1968, Williams et al. 1997).

The disruptive forces that a droplet experiences during simple shear flow are determined by the shear stress ($G\eta_C$) which acts upon the droplet, and so the Weber number is given by (Walstra 1993b):*

---

* The reason that a factor of 4 appears in Equation 6.3, whereas a factor of 2 appears in Equation 6.2, is because only half of the applied shear force goes to deforming the droplet; the remainder causes the droplet to rotate and is therefore not responsible for droplet disruption.

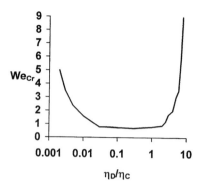

FIGURE 6.4 Dependence of the critical Weber number on the viscosity of the dispersed and continuous phases under simple shear flow conditions. (Adapted from Schubert and Armbruster 1992.)

$$We = \frac{\text{shear forces}}{\text{interfacial forces}} = \frac{G\eta_C d}{2\gamma} \qquad (6.2)$$

Here, G is the shear rate and η_C is the viscosity of the continuous phase. For a given system, it is possible to define a critical Weber number ($We_{critical}$), which is the value of We where the droplets are just stable to disruption. If an emulsion has a Weber number above this critical value (i.e., high shear rates or large droplets), then the droplets will be broken up; otherwise they will remain intact (Figure 6.4). The critical Weber number of an emulsion in the absence of emulsifier depends principally on the ratio of the viscosities of the dispersed and continuous phases (Janssen et al. 1994, Karbstein and Schubert 1995, Williams et al. 1997). $We_{critical}$ has a minimum value when η_D/η_C is between about 0.1 and 1 and increases significantly as the viscosity ratio decreases below about 0.05 or increases above about 5 (Figure 6.4). The behavior of droplets during the disruption process has been widely studied and is now well understood (Bentley and Leal 1986, Stone 1994, Janssen et al. 1994). Droplets are resistant to breakup at low viscosity ratios (<0.05) because they are able to become extremely elongated before any disruption will occur. They are resistant to breakup at high viscosity ratios (>5) because they do not have sufficient time to become deformed before the flow field causes them to rotate to a new orientation and therefore alter the distribution of disruptive stresses acting on them. At intermediate viscosity ratios, the droplets tend to form a dumbbell shape just prior to breaking up.

In the presence of emulsifiers, such as small-molecule surfactants or proteins, the behavior of droplets in flow fields has been found to be different from that in the absence of emulsifiers, which has been attributed to their influence on the rheology of the interfacial membrane (Lucassen-Reyanders and Kuijpers 1992, Janssen et al. 1994, Williams et al. 1997). Droplets are more difficult to disrupt than would be expected from their equilibrium interfacial tension, because the emulsifier imparts rheological properties to the droplet interface, which increases its resistance to tangential stresses.

A food manufacturer usually wants to produce an emulsion which contains droplets below some specified size, and so it is important to establish the factors which determine the size of the droplets produced during homogenization. For simple shear flow, the following relationship gives a good description of the maximum size of droplets which can persist in an emulsion during homogenization under steady-state conditions (Walstra 1983, 1993b):

$$d_{max} = \frac{2\gamma We_{critical}}{G\eta_C} \qquad (6.3)$$

Any droplets larger than d_{max} will be disrupted, whereas any smaller droplets will remain intact. Equation 6.3 indicates that the size of the droplets produced during homogenization decreases as the interfacial tension (γ) decreases, as the shear rate increases, or as the viscosity of the continuous phase increases. It also indicates that a higher shear rate is required to decrease the droplet size when the viscosity of the continuous phase is low. It is for this reason that homogenizers which rely principally on simple shear flow conditions, such as colloid mills, are not suitable for generating emulsions with small droplet sizes when the continuous phase has a low viscosity (Walstra 1983). For this type of system, it is better to use homogenizers which utilize turbulent or cavitational conditions to break up the droplets.

Turbulent Flow Conditions. Turbulence occurs when the flow rate of a fluid exceeds some critical value, which is largely determined by its viscosity (Curle and Davies 1968; Walstra 1983, 1993b; Phipps 1985). Turbulence is characterized by rapid and chaotic fluctuations in the velocity of the fluid with time and location. The disruption of droplets under turbulent flow conditions is caused by the extremely large shear and pressure gradients associated with the eddies generated in the fluid (Walstra 1993b). An eddy is a region within a fluid where there is a close correlation between the fluid velocity of the different elements (Curle and Davies 1968). Normally, a range of different-sized eddies is formed within a liquid during turbulence. The shear and pressure gradients associated with these eddies increase as their size decreases (Walstra 1983). As a consequence, large eddies are believed to be relatively ineffective at disrupting emulsion droplets. Very small eddies are also believed to be ineffective at breaking up droplets because they generate such high shear stresses that most of their energy is dissipated through viscous losses rather than through droplet disruption. For these reasons, intermediate-sized eddies are thought to be largely responsible for droplet disruption under turbulent flow conditions (Walstra 1983, 1993b). When a droplet is in the vicinity of one of these intermediate-sized eddies, it is deformed and disrupted because of the large shear gradient acting across it (Gopal 1968). For isotropic turbulent conditions, the Weber number is given by (Karbstein and Schubert 1995):

$$\text{We} = \frac{\text{turbulent forces}}{\text{interfacial forces}} = \frac{C\rho_C^{1/3}\varepsilon^{2/3}d^{5/3}}{4\gamma} \tag{6.4}$$

where C is a constant related to the critical Weber number, ρ_C is the density of the continuous phase, and ε is the power density. Under isotropic turbulent conditions, the maximum size of droplets that can persist during homogenization once a steady state has been reached has been shown to be given by the following relationship (Walstra 1983):

$$d_{max} = \frac{C'\gamma^{3/5}}{\varepsilon^{2/5}\rho_C^{1/5}} \tag{6.5}$$

where C' is a constant which depends on the characteristic dimensions of the system. This equation indicates that the size of the droplets produced under turbulent conditions decreases as the power density increases, the interfacial tension decreases, or the density of the continuous phase increases. It also suggests that the viscosity of the liquids does not influence the droplet size. Nevertheless, a number of experimental studies have shown that the viscosities of the dispersed and continuous phases do influence the maximum droplet size that can persist during homogenization, with a minimum in d_{max} when η_D/η_C is between about 0.1 and 5 (Braginsky and Belevitskaya 1996). The dependence of the droplet size produced during homogenization on the viscosity ratio under turbulent flow conditions is therefore similar in form to that produced under laminar flow conditions (Figure 6.4). It is therefore possible to reduce the size of the droplets produced during homogenization by ensuring that the viscosity ratio falls within the optimum range for droplet breakup ($0.1 < \eta_D/\eta_C < 5$), which could be achieved by varying the temperature or by adding thickening agents. The viscosity may also influence the droplet size if it is large enough to suppress turbulence. For example, an increase in viscosity due to the presence of thickening agents or high concentrations of droplets may be sufficient to prevent turbulent flow conditions and therefore lead to inefficient homogenization (Walstra 1993b).

An emulsion does not normally remain in a homogenizer long enough for steady-state conditions to be attained, and so Equation 6.5 is not strictly applicable. In practice, the size of the droplets in an emulsion decreases as the length of time they spend in the disruption zone of a homogenizer increases, until eventually a constant value is reached (Karbstein and Schubert 1995). This is because droplets take a finite time to be deformed, and so the turbulent forces must act over a period which is sufficiently longer than this time if all the droplets are to be effectively disrupted (Walstra 1993b). The deformation time is proportional to the viscosity of a droplet; therefore, the more viscous the dispersed phase, the less likely is droplet breakup within a specified time. Emulsions produced under turbulent flow conditions are always polydisperse because of the distribution of eddy sizes in the fluid. In fact, the statistical theories used to derive the above equations indicate that droplets formed under turbulent conditions should follow a log-normal distribution, which is often observed in practice (Gopal 1968).

Cavitational Flow Conditions. Cavitation occurs in fluids which are subjected to rapid changes in pressure and is particularly important in ultrasonic and high-pressure valve homogenizers (Gopal 1968, Phipps 1985). A fluid contracts when the pressure acting on it increases and expands when the pressure decreases. When the instantaneous pressure that a fluid experiences falls below some critical value, a cavity is formed (Lickiss and McGrath 1996). As the fluid continues to expand, the cavity grows and some of the surrounding liquid evaporates and moves into it. During a subsequent compression, the cavity catastrophically collapses, generating an intense shock wave which propagates into the surrounding fluid and causes any droplets in its immediate vicinity to be deformed and disrupted. Extremely high temperatures and pressures are associated with these shock waves, but they are of very short duration and highly localized, so that little damage is usually done to the vessel containing the fluid. Nevertheless, over time, cavitational effects are known to cause significant damage to the surfaces of high-pressure valve homogenizers and ultrasonic transducers, which become "pitted" (Gopal 1968, Phipps 1985). Cavitation only occurs in fluids when the intensity of the fluctuating pressure field exceeds a critical value, known as the *cavitational threshold*. This threshold is high in pure liquids, but is reduced when cavitational nuclei, such as gas bubbles or impurities, are present. The cavitational threshold also depends on the frequency of the pressure fluctuations, decreasing with decreasing frequency (Gopal 1968).

Homogenization of Nonideal Liquids. The equations given above are only strictly applicable to homogenization of ideal (Newtonian) liquids (Chapter 8). In practice, many liquids used in the food industry exhibit nonideal behavior, which can have a pronounced influence on the efficiency of droplet disruption and therefore on the size of the droplets produced during homogenization (Walstra 1983, 1993b). Many biopolymers used to thicken or stabilize emulsions exhibit pronounced shear thinning behavior (Chapters 4 and 8). As a consequence, the viscosity used in the above equations should be that which the droplet experiences at the shear rates that occur during homogenization, rather than that which is measured in a viscometer at low shear rates.

6.3.1.3.   The Role of the Emulsifier in Droplet Disruption

The ease with which a droplet can be disrupted during homogenization increases as the interfacial tension decreases (Equation 6.1). Thus it is possible to produce droplets with smaller sizes by homogenizing in the presence of an emulsifier which reduces the interfacial tension (Walstra 1993b). For example, adding an emulsifier that decreases the interfacial tension from 50 to 5 mN m^{-1} should decrease the size of the droplets produced under laminar flow conditions by an order of magnitude (Schubert and Armbruster 1995). Nevertheless, a number of other factors also determine the effectiveness of emulsifiers at reducing the droplet size. First, the rate at which an emulsifier adsorbs to the surface of the droplets during homogenization must be considered (Walstra 1983). Immediately after their formation, droplets have a low concentration of emulsifier adsorbed to their surface and are therefore more difficult to disrupt because of the relatively high interfacial tension. With time, a greater amount of emulsifier accumulates at the surface, which decreases the interfacial tension and therefore facilitates droplet disruption. Thus, the quicker the emulsifier adsorbs to the surface of the droplets during homogenization, the smaller the droplets produced. Second, the ability of emulsifiers to enhance the interfacial rheology of emulsion droplets hampers the breakup of droplets, which leads to larger droplet sizes than those expected from the equilibrium interfacial tension (Lucassen-Reyanders and Kuijpers 1992, Janssen et al. 1994, Williams et al. 1997). These two effects partly account for the poor correlation between droplet size and equilibrium interfacial tension reported in the literature (Walstra 1983).

6.3.2.   Droplet Coalescence

Emulsions are highly dynamic systems in which the droplets continuously move around and frequently collide with each other (Chapter 7). Droplet–droplet collisions are particularly rapid during homogenization because of the intense mechanical agitation of the emulsion. If droplets are not protected by a sufficiently strong emulsifier membrane, they tend to coalesce with one another during a collision (Walstra 1993b). Immediately after the disruption of an emulsion droplet, there is insufficient emulsifier present to completely cover the newly formed surface, and therefore the new droplets are more likely to coalescence with their neighbors. To prevent coalescence, it is necessary to form a sufficiently concentrated emulsifier membrane around the droplets before they have time to collide with their neighbors. The size of the droplets produced during homogenization therefore depends on the time taken for the emulsifier to be adsorbed to the surface of the droplets ($\tau_{adsorption}$) relative to the time between droplet–droplet collisions ($\tau_{collision}$). These times depend on the flow profile that the droplets experience, as well as the nature of the emulsifier used. Estimates of the adsorption and collision times have been established by Walstra (1983) for laminar and turbulent flow conditions (Table 6.1).

The equations given in Table 6.1 are only strictly applicable to dilute emulsions; nevertheless, they do give some useful insights into the factors which influence homogenization. Ideally, a food manufacturer wants to minimize droplet coalescence during homogenization by ensuring that $\tau_{adsorption}/\tau_{collision} \ll 1$. For both laminar and turbulent flow conditions, this ratio decreases as the dispersed-phase volume fraction decreases, the size of the droplets increases, the excess surface concentration decreases, and the concentration of emulsifier increases. Under laminar flow conditions, the shear rate has no effect on $\tau_{adsorption}/\tau_{collision}$, because the decrease in adsorption time due to increasing G is counterbalanced by the decrease in collision time. Under turbulent flow conditions, increasing the power input decreases $\tau_{adsorption}/\tau_{collision}$, because the decrease in adsorption time is greater than the decrease in the collision time. The above equations assume that every encounter between an emulsifier and droplet leads to adsorption and that every encounter between two droplets

TABLE 6.1

Equations for Calculating the Adsorption and Collision Times of Droplets in Emulsions Under Laminar and Turbulent Flow Conditions

Laminar flow	Turbulent flow
$\tau_{absorption} \approx \dfrac{20\Gamma}{dm_c G}$	$\tau_{absorption} \approx \dfrac{10\Gamma}{dm_c}\sqrt{\dfrac{\eta_C}{\varepsilon}}$
$\tau_{collision} = \dfrac{\pi}{8G\phi}$	$\tau_{collision} = \dfrac{1}{15\phi}\sqrt[3]{\dfrac{d^2\rho_C}{\varepsilon}}$

Note: Γ is the excess surface concentration of the emulsifier, G is the shear rate, ε is the power density, ρ_C is the density of the continuous phase, d is the droplet diameter, ϕ is the dispersed-phase volume fraction, γ is the interfacial tension, and m_c is the concentration of the emulsifier (mol m^{-3}).

Source: Walstra, 1983, 1993.

leads to coalescence. In practice, these mechanisms are not usually 100% efficient because of repulsive hydrodynamic and colloidal interactions (Hunter 1986), as well as the Gibbs–Marangoni effect (Walstra 1993b).

The importance of emulsifier adsorption kinetics on the size of the droplets produced during homogenization has been demonstrated experimentally (Schubert and Armbruster 1992). Under the same homogenization conditions, it has been shown that emulsifiers which adsorb rapidly produce smaller droplet sizes than those which adsorb slowly. Most food emulsifiers do not adsorb rapidly enough to completely prevent droplet coalescence, and so the droplet size achieved during homogenization is greater than that which is theoretically possible (Stang et al. 1994).

In addition to the factors already mentioned, the tendency for droplets to coalesce during (or shortly after) homogenization depends on the effectiveness of the interfacial membrane in resisting coalescence during a droplet–droplet encounter. The resistance of an interfacial membrane to coalescence depends on the concentration of emulsifier molecules present, as well as their structural and physicochemical properties (e.g., dimensions, electrical charge, packing, and interactions) (Chapters 4 and 7).

6.3.3. The Role of the Emulsifier

The above discussion has highlighted two of the most important functions of emulsifiers during the homogenization process:

1. They decrease the interfacial tension between the oil and water phases, thereby reducing the amount of energy required to deform and disrupt the droplets.
2. They form a protective coating around the droplets which prevents them from coalescing with each other.

The size of the droplets produced during homogenization therefore depends on a number of different characteristics of an emulsifier: (1) the ratio of emulsifier to dispersed phase — there must be sufficient emulsifier present to completely cover the surfaces of the droplets formed; (2) the time required for the emulsifier to move from the bulk phase to the droplet surface — the faster the adsorption time, the smaller the droplet size; (3) the probability that an emulsifier molecule will be adsorbed to the surface of a droplet during an encounter between it and the droplet — the greater the adsorption efficiency, the smaller the droplet size; (4) the amount that the emulsifier reduces the interfacial tension — the greater the amount, the smaller the droplet size; and (5) the effectiveness of the emulsifier membrane in protecting the droplets against coalescence — the better the protection, the smaller the droplet size.

6.4. HOMOGENIZATION DEVICES

A number of different types of homogenization device have been developed to produce food emulsions. Each of these devices has its own advantages and disadvantages and range of materials where it is most suitably applied. The choice of a particular homogenizer depends on whether the emulsion is being made in a factory or in a laboratory, the equipment available, the volume of material to be homogenized, the throughput, the nature of the starting materials, the desired droplet size distribution, the required physicochemical properties of the final product, and the cost of purchasing and running the equipment. The most important types of homogenizer used in the food industry are discussed below.

6.4.1. High-Speed Blenders

High-speed blenders are the most commonly used method for directly homogenizing oil and aqueous phases in the food industry (Brennan et al. 1981, Loncin and Merson 1979, Fellows 1988). The liquids to be homogenized are placed in a suitable vessel (Figure 6.5), which may contain as little as a few cubic centimeters or as much as several cubic meters of liquid, and are then agitated by a stirrer which rotates at high speeds (typically 20 to 2000 rev min^{-1}). The rapid rotation of the blade generates a combination of longitudinal, rotational, and radial

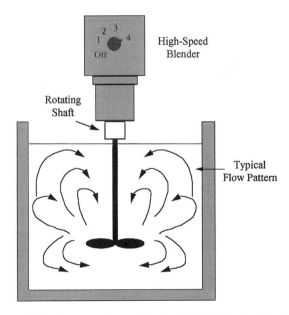

FIGURE 6.5   High-speed blenders are often used in the food industry to directly homogenize oil and aqueous phases.

TABLE 6.2
Comparison of Different Types of Homogenizer

	Throughput	Relative energy efficiency	Minimum droplet size	Sample viscosity
High-speed blender	Batch	Low	2 μm	Low to medium
Colloid mill	Continuous	Intermediate	1 μm	Medium to high
High-pressure homogenizer	Continuous	High	0.1 μm	Low to medium
Ultrasonic probe	Batch	Low	0.1 μm	Low to medium
Ultrasonic jet homogenizer	Continuous	High	1 μm	Low to medium
Microfluidization	Continuous	High	<0.1 μm	Low to medium
Membrane processing	Batch or continuous	High	0.3 μm	Low to medium

velocity gradients in the liquids which disrupts the interface between the oil and water, causes the liquids to become intermingled, and breaks the larger droplets into smaller ones (Fellows 1988). Efficient homogenization is achieved when the horizontal and vertical flow profiles distribute the liquids evenly throughout the vessel, which can be facilitated by fixing baffles to the inside walls of the vessel (Gopal 1968). The design of the stirrer also determines the efficiency of the homogenization process; a number of different types are available for different situations (e.g., blades, propellers, and turbines) (Fellows 1988). Blending generally leads to an increase in the temperature of an emulsion because some of the mechanical energy is converted into heat due to viscous dissipation. If any of the ingredients in the emulsion are sensitive to heat, it may be necessary to control the temperature of the vessel during homogenization. High-speed blenders are particularly useful for preparing emulsions with low or intermediate viscosities (Table 6.2). The droplet size usually decreases as the homogenization time or the rotation speed of the stirrer is increased, until a lower limit is achieved which depends on the nature and concentration of the ingredients used. Typically, the droplets produced by a high-speed blender range between about 2 and 10 μm in diameter.

6.4.2. Colloid Mills

Colloid mills are widely used in the food industry to homogenize medium- and high-viscosity liquids (Loncin and Merson 1979, Walstra 1983, Fellows 1988). A variety of different designs of colloid mill are commercially available, but they all operate on fairly similar physical principles (Figure 6.6). The liquids to be homogenized are fed into the colloid mill in the form of a coarse emulsion, rather than as separate oil and aqueous phases, because the device is much more efficient at reducing the size of the droplets in a preexisting emulsion (secondary homogenization) than at homogenizing two separate phases (primary homogenization). The coarse emulsion is usually produced directly from the oil and aqueous phases using a high-speed blender. It is then fed into the homogenizer and flows through a narrow gap between two disks: the *rotor* (a rotating disk) and the *stator* (a static disk). The rapid rotation of the rotor generates a shear stress in the gap which causes the larger droplets to be broken down into smaller ones. The intensity of the shear stresses can be altered by varying the thickness of the gap between the rotor and stator (from about 50 to 1000 μm), varying the rotation speed (from about 1000 to 20,000 rev min^{-1}), or by using disks that have roughened surfaces or interlocking teeth (Gopal 1968). Droplet disruption can also be enhanced by increasing the length of time the emulsion spends in the colloid mill, either by decreasing the flow rate or by passing the emulsion through the device a number of times. Typically, the flow rate can be varied between about 4 and 20,000 l h^{-1}. It should be noted

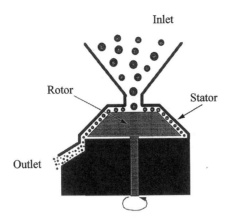

FIGURE 6.6   Colloid mills are mainly used in the food industry to homogenize intermediate- and high-viscosity materials.

that many of the factors which increase the effectiveness of droplet disruption also increase the manufacturing costs (by increasing energy costs or reducing product throughput). Food manufacturers must therefore select the rotation speed, gap thickness, rotor/stator type, and throughput which give the best compromise between droplet size and manufacturing costs. It is usually necessary to have some form of cooling device as part of a colloid mill in order to offset the increase in temperature caused by viscous dissipation losses. Colloid mills are more suitable for homogenizing intermediate- and high-viscosity fluids (such as peanut butter, fish, or meat pastes) than high-pressure valve or ultrasonic homogenizers (Table 6.2). Typically, they can be used to produce emulsions with droplet diameters between about 1 and 5 μm.

6.4.3.   High-Pressure Valve Homogenizers

High-pressure valve homogenizers are the most commonly used method of producing fine emulsions in the food industry. Like colloid mills, they are more effective at reducing the size of the droplets in a preexisting emulsion than at creating an emulsion directly from two separate liquids (Pandolfe 1991, 1995). A coarse emulsion is usually produced using a high-speed blender and is then fed directly into the input of the high-pressure valve homogenizer. The homogenizer has a pump which pulls the coarse emulsion into a chamber on its back-stroke and then forces it through a narrow valve at the end of the chamber on its forward stroke (Figure 6.7). As the coarse emulsion passes through the valve, it experiences a combination of intense shear, cavitational, and turbulent flow conditions which cause the larger droplets to be broken down into smaller ones (Phipps 1985). A variety of different types of valve have been designed for different types of application. Most commercial homogenizers use spring-loaded valves so that the gap through which the emulsion passes can be varied (typically between about 15 and 300 μm). Decreasing the gap size increases the pressure drop across the valve, which causes a greater degree of droplet disruption and smaller droplets to be produced. On the other hand, narrowing the gap increases the energy input required to form an emulsion, thereby increasing manufacturing costs. Experiments have shown that there is an approximately linear relationship between the logarithm of the homogenization pressure (P) and the logarithm of the droplet diameter (d) (i.e., $\log d \propto \log P$) (Walstra 1983, Phipps 1985). Thus a food manufacturer is able to develop empirical equations which can be used to predict the homogenization pressure required to produce an emulsion with a given droplet size. The throughputs of industrial homogenizers typically vary

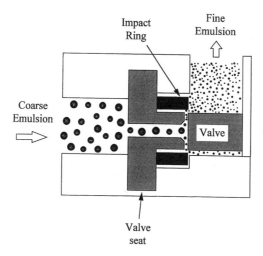

FIGURE 6.7   High-pressure valve homogenizers are used to produce emulsions with fine droplet sizes.

between about 100 to 20,000 l h^{-1}, whereas homogenization pressures vary between about 3 and 20 MPa.

Some commercial devices use a "two-stage" homogenization process, in which the emulsion is forced through two consecutive valves. The first valve is at high pressure and is responsible for breaking up the droplets, whereas the second valve is at low pressure and is responsible for disrupting any "flocs" which are formed during the first stage (Phipps 1985).

High-pressure valve homogenizers can be used to produce a wide variety of different food products, although they are most suitable for low- and intermediate-viscosity materials, particularly when a small droplet size is required. If the oil and aqueous phases have been blended prior to homogenization, it is often possible to create an emulsion with submicron particles using a single pass through the homogenizer (Pandolfe 1995). If very fine emulsion droplets are required, it is usually necessary to pass the emulsion through the homogenizer a number of times. Emulsion droplets with diameters as small as 0.1 μm can be produced using this method. The temperature rise in a high-pressure valve homogenizer is usually fairly small, but it can become appreciable if the emulsion is recirculated or extremely high pressures are used. In these cases, it may be necessary to keep the emulsion cool by using a water-jacketed homogenization chamber.

6.4.4. Ultrasonic Homogenizers

The possibility of using ultrasound to form emulsions was realized in the early part of this century, and ultrasonic homogenizers have been widely used for this purpose in both industry and research since then (McCarthy 1964, Gopal 1968). This type of homogenizer utilizes high-intensity ultrasonic waves which generate intense shear and pressure gradients within a material that disrupt the droplets mainly due to cavitational effects (Section 6.3.1.2).

A number of methods are available for generating high-intensity ultrasonic waves, but only two are commonly used in the food industry: piezoelectric transducers and liquid jet generators (Gopal 1968). Piezoelectric transducers are used in the benchtop ultrasonic homogenizers that are found in many research laboratories. They are ideal for preparing small volumes of emulsion (a few to a few hundred cubic centimeters), which is an important consideration in research laboratories because the chemicals used are often expensive. An ultrasonic transducer consists of a piezoelectric crystal contained within a protective metal casing,

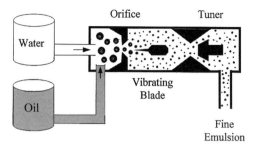

FIGURE 6.8 Liquid jet generators are commonly used in the food industry for the continuous production of emulsions.

which is usually tapered at the end. A high-intensity electrical wave is applied to the transducer, which causes the piezoelectric crystal to rapidly oscillate and generate an ultrasonic wave. The ultrasonic wave is directed toward the tip of the transducer, where it radiates into the surrounding liquids and generates intense pressure and shear gradients (mainly due to cavitational effects) which cause the liquids to be broken up into smaller fragments and intermingle with one another. The fact that the ultrasonic energy is focused on a small volume of the sample near the tip of the ultrasonic transducer means that it is important to have good agitation in the sample container. In small vessels, this is achieved by the fluid flow induced by the ultrasonic field itself, but in large vessels, it is often necessary to have additional agitation to ensure effective mixing and homogenization of all of the sample. To create a stable emulsion, it is usually necessary to irradiate a sample with ultrasound for periods ranging from a few seconds to a few minutes. Continuous application of ultrasound to a sample can cause appreciable heating, and so it is often advantageous to apply the ultrasound in a number of short bursts.

Ultrasonic jet homogenizers are used mainly for the industrial preparation of food emulsions (Figure 6.8). A stream of fluid is made to impinge on a sharp-edged blade, which causes the blade to rapidly vibrate, thus generating an intense ultrasonic field that breaks up any droplets in its immediate vicinity due to a combination of cavitation, shear, and turbulence (Gopal 1968). The major advantages of this device are that it can be used for the continuous production of emulsions, it can generate very small droplets, and it is more energy efficient than high-pressure valve homogenizers (i.e., less energy is required to produce droplets of the same size). Even so, the vibrating blade is prone to erosion because of the ultrasonic field, which means that it has to be replaced frequently. Fluid flow rates between 1 and 5000 l min^{-1} are possible using this technique.

The principal factors determining the efficiency of ultrasonic homogenizers are the intensity, duration, and frequency of the ultrasonic waves (Gopal 1968). Below a frequency of about 16 kHz, ultrasonic waves become audible and are therefore objectionable to users. Emulsions can be formed using ultrasonic waves with frequencies as high as 5 MHz, but the homogenization efficiency decreases with increasing frequency. For these reasons, most commercial devices use ultrasonic waves with frequencies between about 20 and 50 kHz. The size of the droplets produced during homogenization can be decreased by increasing either the intensity of the ultrasonic radiation or the length of time it is applied.

6.4.5. Microfluidization

Microfluidization is a technique which is capable of creating emulsions with extremely small droplet sizes directly from the individual oil and aqueous phases (Dickinson and Stainsby 1988). Separate streams of the oil and aqueous phases are accelerated to a high velocity and

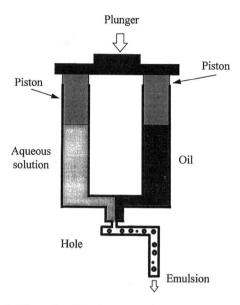

FIGURE 6.9 In a microfluidizer, the oil and aqueous phases are brought together at a high velocity, which causes the intermingling of the liquids and the disruption of the droplets.

then made to simultaneously impinge on a surface, which causes them to be intermingled with each other and broken up into droplets (Figure 6.9). Extremely small emulsion droplets can be produced by recirculating an emulsion through the microfluidizer a number of times (Strawbridge et al. 1995). The *jet homogenizer* developed by Dickinson and co-workers is an example of a laboratory-scale instrument based on the principle of microfluidization which can be used to create emulsions when the volume of starting materials is limited or expensive (Dickinson and Stainsby 1988).

6.4.6. Membrane Homogenizers

An emulsion is formed by forcing one immiscible liquid into another through a glass membrane which contains a uniform pore size (Figure 6.10). The size of the droplets formed

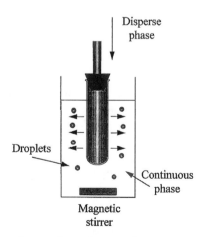

FIGURE 6.10 Batch version of a membrane homogenizer.

depends on the diameter of the pores in the membrane and the interfacial tensions between the oil and water phases (Kandori 1995). Membranes can be manufactured with different pore diameters so that emulsions with different droplet sizes can be produced (Kandori 1995). The membrane must be sufficiently strong so that it is not broken by the pressures applied to the fluids during homogenization. The membrane technique can be used as either a batch or a continuous process. In the batch process, droplets are formed by forcing the dispersed phase through a cylindrical membrane which is dipped into a vessel containing the continuous phase (Figure 6.10). In the continuous process, the homogenizer consists of a cylindrical membrane through which the continuous phase flows, which is located within a tube through which the dispersed phase flows. The dispersed phase is pressurized so that it is forced through the membrane, where it forms small droplets in the continuous phase.

An increasing number of applications for the membrane homogenizer are being identified, and the technique can now be purchased commercially for preparing emulsions in the laboratory (Kandori 1995). These instruments can be used to produce oil-in-water, water-in-oil, and multiple emulsions. The major advantages of membrane homogenizers are their ability to produce emulsions with very narrow droplet size distributions and the fact that they are highly energy efficient because much less energy is lost due to viscous dissipation. Emulsions with mean droplet sizes between about 0.3 and 10 μm can be produced using this technique.

6.4.7. Homogenization Efficiency

The energy efficiency of a homogenizer (E_H) can be calculated by comparing the minimum amount of energy theoretically required to form an emulsion (ΔE_{min}) with the actual amount of energy that is expended during homogenization (ΔE_{total}):

$$E_H = \frac{\Delta E_{min}}{\Delta E_{total}} \times 100 \qquad (6.6)$$

The minimum amount of energy required to form an emulsion is equal to that needed to increase the interfacial area between the oil and water phases: $\Delta E_{min} = \Delta A \gamma$, where ΔA is the increase in interfacial area and γ is the interfacial tension. For a typical oil-in-water emulsion, ΔE_{min} has a value of about 3 kJ m^{-3}, assuming that $\phi = 0.1$, $r = 1$ μm, and $\gamma = 10$ mN m^{-1} (Walstra 1983). The actual amount of energy required to form an emulsion depends on the type of homogenizer used and the operating conditions. For a high-pressure valve homogenizer, ΔE_{total} is typically about 10,000 kJ m^{-3}, and so the homogenization efficiency is less than 0.1% (Walstra 1983). The reason that homogenization is such an inefficient process is because the disruption of small droplets requires the generation of extremely high pressure gradients. These pressure gradients must be large enough to overcome the Laplace pressure gradient ($\approx 2\gamma/r^2$), which is about 2×10^{10} Pa m^{-1} for a 1-μm emulsion droplet. The pressure gradient due to shear forces acting across a droplet is given by $G\eta_c/r$, which indicates that the shear rate must exceed about 2×10^7 s^{-1} in order to disrupt the droplets. The movement of liquids at such high shear rates leads to large amounts of energy dissipation because of frictional losses. The conversion of mechanical work into heat also accounts for the increase in temperature observed during homogenization. To improve the energy efficiency of a homogenizer, it is necessary to either reduce the pressure gradient required to disrupt the droplets (e.g., by reducing the interfacial tension) or minimize the viscous dissipation caused by fluid flow.

6.4.8. Comparison of Homogenizers

The choice of a homogenizer for a particular application depends on a number of factors, including the volume of sample to be homogenized, the desired throughput, energy consumption, the physicochemical properties of the component phases, the desired droplet size distribution, the equipment available, initial costs, and running costs. After choosing the most suitable type of homogenizer, one must select the optimum operating conditions for that particular device, such as flow rate, pressure, gap thickness, temperature, homogenization time, and rotation speed.

The conversion of separate oil and aqueous phases into an emulsion containing fairly large droplets (>2 μm) is normally achieved using a high-speed blender. For industrial applications using high-viscosity fluids ($0.1 < \eta_C < 1$ Pa · s), a colloid mill is the most efficient type of homogenizer, but for lower viscosity liquids, either the high-pressure valve homogenizer or ultrasonic jet homogenizer is more suitable. For fundamental studies, one often needs small sample volumes; therefore, specially designed laboratory homogenizers are available, which are either scaled-down versions of the industrial equipment or instruments specifically designed for use in the laboratory. For studies where the ingredients are limited or expensive, an ultrasonic transducer could be used because it can produce small sample sizes. For studies where it is important to have monodisperse emulsions, it may be advantageous to use a membrane homogenizer.

The selection of an appropriate homogenizer for a particular application usually involves close cooperation between the food processor and the manufacturer of the equipment. The food processor must first specify the desired throughput, pressure, temperature, particle size, hygiene requirements, and properties of the sample. The equipment manufacturer will then be able to recommend a piece of equipment that is most suitable for the specific product (e.g., size, valve design, flow rates, and construction materials). It is good practice for food processors to test a number of homogenizers from different manufacturers under the conditions that will be used in the factory prior to making a purchase.

6.5. FACTORS WHICH INFLUENCE DROPLET SIZE

The size of the droplets produced during homogenization is important because it determines the stability, appearance, and texture of the final product. To create a product with specific properties, it is therefore necessary to ensure that the majority of the droplets fall within some preestablished size range. For this reason, it is important for food scientists to be aware of the major factors which influence droplet size.

6.5.1. Emulsifier Type and Concentration

For a fixed concentration of oil, water, and emulsifier, there is a maximum interfacial area which can be completely covered by the emulsifier. As homogenization proceeds, the size of the droplets decreases and the interfacial area increases. Once the droplets fall below a certain size, there is insufficient emulsifier present to completely cover their surface, and so they tend to coalesce with their neighbors. The minimum size of stable droplets that can be produced during homogenization (assuming monodisperse droplets*) is therefore governed by the type and concentration of emulsifier present:

$$r_{min} = \frac{3 \cdot \Gamma_{sat} \cdot \phi}{c_S} \qquad (6.7)$$

* For a polydisperse emulsion, the radius used in Equation 6.6 should be the volume–surface mean radius (Chapter 1).

where Γ_{sat} is the excess surface concentration of the emulsifier at saturation (in kg m^{-2}), ϕ is the dispersed-phase volume fraction, and c_S is the concentration of emulsifier in the emulsion (in kg m^{-3}). Equation 6.7 indicates that the minimum droplet size can be decreased by increasing the emulsifier concentration, decreasing the droplet concentration, or using an emulsifier with a smaller Γ_{sat}. For a 10% oil-in-water emulsion containing 1% of emulsifier, the minimum droplet radius is about 60 nm (assuming $\Gamma_{sat} = 2 \times 10^{-6}$ kg m^{-2}). In practice, there are a number of factors which mean that the droplet size produced during homogenization is greater than the theoretical minimum.

In order to attain the theoretical minimum droplet size, a homogenizer must be capable of generating a pressure gradient that is large enough to disrupt any droplets that are greater than r_{min} (i.e., $>2\gamma/r_{min}^2$) (Section 6.4.7). Some types of homogenizer are not capable of generating such high pressure gradients and are therefore not suitable for producing emulsions with small droplet sizes, even though there may be sufficient emulsifier present (Walstra 1983). The emulsion must also spend sufficient time within the homogenization zone for all of the droplets to be completely disrupted. If an emulsion passes through a homogenizer too rapidly, some of the droplets may not be disrupted. Even if a homogenizer is capable of producing small droplets, the emulsifier molecules must adsorb rapidly enough to form a protective interfacial layer around the droplets which prevents them from coalescing with their neighbors.

The emulsifier also influences the droplet size by reducing the interfacial tension between the oil and aqueous phases and therefore facilitating droplet disruption. Consequently, the more rapidly an emulsifier adsorbs, and the greater the reduction of the interfacial tension, the smaller the droplets that can be produced at a certain energy input.

Many different types of emulsifier can be used in the food industry, and each of these exhibits different characteristics during homogenization (e.g., the speed at which they adsorb, the maximum reduction in interfacial tension, and the effectiveness of the interfacial membrane in preventing droplet coalescence). Factors which influence the adsorption kinetics of emulsifiers, and their effectiveness at reducing interfacial tension, were discussed in Chapter 5, and factors which affect the stability of droplets against coalescence will be covered in Chapter 7. A food manufacturer must select the most appropriate emulsifier for each type of food product, taking into account its performance during homogenization, solution conditions, cost, availability, legal status, ability to provide long-term stability, and the desired physicochemical properties of the product.

6.5.2. Energy Input

The size of the droplets in an emulsion can be reduced by increasing the amount of energy supplied during homogenization (as long as there is sufficient emulsifier to cover the surfaces of the droplets formed). The energy input can be increased in a number of different ways depending on the nature of the homogenizer. In a high-speed blender, the energy input can be enhanced by increasing the rotation speed or the length of time that the sample is blended. In a high-pressure valve homogenizer, it can be enhanced by increasing the homogenization pressure or recirculating the emulsion through the device. In a colloid mill, it can be enhanced by using a narrower gap between the stator and rotator, increasing the rotation speed, by using disks with roughened surfaces, or by passing the emulsion through the device a number of times. In an ultrasonic homogenizer, the energy input can be enhanced by increasing the intensity of the ultrasonic wave or by sonicating for a longer time. In a microfluidizer, the energy input can be enhanced by increasing the velocity at which the liquids are brought into contact with each other or by recirculating the emulsion. In a membrane homogenizer, the energy input can be enhanced by increasing the pressure at which the liquid is forced through the membrane. Under a given set of homogenization conditions (energy input, temperature, composition), there is a certain size below which the emulsion droplets cannot be reduced with repeated homogenization, and therefore homogenizing the system any longer would be inefficient.

Increasing the energy input usually leads to an increase in manufacturing costs, and therefore a food manufacturer must establish the optimum compromise between droplet size, time, and cost. The energy input required to produce an emulsion containing droplets of a given size depends on the energy efficiency of the homogenizer used (Walstra 1983).

Under most circumstances, there is a decrease in droplet size as the energy input is increased. Nevertheless, there may be occasions when increasing the energy actually leads to an increase in droplet size because the effectiveness of the emulsifier is reduced by excessive heating or exposure to high pressures. This could be particularly important for protein-stabilized emulsions, because the molecular structure and functional properties of proteins are particularly sensitive to changes in their environmental conditions. For example, globular proteins, such as β-lactoglobulin, are known to unfold and aggregate when they are heated above a certain temperature, which reduces their ability to stabilize emulsions (Section 4.6).

6.5.3. Properties of Component Phases

The composition and physicochemical properties of both the oil and aqueous phases influence the size of the droplets produced during homogenization (Phipps 1985). Variations in the type of oil or aqueous phase will alter the viscosity ratio (η_D/η_C) which determines the minimum size that can be produced under steady-state conditions (Section 6.3). Different oils have different interfacial tensions when placed in contact with water because they have different molecular structures or because they contain different amounts of surface-active impurities, such as free fatty acids, monoacylglycerols, or diacylglycerols. These surface-active lipid components tend to accumulate at the oil–water interface and lower the interfacial tension, thus lowering the amount of energy required to disrupt a droplet.

The aqueous phase of an emulsion may contain a wide variety of components, including minerals, acids, bases, biopolymers, sugars, alcohols, ice crystals, and gas bubbles. Many of these components will alter the size of the droplets produced during homogenization because of their influence on rheology, interfacial tension, coalescence stability, or adsorption kinetics. For example, the presence of low concentrations of short-chain alcohols in the aqueous phase of an emulsion reduces the size of the droplets produced during homogenization because of the reduction in interfacial tension (Banks and Muir 1988). The presence of biopolymers in an aqueous phase has been shown to increase the droplet size produced during homogenization due to their ability to suppress the formation of small eddies during turbulence (Walstra 1983). Protein-stabilized emulsions cannot be produced close to the isoelectric point of a protein or at high electrolyte concentrations because the proteins are susceptible to aggregation. A knowledge of the composition of both the oil and aqueous phases of an emulsion and the role that each component plays during homogenization is therefore important when optimizing the size of the droplets produced by a homogenizer.

Experiments have shown that the smallest droplet size that can be achieved using a high-pressure valve homogenizer increases as the dispersed-phase volume fraction increases (Phipps 1985). There are a number of possible reasons for this: (1) increasing the viscosity of an emulsion may suppress the formation of eddies responsible for breaking up droplets; (2) if the emulsifier concentration is kept constant, there may be an insufficient amount present to completely cover the droplets; and (3) the rate of droplet coalescence is increased.

6.5.4. Temperature

Temperature influences the size of the droplets produced during homogenization in a number of ways. The viscosity of both the oil and aqueous phases is temperature dependent, and therefore the minimum droplet size that can be produced may be altered because of a variation in the viscosity ratio (η_D/η_C) (Section 6.3). Heating an emulsion usually causes a slight reduction in the interfacial tension between the oil and water phases, which would be expected to facilitate the production of small droplets (Equation 6.1). Certain types of emulsifiers lose their ability to stabilize emulsion droplets against aggregation when they are heated above a certain temperature. For example, when small-molecule surfactants are heated close to their phase inversion temperature, they are no longer effective in preventing droplet coalescence, or when globular proteins are heated above a critical temperature, they unfold and aggregate (Chapter 4). Alterations in temperature also influence the competitive adsorption of surface-active components, thereby altering interfacial composition (Dickinson and Hong 1994).

The temperature is also important because it determines the physical state of the lipid phase (Section 4.2). It is practically impossible to homogenize a fat that is either completely or substantially solid because it will not flow through a homogenizer or because of the huge amount of energy required to break up the fat crystals into small particles. There are also problems associated with the homogenization of oils that contain even small amounts of fat crystals because of partial coalescence (Chapter 7). The crystals from one droplet penetrate another droplet, leading to the formation of a "clump." Extensive clump formation leads to the generation of large particles and to a dramatic increase in the viscosity, which would cause a homogenizer to become blocked. For this reason, it is usually necessary to warm a sample prior to homogenization to ensure that the lipid phase is completely liquid. For example, milk fat is usually heated to about 40°C to melt all the crystals prior to homogenization (Phipps 1985).

6.6. DEMULSIFICATION

Demulsification is the process whereby an emulsion is converted into the separate oil and aqueous phases from which it was comprised and is therefore the opposite of homogenization (Lissant 1983, Menon and Wasan 1985, Hunter 1989). Demulsification is important in a number of technological processes in the food industry (e.g., oil recovery or the separation of lipid and aqueous phases). Demulsification is also important in research and development because it is often necessary to divide an emulsion into the separate oil and aqueous phases so that their composition or properties can be characterized. For example, the oil phase must often be extracted from an emulsion in order to determine the extent of lipid oxidation (Coupland et al. 1996) or to measure the partition coefficient of a food additive (Huang et al. 1997). Demulsification is achieved by causing the droplets to come into close contact with each other and then to coalesce. As this process continues, it eventually leads to the complete separation of the oil and aqueous phases. A knowledge of the physical principles of demulsification requires an understanding of the factors that determine the stability of emulsions to flocculation and coalescence (Chapter 7).

A variety of different types of emulsifier are used in the food industry to stabilize droplets against flocculation and coalescence (Chapter 4). Each type of emulsifier relies on different physicochemical mechanisms to prevent droplet aggregation, including electrostatic, steric, hydration, and thermal fluctuation interactions (Chapter 3). The selection of the most appropriate demulsification technique for a given emulsion therefore depends on a knowledge of the type of emulsifier used to stabilize the system and the mechanisms by which it provides stability.

6.6.1. Nonionic Surfactants

Nonionic surfactants usually stabilize emulsion droplets against flocculation through a combination of polymeric steric, hydration, and thermal fluctuation interactions (Section 4.5). Nevertheless, the interfacial membranes formed by nonionic surfactants are often unstable to rupture when the droplets are brought into close contact. Demulsification can therefore be achieved by altering the properties of an emulsion so that the droplets come into close contact for prolonged periods. This can often be achieved by heating an emulsion so that the polar head groups of the surfactant molecules become dehydrated, because this reduces the hydration repulsion between the droplets and allows them to come closer together. In addition, the optimum curvature of the surfactant monolayer tends toward zero as the size of the head group decreases, which increases the likelihood of coalescence (Section 4.5). This demulsification technique cannot be used for some emulsions stabilized by nonionic surfactants because the phase inversion temperature is much greater than 100°C or because heating may cause degradation or loss of one of the components being analyzed. In these cases, it is necessary to induce demulsification using alternative methods.

The addition of medium-chain alcohols has also been found to be effective in promoting demulsification in some systems (Menon and Wasan 1985). There are two possible explanations for this behavior: (1) the alcohol displaces some of the surfactant molecules from the interface and forms an interfacial membrane which provides little protection against droplet aggregation or (2) the alcohol molecules are able to get between the tails of the surfactant molecules at the interface, thereby increasing the optimum curvature of the interface toward zero and increasing the likelihood of coalescence (Section 4.5). In some emulsions, it is possible to promote droplet coalescence by adding a strong acid which cleaves the head groups of the surfactants from their tails so that the polar head group moves into the aqueous phase and the nonpolar tail moves into the droplet, thereby providing little protection against droplet coalescence.

6.6.2. Ionic Surfactants

Ionic surfactants stabilize droplets against coalescence principally by electrostatic repulsion (Chapter 3). Like nonionic surfactants, the membranes formed by ionic surfactants are not particularly resistant to rupture once the droplets are brought into close contact (Evans and Wennerstrom 1994). The most effective method of inducing droplet coalescence in these systems is therefore to reduce the magnitude of the electrostatic repulsion between the droplets (Menon and Wasan 1985). This can be achieved by adding electrolyte to the aqueous phase of the emulsion so as to screen the electrostatic interactions. Sufficient electrolyte must be added so that the energy barrier between the droplets becomes of the order of the thermal energy or less (Chapter 3). This process can most easily be achieved using multivalent ions, because they are more effective at screening electrostatic interactions at low concentrations than monovalent ions. Alternatively, the pH may be altered so that the surfactant loses its charge, which depends on the dissociation constant of the ionizable groups. Electromechanical methods can also be used to promote demulsification. An electric field is applied across an emulsion, which causes the charged droplets to move toward the oppositely charged electrode. A semipermeable membrane is placed across the path of the droplets, which captures the droplets but allows the continuous phase to pass through. The droplets are therefore forced against the membrane until their interfacial membranes are ruptured and they coalesce.

6.6.3. Biopolymer Emulsifiers

Biopolymers principally stabilize droplets against coalescence through a combination of electrostatic and polymeric steric interactions. In addition, they tend to form thick viscoelastic membranes that are highly resistant to rupture. There are two different strategies that can be use to induce droplet coalescence in this type of system:

1. The biopolymer can be digested by strong acids or enzymes so that it is broken into small fragments that are either not surface active or do not form a sufficiently strong membrane.
2. The biopolymers are displaced from the interface by small-molecule surfactants, and then the droplets are destabilized using one of the methods described above. Some proteins are capable of forming an interfacial membrane in which the molecules are covalently bound to each other through disulfide bonds. In order to displace these proteins, it may be necessary to cleave the disulfide bonds prior to displacing the proteins (e.g., by adding mercaptoethanol).

The Gerber and Babcock methods of determining the total fat content of milk are examples of the first of these strategies, while the detergent method is an example of the second (Pike 1994).

6.6.4. General Methods of Demulsification

A variety of physical techniques are available which can be used to promote demulsification in most types of emulsions. In all of the demulsification processes mentioned above, the separation of the oil phase from the aqueous phase can be facilitated by centrifuging the emulsion after the coalescence process has been initiated. In some emulsions, it is also possible to separate the phases directly by centrifugation at high speeds, without the need for any pretreatment (Menon and Wasan 1985). Centrifugation forces the droplets to one end of the container, which causes their interfacial membranes to become ruptured and therefore leads to phase separation.

Demulsification can also be achieved using various types of filtration devices (Menon and Wasan 1985). The emulsion is passed through a filter which adsorbs emulsion droplets. When a number of these adsorbed droplets come into close contact, they merge together to form a single large droplet which is released back into the aqueous phase. As the emulsion passes through the filter, this process continues until eventually the oil and water phases are completely separated from each another.

6.6.5. Selection of Most Appropriate Demulsification Technique

In addition to depending on the type of emulsifier present, the choice of an appropriate demulsification technique also depends on the sensitivity of the other components in the system to the separation process. For example, if one is monitoring lipid oxidation or trying to determine the concentration of an oil-soluble volatile component, it is inadvisable to use a demulsification technique that requires excessive heating. On the other hand, if the sample contains a lipid phase that is crystalline, it is usually necessary to warm the sample to a temperature where all the fat melts prior to carrying out the demulsification procedure.

6.7. FUTURE DEVELOPMENTS

Homogenization is an extremely important step in the production of emulsion-based food products. The efficiency of this process has a large impact on the bulk physicochemical and sensory properties of the final product. The progress that has already been made in identifying the factors which influence homogenization was reviewed in this chapter. Nevertheless, a great deal of research is still required before this process can be fully understood because of the inherent complexity of the physicochemical processes involved. The existing theories need to be extended to give a more realistic description of emulsion formation in commercial homogenization devices. In addition, systematic experiments using well-characterized emulsions and homogenization devices need to be carried out. Our understanding of this area will also be advanced due to the valuable insights provided by computer simulations. Computer models have been developed to predict the size distribution of droplets produced in homogenizers (Lachaise et al. 1996). These models can take into account the competition between droplet disruption and droplet coalescence mechanisms, as well as the influence of emulsifier adsorption kinetics on these processes.

Before ending this chapter, it is important to mention that homogenization is only one step in the formation of a food emulsion. A number of other processing operations usually come before or after homogenization, including chilling, freezing, pasteurization, drying, mixing, churning, and whipping. The quality of the final product is determined by the effect that each of these processing operations has on the properties of the food. Homogenization efficiency may be influenced by the effectiveness of any of the preceding processing operations, and it may alter the effectiveness of any of the following processing operations. Thus it is important to establish the interrelationship between the various food-processing operations on the final properties of a product.

7 Emulsion Stability

7.1. INTRODUCTION

The term "emulsion stability" refers to the ability of an emulsion to resist changes in its properties over time: the more stable the emulsion, the more slowly its properties change. An emulsion may become unstable due to a number of different types of physical and chemical processes.* Physical instability results in an alteration in the spatial distribution or structural organization of the molecules, whereas chemical instability results in an alteration in the chemical structure of the molecules. Creaming, flocculation, coalescence, partial coalescence, phase inversion, and Ostwald ripening are examples of physical instability (Dickinson and Stainsby 1982, Dickinson 1992, Walstra 1996a,b), whereas oxidation and hydrolysis are common examples of chemical instability (Fennema 1996a). In practice, two or more of these mechanisms may operate in concert. It is therefore important for food scientists to identify the relative importance of each mechanism, the relationship between them, and the factors which influence them, so that effective means of controlling the stability and physicochemical properties of emulsions can be established.

The length of time that an emulsion must remain stable depends on the nature of the food product (Dickinson 1992). Some food emulsions are formed as intermediate steps during a manufacturing process and therefore only need to remain stable for a few seconds, minutes, or hours (e.g., cake batter, ice cream mix, margarine premix), whereas others must remain stable for days, months, or even years prior to consumption, (e.g., mayonnaise, salad dressings, and cream liqueurs). On the other hand, the production of some foods involves a controlled *destabilization* of an emulsion during the manufacturing process (e.g., margarine, butter, whipped cream, and ice cream) (Mulder and Walstra 1974, Boode 1992). One of the major objectives of emulsion scientists working in the food industry is to establish the factors which determine the stability of each type of food product, as well as to elucidate general principles which can be used to predict the behavior of new products or processes. In practice, it is very difficult to quantitatively predict the stability of food emulsions from first principles because of their compositional and structural complexity. Nevertheless, an appreciation of the origin and nature of the various destabilization mechanisms is still an invaluable tool for controlling and improving emulsion stability. Because of the difficulties in theoretically predicting emulsion stability, food scientists often rely on the use of analytical techniques to experimentally monitor changes in emulsion properties over time. By using a combination of theoretical understanding and experimental measurements, food manufacturers are able to predict the effectiveness of different ingredients and processing conditions on the stability and properties of food emulsions.

The rate at which an emulsion breaks down, and the mechanism by which this process occurs, depends on its composition and microstructure, as well as on the environmental conditions it experiences during its lifetime (e.g., temperature variations, mechanical agita-

---

* It should be noted that the properties of emulsions may also change with time due to microbiological changes (e.g., the growth of specific types of bacteria or mold).

tion, and storage conditions). In this chapter, we examine the physicochemical basis of each of the major destabilization mechanisms, as well as discuss the major factors which influence them, methods of controlling them, and experimental techniques for monitoring them. This information is particularly useful for food manufacturers who need to formulate emulsions with enhanced shelf life or promote emulsion instability in a controlled fashion.

7.2.   ENERGETICS OF EMULSION STABILITY

In considering the "stability" of an emulsion, it is extremely important to distinguish between its *thermodynamic stability* and its *kinetic stability* (Dickinson 1992). Thermodynamics tells us whether or not a given process will occur, whereas kinetics tells us the rate at which it will proceed if it does occur (Atkins 1994). All food emulsions are thermodynamically unstable systems and will eventually break down if they are left long enough. For this reason, it is differences in kinetic stability which are largely responsible for the diverse range of properties exhibited by different food emulsions.

7.2.1.   Thermodynamics

The thermodynamic instability of an emulsion is readily demonstrated if one agitates a sealed vessel containing pure oil and pure water and then observes the change in the appearance of the system with time. The optically opaque emulsion which is initially formed by agitation breaks down over time until a layer of oil is observed on top of a layer of water (Figure 6.1).

The origin of this thermodynamic instability can be illustrated by comparing the free energy of a system consisting of an oil and an aqueous phase before and after emulsification (Hunter 1989). To simplify this analysis, we will initially assume that the oil and water have similar densities so that no creaming or sedimentation occurs. As a consequence, the final state consists of a single large droplet suspended in the continuous phase (Figure 7.1), rather than a layer of oil on top of a layer of water (see Figure 6.1). In its initial state, prior to emulsification, the free energy is given by:

$$G^i = G_O^i + G_W^i + G_I^i - TS_{config}^i \tag{7.1}$$

and in its final state, after emulsification, it is given by:

$$G^f = G_O^f + G_W^f + G_I^f - TS_{config}^f \tag{7.2}$$

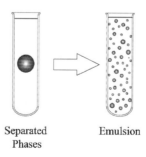

Separated   Emulsion
Phases

FIGURE 7.1   The formation of an emulsion is thermodynamically unfavorable because of the increase in surface area between the oil and water phases.

where G_O, G_W, and G_I are the free energies of the oil phase, water phase, and the oil–water interface, respectively; T is the absolute temperature; and S is the configurational entropy of the droplets in the system. The superscripts i and f refer to the initial and final states of the system. The free energies of the bulk oil and water phases remain constant before and after homogenization: $G_O^i = G_O^f$ and $G_W^i = G_W^f$, and so the difference in free energy between the initial and final states is given by (Hunter 1989):

$$\Delta G_{\text{formation}} = G^f - G^i = G_I^f - G_I^i - (TS_{\text{config}}^f - TS_{\text{config}}^i) = \Delta G_I - T\Delta S_{\text{config}} \quad (7.3)$$

By definition, the difference in interfacial free energy between the initial and final states (ΔG_I) is equal to the increase in surface area between the oil and aqueous phases (ΔA) multiplied by the interfacial tension (γ): $\Delta G_I = \gamma \Delta A$. Hence,

$$\Delta G_{\text{formation}} = \gamma \Delta A - T\Delta S_{\text{config}} \quad (7.4)$$

The change in interfacial free energy ($\gamma \Delta A$) is always positive, because the interfacial area increases after homogenization, and therefore it opposes emulsion formation. On the other hand, the configurational entropy term ($-T\Delta S_{\text{config}}$) is always negative, because the number of arrangements accessible to the droplets in the emulsified state is much greater than in the nonemulsified state, and therefore it favors emulsion formation. An expression for the configurational entropy can be derived from a statistical analysis of the number of configurations emulsion droplets can adopt in the initial and final states (Hunter 1989):

$$\Delta S_{\text{config}} = -\frac{nk}{\phi} [\phi \ln \phi + (1 - \phi) \ln(1 - \phi)] \quad (7.5)$$

where k is Boltzmann's constant, n is the number of droplets, and ϕ is the dispersed-phase volume fraction. In most food emulsions, the configurational entropy is much smaller than the interfacial free energy and can be ignored (Hunter 1989). As an example, consider a 10% oil-in-water emulsion containing 1-μm droplets ($\gamma = 0.01$ N m^{-1}). The interfacial free energy term ($\gamma \Delta A$) is about 3 kJ/m^3 of emulsion, whereas the configurational entropy term ($T\Delta S$) is about 3×10^{-7} kJ/m^3.

The overall free energy change associated with the creation of a food emulsion can therefore be represented by the following expression:

$$\Delta G_{\text{formation}} = \gamma \Delta A \quad (7.6)$$

Thus the formation of a food emulsion is always thermodynamically unfavorable, because of the increase in interfacial area after emulsification. It should be noted that the configurational entropy term can dominate the interfacial free energy term in emulsions in which the interfacial tension is extremely small and that these systems are therefore thermodynamically stable (Hunter 1989). This type of thermodynamically stable system is usually referred to as a *microemulsion*, to distinguish it from thermodynamically unstable (macro)emulsions.

In practice, the oil and water phases normally have different densities, and so it is necessary to include a free energy term which accounts for gravitational effects (i.e., the tendency for the liquid with the lowest density to move to the top of the emulsion). This term contributes to the thermodynamic instability of emulsions and accounts for the observed creaming or sedimentation of droplets (Section 7.3.2).

7.2.2. Kinetics

The free energy change associated with emulsion formation determines whether or not an emulsion is thermodynamically stable, but it does not give any indication of the rate at which the properties of an emulsion change with time, the type of changes which occur, or the physical mechanism(s) responsible for these changes. Information about the time dependence of emulsion stability is particularly important to food scientists who need to create food products which retain their desirable properties for a sufficiently long time under a variety of different environmental conditions. For this reason, food scientists are usually more interested in the *kinetic stability* of emulsions, rather than their thermodynamic stability.

The importance of kinetic effects can be highlighted by comparing the long-term stability of emulsions with the same composition but with different droplet sizes. An emulsion which contains small droplets usually has a longer shelf life (greater kinetic stability) than one which contains large droplets, even though it is more thermodynamically unstable (because it has a larger interfacial area, ΔA).

Despite the fact that food emulsions exist in a thermodynamically unstable state, many of them remain kinetically stable (*metastable*) for months or even years. What is the origin of this kinetic stability? Conceptually, the kinetic stability of an emulsion can be attributed to an activation energy (ΔG^*), which must be overcome before it can reach its most thermodynamically favorable state (Figure 7.2). An emulsion which is kinetically stable must have an activation energy which is significantly larger than the thermal energy of the system (kT). For most emulsions, an activation energy of about 20 kT is sufficient to provide long-term stability (Friberg 1997). In reality, emulsions have a number of different metastable states, and each of these has its own activation energy. Thus, an emulsion may move from one metastable state to another before finally reaching the most thermodynamically stable state. A change from one of these metastable states to another may be sufficient to have a deleterious effect on food quality.

The kinetic stability of emulsions can only be understood with reference to their dynamic nature. The droplets in an emulsion are in a continual state of motion and frequently collide into one another, due to their Brownian motion, gravity, or applied external forces. Whether droplets move apart, remain loosely associated with each other, or fuse together after a collision depends on the nature of the interactions between them. The kinetic stability of emulsions is therefore largely determined by the dynamics and interactions of the droplets

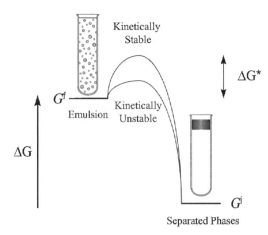

FIGURE 7.2 Emulsions are thermodynamically unstable systems, but may exist in a metastable state and therefore be kinetically stable.

they contain. Consequently, a great deal of this chapter will be concerned with the nature of the interactions between droplets and the factors which determine their movement in emulsions.

Earlier it was mentioned that if pure oil and pure water are agitated together, a temporary emulsion is formed which rapidly reverts back to its individual components. This is because there is a very low activation energy between the emulsified and unemulsified states. To create an emulsion which is kinetically stable for a reasonably long period of time, it is necessary to have either an emulsifier or a thickening agent present that produces an activation energy that is sufficiently large to prevent instability. An emulsifier adsorbs to the surface of freshly formed droplets and forms a protective membrane which prevents them from merging together, while a thickening agent increases the viscosity of the continuous phase so that droplets collide with one another less frequently (Chapter 4). The role of emulsifiers and thickening agents in emulsion stability will therefore be another common theme of this chapter.

7.3. GRAVITATIONAL SEPARATION

In general, the droplets in an emulsion have a different density to that of the liquid which surrounds them, and so a net gravitational force acts upon them (Dickinson and Stainsby 1982, Hunter 1989, Dickinson 1992, Walstra 1996a,b). If the droplets have a lower density than the surrounding liquid, they have a tendency to move upward, which is referred to as *creaming* (Figure 7.3). Conversely, if they have a higher density than the surrounding liquid, they tend to move downward, which is referred to as *sedimentation*. The densities of most edible oils (in their liquid state) are lower than that of water, and so there is a tendency for oil to accumulate at the top of an emulsion and water at the bottom. Thus, droplets in an oil-in-water emulsion tend to cream, whereas those in a water-in-oil emulsion tend to sediment.

Gravitational separation is usually regarded as having an adverse effect on the quality of food emulsions. A consumer expects to see a product which appears homogeneous, and therefore the separation of an emulsion into an optically opaque droplet-rich layer and a less opaque droplet-depleted layer is undesirable. The textural attributes of a product are also adversely affected by gravitational separation, because the droplet-rich layer tends to be more viscous than expected, whereas the droplet-depleted layer tends to be less viscous. The taste and mouthfeel of a portion of food therefore depend on the location from which it was taken from the creamed emulsion. A sample selected from the top of an oil-in-water emulsion that has undergone creaming will seem too "rich" because of the high fat content, whereas a sample selected from the bottom will seem too "watery" because of the low fat content. Gravitational separation is also a problem because it causes droplets to come into close

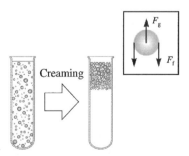

FIGURE 7.3 Food emulsions are prone to creaming because of the density difference between the oil and water phases. Inset: the forces acting on an emulsion droplet.

contact for extended periods, which can lead to enhanced flocculation or coalescence and eventually to oiling off, which is the formation of a layer of pure oil on top of the emulsion. When a food manufacturer is designing an emulsion-based product, it is therefore important to control the rate at which gravitational separation occurs.

Each food product is unique, containing different types of ingredients and experiencing different environmental conditions during its processing, storage, and consumption. As a consequence, the optimum method of controlling gravitational separation varies from product to product. In this section, we consider the most important factors which influence gravitational separation, as well as strategies for controlling it.

7.3.1. Physical Basis of Gravitational Separation

7.3.1.1. Stokes' Law

The rate at which an isolated spherical particle creams in an ideal liquid is determined by the balance of forces which act upon it (Figure 7.3 inset). When a particle has a lower density than the surrounding liquid, an upward gravitational force acts upon it (Hiemenz 1986, Hunter 1989):

$$F_g = -\frac{4}{3}\pi r^3 (\rho_2 - \rho_1)g \tag{7.7}$$

where r is the radius of the particle, g is the acceleration due to gravity, ρ is the density, and the subscripts 1 and 2 refer to the continuous and dispersed phases, respectively. As the particle moves upward through the surrounding liquid, it experiences a hydrodynamic frictional force that acts in the opposite direction and therefore retards its motion:

$$F_f = 6\pi\eta_1 rv \tag{7.8}$$

where v is the creaming velocity and η is the shear viscosity. The particle rapidly reaches a constant velocity, where the upward force due to gravity balances the downward force due to friction (i.e., $F_g = F_f$). By combining Equations 7.7 and 7.8, we obtain the Stokes' law equation for the creaming rate of an isolated spherical particle in a liquid:

$$v_{\text{Stokes}} = -\frac{2gr^2(\rho_2 - \rho_1)}{9\eta_1} \tag{7.9}$$

The sign of v determines whether the droplet moves upward (+) or downward (−). To a first approximation, the stability of a food emulsion to creaming can be estimated using Equation 7.9. For example, an oil droplet ($\rho_2 = 910$ kg m^{-3}) with a radius of 1 μm suspended in water ($\eta_1 = 1$ mPa s, $\rho_1 = 1000$ kg m^{-3}) will cream at a rate of about 17 mm/day. Thus one would not expect an emulsion containing droplets of this size to have a particularly long shelf life. As a useful rule of thumb, an emulsion in which the creaming rate is less than about 1 mm/day can be considered to be stable toward creaming (Dickinson 1992).

In the rest of this section, we shall mainly consider creaming, rather than sedimentation, because it is more common in food systems. Nevertheless, the same physical principles are important in both cases, and the methods of controlling them are similar. In the initial stages of creaming, the droplets move upward and a droplet-depleted layer is observed at the bottom of the container (Figure 7.4). When the droplets reach the top of the emulsion, they cannot move upward any further, and so they pack together to form a "creamed layer." The

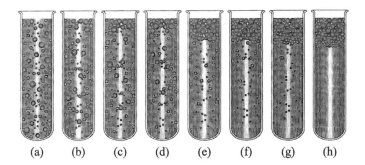

(a)   (b)   (c)   (d)   (e)   (f)   (g)   (h)

FIGURE 7.4 Time dependence of the creaming process in emulsions. The droplets move upward until they cannot move any further and form a "creamed" layer.

thickness of the creamed layer depends on the effectiveness of the droplet packing. Droplets may pack tightly or loosely depending on their polydispersity and the nature of the interactions between them. Tightly packed droplets tend to form a thin creamed layer, whereas loosely packed droplets form a thick creamed layer. Many of the factors which determine the packing of droplets in a creamed layer also determine the structure of flocs (see Section 7.4). The droplets in a creamed emulsion can often be redispersed by mild agitation, provided they are not too strongly attracted to each other or that coalescence has not occurred.

7.3.1.2. Modifications of Stokes' Law

Stokes' law can only be strictly used to calculate the velocity of an isolated rigid spherical particle suspended in an ideal liquid of infinite extent (Hiemenz 1986, Dickinson 1992). In practice, there are often large deviations between the creaming velocity predicted by Stokes' law and experimental measurements of creaming in food emulsions, because many of the assumptions used in deriving Equation 7.9 are invalid. Some of the most important factors that alter the creaming rate in food emulsions are considered below.

Droplet Fluidity. Stokes' equation assumes there is no slip at the interface between the droplet and the surrounding fluid, which is only strictly true for solid particles. The liquid within a droplet can move when a force is applied to the surface of the droplet; thus the frictional force that opposes the movement of a droplet is reduced, which causes an increase in the creaming velocity (Dickinson and Stainsby 1982):

$$v = v_{Stokes} \frac{3(\eta_2 + \eta_1)}{(3\eta_2 + 2\eta_1)} \tag{7.10}$$

This expression reduces to Stokes' equation when the viscosity of the droplet is much greater than that of the continuous phase ($\eta_2 \gg \eta_1$). Conversely, when the viscosity of the droplet is much less than that of the continuous phase ($\eta_2 \ll \eta_1$), the creaming rate is 1.5 times faster than that predicted by Equation 7.9. In practice, the droplets in most food emulsions can be considered to act like rigid spheres because they are surrounded by a viscoelastic interfacial layer which prevents the fluid within them from moving (Walstra 1996a,b).

Nondilute Systems. The creaming velocity of droplets in concentrated emulsions is less than that in dilute emulsions because of hydrodynamic interactions between the droplets (Hunter

1989, Walstra 1996b). As an emulsion droplet moves upward due to gravity, an equal volume of continuous phase must move downward to compensate. Thus there is a net flow of continuous phase downward, which opposes the upward movement of the droplets and therefore decreases the creaming velocity. In fairly dilute emulsions (i.e., <2% droplets), the creaming velocity of droplets is given by (Hunter 1989):

$$v = v_{\text{Stokes}}(1 - 6.55\phi)$$
(7.11)

In more concentrated emulsions, a number of other types of hydrodynamic interaction also reduce the creaming velocity of the droplets. These hydrodynamic effects can be partly accounted for by using a value for the viscosity of a concentrated emulsion, rather than that of the continuous phase, in Equation 7.9 (see Section 8.4).

A semiempirical equation that has been found to give relatively good predictions of the creaming behavior of concentrated emulsions has been developed (Hunter 1989):

$$v = v_{\text{Stokes}}\left(1 - \frac{\phi}{\phi_c}\right)^{k\phi_c}$$
(7.12)

Here, ϕ_c and k are parameters which depend on the nature of the emulsion. This equation predicts that the creaming velocity decreases as the droplet concentration increases, until creaming is completely inhibited once a critical dispersed-phase volume fraction (ϕ_c) has been exceeded (Figure 7.5). In general, the value of ϕ_c depends on the packing of the droplets within an emulsion, which is governed by their polydispersity and colloidal interactions. Polydisperse droplets are able to fill the available space more effectively than monodisperse droplets because the small droplets can fit into the gaps between the larger ones (Das and Ghosh 1990), and so ϕ_c is increased. When the droplets are strongly attracted to each other, they can form a particle gel at relatively low droplet concentrations, which prevents any droplet movement (Figure 7.6). When the droplets are strongly repelled from each other, their effective size increases, which also causes complete restriction of their movement at lower values of ϕ_c.

Polydispersity.   Food emulsions contain a range of different droplet sizes, and the larger droplets cream more rapidly than the smaller droplets, so that there is a distribution of creaming rates within an emulsion (Figure 7.4). As the larger droplets move upward more

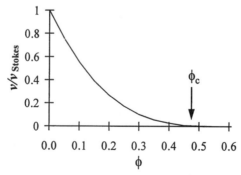

FIGURE 7.5 Prediction of the creaming rate as a function of dispersed-phase volume fraction in concentrated emulsions (Equation 7.11). Notice that creaming is effectively prevented when the emulsion droplets become closely packed.

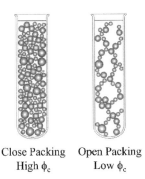

Close Packing   Open Packing
High ϕ_c Low ϕ_c

FIGURE 7.6 Creaming is prevented at lower dispersed-phase volume fractions when the droplets in the flocs are more loosely packed.

rapidly, they collide with smaller droplets (Melik and Fogler 1988, Dukhin and Sjoblom 1996). If the droplets aggregate after a collision, they will cream at a faster rate than either of the isolated droplets. Detailed information about the evolution of the droplet concentration throughout an emulsion can be obtained by using computer simulations which take into account polydispersity (Davis 1996, Tory 1996). For many purposes, it is only necessary to have an average creaming velocity, which can be estimated by using a mean droplet radius (r_{54}) in the Stokes equation (Dickinson 1992, Walstra 1996b).

Droplet Flocculation. In many food emulsions, the droplets aggregate to form flocs (Section 7.4). The size and structure of the flocs within an emulsion have a large influence on the rate at which the droplets cream (Bremer 1992, Bremer et al. 1993, Walstra 1996a, Pinfield et al. 1997). At low or intermediate droplet concentrations, where flocs do not substantially interact with one another, flocculation tends to increase the creaming velocity because the flocs have a larger effective size than the individual droplets (which more than compensates for the fact that the density difference between the flocs and the surrounding liquid is reduced). In concentrated emulsions, flocculation retards creaming because a three-dimensional network of aggregated flocs is formed, which prevents the individual droplets from moving (Figure 7.6). The droplet concentration at which creaming is prevented depends on the structure of the flocs formed. A network can form at lower dispersed-phase volume fractions when the droplets in a floc are more loosely packed, and therefore creaming is prevented at lower droplet concentrations (Figure 7.6). These loosely packed flocs tend to form when there is a strong attraction between the droplets (Section 7.4.3).

Non-Newtonian Rheology of Continuous Phase. The continuous phase of many food emulsions is non-Newtonian (i.e., the viscosity depends on shear rate or has some elastic characteristics) (Chapter 8). As a consequence, it is important to consider which is the most appropriate viscosity to use in Stokes' equation (van Vliet and Walstra 1989, Walstra 1996a). Biopolymers, such as modified starches or gums, are often added to oil-in-water emulsions to increase the viscosity of the aqueous phase (Section 4.6). Many of these biopolymer solutions exhibit shear-thinning behavior; that is, they have a high viscosity at low shear rates which decreases dramatically as the shear rate is increased (Chapter 8). This property is important because it means that the droplets are prevented from creaming, but that the food emulsion still flows easily when poured from a container (Dickinson 1992). Creaming usually occurs when an emulsion is at rest, and therefore it is important to know the apparent viscosity that a droplet experiences as it moves through the continuous phase under these conditions. Typically, the shear rate that a droplet experiences as it creams is between about 10^{-4} to 10^{-7} s^{-1} (Dickinson 1992). Solutions of thickening agents have extremely high apparent shear viscosities at these low shear rates, which means that the droplets cream extremely slowly (Walstra 1996a).

Some aqueous biopolymer solutions have a yield stress (τ_B), below which the solution acts like an elastic solid and above which it acts like a viscous fluid (van Vliet and Walstra 1989). In these systems, droplet creaming is effectively eliminated when the yield stress of the solution is larger than the stress exerted by a droplet as it moves through the continuous phase, i.e., $\tau_B \geq 2r(\rho_1 - \rho_2)g$ (Dickinson 1992). Typically, a value of about 10 mPa is required to prevent emulsion droplets of a few micrometers from creaming. Similar behavior is observed in water-in-oil emulsions that contain a network of aggregated fat crystals which has a sufficiently high yield stress to prevent the water droplets from sedimenting.

The above discussion highlights the importance of carefully defining the rheological properties of the continuous phase. For this reason, it is good practice to measure the viscosity of the continuous phase over the range of shear rates that an emulsion droplet is likely to experience during processing, storage, and handling, which may be as wide as 10^{-7} to 10^3 s^{-1} (Dickinson 1992).

Electrical Charge. Charged emulsion droplets tend to move more slowly than uncharged droplets for two reasons (Dickinson and Stainsby 1982, Walstra 1986a). First, repulsive electrostatic interactions between similarly charged droplets mean that they cannot get as close together as uncharged droplets. Thus, as a droplet moves upward, there is a greater chance that its neighbors will be caught in the downward flow of the continuous phase. Second, the cloud of counterions surrounding a droplet moves less slowly than the droplet itself, which causes an imbalance in the electrical charge which opposes the movement of the droplet.

Fat Crystallization. Many food emulsions contain a lipid phase that is either partly or wholly crystalline (Mulder and Walstra 1974, Boode 1992, Dickinson and McClements 1995). In oil-in-water emulsions, the crystallization of the lipid phase affects the overall creaming rate because solid fat ($\rho \approx 1200$ kg m^{-3}) has a higher density than liquid oil ($\rho \approx 910$ kg m^{-3}). The density of a droplet containing partially crystallized oil is given by $\rho_{droplet} = \phi_{SFC} \rho_{solid} + (1 - \phi_{SFC}) \rho_{liquid}$, where ϕ_{SFC} is the solid fat content At a solid fat content of about 30%, an oil droplet has a similar density as water and will therefore neither cream nor sediment. At lower solid fat contents, the droplets cream, and at higher solid fat contents they sediment. This accounts for the more rapid creaming of milk fat globules at 40°C, where they are completely liquid, compared to at 20°C, where they are partially solid (Mayhill and Newstead 1992).

As mentioned above, crystallization of the fat in a water-in-oil emulsion may lead to the formation of a three-dimensional network of aggregated fat crystals, which prevents the water droplets from sedimenting (e.g., in butter and margarine) (Dickinson and Stainsby 1982). In these systems, there is a critical solid fat content necessary for the formation of a network, which depends on the morphology of the fat crystals (Walstra 1987). The importance of network formation is illustrated by the effect of heating on the stability of margarine. When margarine is heated above a temperature where most of the fat crystals melt, the network breaks down and the water droplets sediment, leading to the separation of the oil and aqueous phases.

Adsorbed Layer. The presence of a layer of adsorbed emulsifier molecules at the surface of an emulsion droplet affects the creaming rate in a number of ways. First, it increases the effective size of the emulsion droplet, and therefore the creaming rate is increased by a factor of $(1 + \delta/r)^2$, where δ is the thickness of the adsorbed layer. Typically, the thickness of an adsorbed layer is between about 2 and 10 nm, and therefore this effect is only significant for very small emulsion droplets (<0.1 μm). Second, the adsorbed layer may alter the effective density of the dispersed phase, ρ_2 (Tan 1990). The effective density of the dispersed phase when the droplets are surrounded by an adsorbed layer can be calculated using the following relationship, assuming that the thickness of the adsorbed layer is much smaller than the radius of the droplets ($\delta \ll r$):

$$\rho_2 = \frac{\rho_{droplet} + 3(\delta / r)\rho_{layer}}{1 + 3(\delta / r)} \tag{7.13}$$

The density of the emulsifier layer is usually greater than that of either the continuous or dispersed phase, and therefore the adsorption of emulsifier increases the effective density of the dispersed phase (Tan 1990). The density of large droplets ($r \gg \delta$) is approximately the same as that of the bulk dispersed phase, but that of smaller droplets may be altered significantly. It is therefore possible to retard creaming by using a surface-active biopolymer which forms a high-density interfacial layer.

Brownian Motion. Another major limitation of Stokes' equation is that it ignores the effects of Brownian motion on the creaming velocity of emulsion droplets (Pinfield et al. 1994, Walstra 1996a). Gravity favors the accumulation of droplets at either the top (creaming) or bottom (sedimentation) of an emulsion. On the other hand, Brownian motion favors the random distribution of droplets throughout the whole of the emulsion because this maximizes the configurational entropy of the system. The equilibrium distribution of droplets in an emulsion which is susceptible to both creaming and Brownian motion is given by the following equation (ignoring the finite size of the droplets) (Walstra 1996a):

$$\phi(h) = \phi_0 \exp\left(\frac{-4\pi r^3 \Delta\rho g h}{3kT}\right) \tag{7.14}$$

where $\phi(h)$ is the concentration of the droplets at a distance h below the top of the emulsion, and ϕ_0 is the concentration of the droplets at the top of the emulsion. If $\phi(h) = \phi_0$, then the droplets are evenly dispersed between the two locations (i.e., Brownian motion dominates), but if $\phi(h) \ll \phi_0$, the droplets tend to accumulate at the top of the emulsion (i.e., creaming dominates). It has been proposed that if $\phi(h)/\phi_0$ is less than about 0.02, then the influence of Brownian motion on the creaming behavior of emulsions is negligible (Walstra 1996a). This condition is met for droplets with radii greater than about 25 nm, which is nearly always the case in food emulsions.

Complexity of Creaming. The above discussion has highlighted the many factors which need to be considered when predicting the rate at which droplets cream in emulsions. In practice, it is difficult to simultaneously account for all of these factors in a single analytical equation. The most comprehensive method of predicting gravitational separation in emulsions it is to use computer simulations (Pinfield et al. 1994, Tory 1996).

7.3.2.   Methods of Controlling Gravitational Separation

The discussion of the physical basis of gravitational separation in the previous section has highlighted a number of ways of retarding its progress in food emulsions.

7.3.2.1. Minimize Density Difference

The driving force for gravitational separation is the density difference between the droplets and the surrounding liquid: $\Delta\rho = (\rho_2 - \rho_1)$. It is therefore possible to prevent gravitational separation by "matching" the densities of the oil and aqueous phases (Tan 1990). Most naturally occurring edible oils have fairly similar densities ($\approx$910 kg m^{-3}), and therefore food manufacturers have limited flexibility in preventing creaming by changing the type of oil used in their products. Nevertheless, a number of alternative strategies have been developed which enable food manufacturers to match the densities of the dispersed and continuous phases more closely.

Density matching can be achieved by mixing natural oils with brominated vegetable oils (which have a higher density than water), so that the overall density of the oil droplets is similar to that of the aqueous phase (Tan 1990). Nevertheless, the utilization of brominated oils in foods is restricted in many countries, and so food manufacturers only have a limited amount of control over creaming stability using this method. The restrictions on the use of brominated oils has led to the utilization of other types of "weighting agents" which are soluble in the oil phase and increase the density of the oil droplets, such as ester gum, damar gum, and sucrose acetate isobutyrate (Tan 1990). These weighting agents are usually incorporated into the oil phase prior to the homogenization process.

If the droplets in an oil-in-water emulsion are small, it may be possible to prevent gravitational separation by using an emulsifier which forms a relatively thick and dense interfacial layer, because this decreases the density difference between the oil droplets and the surrounding liquid (Section 7.3.1).

In some emulsions, it is possible to control the degree of gravitational separation by varying the solid fat content of the lipid phase. As mentioned in Section 7.3.1, an oil droplet with a solid fat content of about 30% has a density similar to water and will therefore be stable to gravitational separation. The solid fat content of a droplet could be controlled by altering the composition of the lipid phase or by controlling the temperature (Dickinson and McClements 1995). In practice, this procedure is unsuitable for many food emulsions because partially crystalline droplets are susceptible to partial coalescence, which severely reduces their stability (Section 7.6).

7.3.2.2. Reduce Droplet Size

Stokes' law indicates that the velocity at which a droplet moves is proportional to the square of its radius (Equation 7.9). The stability of an emulsion to gravitational separation can therefore be enhanced by reducing the size of the droplets it contains. Homogenization of raw milk is one of the most familiar examples of the retardation of creaming in a food emulsion by droplet size reduction (Swaisgood 1996). A food manufacturer generally aims to reduce the size of the droplets in an emulsion below some critical radius that is known to be small enough to prevent them from creaming during the lifetime of the product. In practice, homogenization leads to the formation of emulsions that contain a range of different sizes, and a certain percentage of the droplets is likely to be above the critical radius and therefore susceptible to gravitational separation. For this reason, a food manufacturer usually specifies the minimum percentage of droplets which can be above the critical radius. For example, cream liqueurs are usually designed so that less than 3% of the droplets have diameters greater than 0.4 μm (Dickinson 1992). Even though a small fraction of the droplets are greater than this size, and therefore susceptible to creaming, this does not cause a major problem because the presence of the droplet-rich layer is obscured by the turbidity produced by the droplets which remain in the droplet-depleted layer.

7.3.2.3. Modify Rheology of Continuous Phase

Increasing the viscosity of the liquid surrounding a droplet (η_1) decreases the velocity at which the droplet moves (Equation 7.9). Thus the stability of an emulsion to gravitational separation can be enhanced by increasing the viscosity of the continuous phase (e.g., by adding a thickening agent) (Section 4.6). Gravitational separation may be completely retarded if the continuous phase contains a three-dimensional network of aggregated molecules or particles which traps the droplets and prevents them from moving. Thus the droplets in oil-in-water emulsions can be completely stabilized against creaming by using biopolymers which form a gel in the aqueous phase, while the droplets in water-in-oil emulsions can be completely stabilized against sedimentation by ensuring there is a network of aggregated fat crystals in the oil phase (van Vliet and Walstra 1989).

7.3.2.4. Increase Droplet Concentration

The rate of gravitational separation can be retarded by increasing the droplet concentration. At a sufficiently high dispersed-phase volume fraction, the droplets are prevented from moving because they are so closely packed together (Figure 7.6). It is for this reason that the droplets in mayonnaise, which has a high dispersed-phase volume fraction, are more stable to creaming than those in salad dressings, which have a lower dispersed-phase volume fraction. Nevertheless, it should be mentioned that it is often not practically feasible to alter the droplet concentration, and therefore one of the alternative methods of preventing creaming should be used.

7.3.2.5. Alter Degree of Droplet Flocculation

The rate of gravitational separation can be controlled by altering the degree of flocculation of the droplets in an emulsion. In dilute emulsions, flocculation causes enhanced gravitational separation because it increases the effective size of the particles. To improve the stability of these systems, it is important to ensure that the droplets are prevented from flocculating (Section 7.4). In concentrated emulsions, flocculation reduces the rate of gravitational separation because the droplets are prevented from moving past one another (Figure 7.6). The critical dispersed-phase volume fraction at which separation is prevented depends on the structural organization of the droplets within the flocs (Section 7.4.3). The stability of concentrated emulsions may therefore be enhanced by altering the nature of the colloidal interactions between the droplets and therefore the structure of the flocs formed.

7.3.3. Experimental Measurement of Gravitational Separation

To theoretically predict the rate at which gravitational separation occurs in an emulsion, it is necessary to have information about the densities of the dispersed and continuous phases, the droplet size distribution, and the rheological properties of the continuous phase. The density of the liquids can be measured using a variety of techniques, including density bottles, hydrometers, and oscillating U-tube density meters (Pomeranz and Meloan 1994). The droplet size distribution can be measured by microscopy, light-scattering, electrical pulse counting, or ultrasonic methods (see Chapter 10). The rheological properties of the continuous phase can be characterized using various types of viscometers and dynamic shear rheometers (see Chapter 8). In principle, it is possible to predict the long-term stability of a food emulsion from a knowledge of these physicochemical properties and a suitable mathematical model. In practice, this approach has limited use because the mathematical models are not currently sophisticated enough to take into account the inherent complexity of food emulsions. For this reason, it is often more appropriate to directly measure the gravitational separation of the droplets in an emulsion.

The simplest method of monitoring gravitational separation is to place an emulsion in a transparent test tube, leave it for a certain length of time, and then measure the height of the interface between the droplet-depleted and droplet-rich layers using a ruler. This procedure

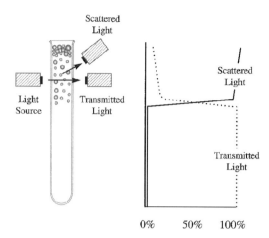

FIGURE 7.7 Light-scattering device for vertically scanning the creaming and sedimentation of droplets in emulsions.

can often be accelerated by centrifuging an emulsion at a fixed speed for a certain length of time (Sherman 1995). Nevertheless, this accelerated creaming test should be used with caution because the factors which determine droplet movement in a gravitational field are often different from those which are important in a centrifugal field. For example, the continuous phase may have a yield stress which is exceeded in a centrifuge but which would never be exceeded under normal storage conditions. The two major problems associated with visually determining the extent of creaming are (1) it is only possible to obtain information about the location of the boundary between the droplet-rich layer and droplet-depleted layers, rather than about the full vertical concentration profile of the droplets, and (2) in some systems, the droplet-depleted layer may be optically opaque, and so it is difficult to clearly locate the interface between it and the droplet-rich layer.

A more sophisticated method of monitoring gravitational separation is to use light scattering (Davis 1996). An emulsion is placed in a vertical glass tube and a monochromatic beam of near-infrared light is directed through it (Figure 7.7). The percentage of transmitted and scattered light is measured as a function of emulsion height by scanning the light beam up and down the sample using a stepper motor. The variation in droplet concentration with emulsion height can then be deduced from the percentage of transmitted and scattered light using a suitable theory or calibration curve. The technique could be used to measure both the size and concentration of the droplets at any height by measuring the angular dependence of the intensity of the scattered light. This device is mainly used to study creaming and sedimentation kinetics in fairly dilute emulsions, because it is not possible to transmit light through concentrated emulsions.

Traditionally, the kinetics of gravitational separation was monitored in concentrated emulsions by physically removing sections of an emulsion from different heights and then analyzing the concentration of droplets in each section (e.g., by measuring the density or by evaporating the water) (Pal 1994). These techniques cause the destruction of the sample being analyzed and therefore cannot be used to monitor creaming in the same sample as a function of time. Instead, a large number of similar samples have to be prepared and each one analyzed at a different time. Recently, a number of analytical methods have been developed to monitor gravitational separation in concentrated emulsions without disturbing the sample (e.g., ultrasonics, dielectric spectroscopy, and nuclear magnetic resonance) (Chapter 10). The ultrasonic device is very similar to the light-scattering technique described above, except that it is based on the propagation of ultrasonic waves through an emulsion rather than electromagnetic waves. An ultrasonic transducer is scanned vertically up and down an emulsion, which enables one to determine both the size and concentration of the droplets as a function of emulsion height (Chapter 10). Nuclear magnetic resonance imaging techniques, which are based on differences in the response of oil and water to the application of a radio-frequency pulse, have also been used to monitor gravitational separation in emulsions (Chapter 10). These techniques enable one to obtain a three-dimensional image of the droplet concentration within an emulsion, but the equipment is expensive to purchase and requires highly skilled operators, which has limited their application (Dickinson and McClements 1995, Soderman and Bailnov 1996).

7.4. FLOCCULATION

The droplets in emulsions are in continual motion because of the effects of thermal energy, gravity, or applied mechanical forces, and as they move about, they frequently collide with their neighbors (Lips et al. 1993, Dukhin and Sjoblom 1996). After a collision, emulsion droplets may either move apart or remain aggregated, depending on the relative magnitude of the attractive and repulsive interactions between them (Chapter 3). Droplets aggregate when there is a minimum in the interdroplet pair potential which is sufficiently deep and which is accessible to the droplets. The two major types of aggregation in food emulsions are flocculation and coalescence (Dickinson and Stainsby 1982; Dickinson 1992; Walstra 1996a,b). Flocculation is the process whereby two or more droplets come together to form an aggregate in which the droplets retain their individual integrity, whereas coalescence is the process whereby two or more droplets merge together to form a single larger droplet. Droplet flocculation is discussed in this section, and droplet coalescence is covered in Section 7.5.

Droplet flocculation may be either advantageous or detrimental to emulsion quality depending on the nature of the food product. Flocculation accelerates the rate of gravitational separation in dilute emulsions, which is undesirable because it reduces their shelf life (Luyten et al. 1993). It also causes a pronounced increase in emulsion viscosity and may even lead to the formation of a gel (Demetriades et al. 1997a,b). Some food products are expected to have a low viscosity, and therefore flocculation is detrimental. In other products, a controlled amount of flocculation may be advantageous because it leads to the creation of a desirable texture. Improvements in the quality of emulsion-based food products therefore depend on a better understanding of the factors which determine the degree of floc formation, the structure of the flocs formed, and the rate at which flocculation proceeds. In addition, it is important to understand the effect that flocculation has on the bulk physicochemical properties of emulsions.

7.4.1. Physical Basis of Flocculation

As flocculation proceeds, there is a decrease in the total number of particles (monomers + aggregates) in the emulsion, which can be described by the following equation:

$$\frac{dn_T}{dt} = -\frac{1}{2} FE \tag{7.15}$$

where dn_T/dt is the flocculation rate, n_T is the total number of particles per unit volume, t is the time, F is the collision frequency, and E is the collision efficiency. A factor of one-half appears in the equation because a collision between *two* particles leads to a reduction of *one* in the total number of particles present. Equation 7.15 indicates that the rate at which flocculation occurs depends on two factors: the frequency of collisions between the droplets and the fraction of collisions which leads to aggregation.

7.4.1.1. Collision Frequency

The collision frequency is the total number of droplet encounters per unit time per unit volume of emulsion. Any factor which increases the collision frequency increases the flocculation rate (provided that it does not also decrease the collision efficiency). Collisions between droplets occur as a result of their movement, which may be induced by Brownian motion, gravitational separation, or applied mechanical forces.

Collisions Due to Brownian Motion. In quiescent systems, the collisions between droplets are mainly a result of their Brownian motion. By considering the diffusion of particles in a dilute suspension, von Smoluchowski was able to derive the following expression for the collision frequency (Dickinson and Stainsby 1982):

$$F_B = 16\pi D_0 rn^2 \tag{7.16}$$

Here, F_B is the collision frequency due to Brownian motion (m^{-3} s^{-1}), D_0 is the diffusion coefficient of a single particle (m^2 s^{-1}), n is the number of particles per unit volume (m^{-3}), and r is the droplet radius (m). For spherical particles, $D_0 = kT/6\pi\eta_1 r$, where η_1 is the viscosity of the continuous phase, k is Boltzmann's constant, and T is the absolute temperature. Hence:

$$F_B = k_B n^2 = \frac{8kTn^2}{3\eta_1} = \frac{3kT\phi^2}{2\eta_1\pi r^6} \tag{7.17}$$

where k_B is a second-order rate constant (m^3 s^{-1}) and ϕ is the dispersed-phase volume fraction. For particles dispersed in water at room temperature, the collision frequency is $\approx 2 \times 10^{18} \phi^2/r^6$, when the radius is expressed in micrometers. Equation 7.17 indicates that the frequency of collisions between droplets can be reduced by decreasing their volume fraction, increasing their size, or increasing the viscosity of the continuous phase. If it is assumed that every collision between two particles leads to aggregation, and that the rate constant is independent of aggregate size, then the flocculation rate is given by $dn_T/dt = -\frac{1}{2}F_B$, which can be integrated to give the following expression for the change in the total number of particles with time (Evans and Wennerstrom 1994):

$$n_T = \frac{n_0}{1 + \frac{1}{2}k_B n_0 t} \tag{7.18}$$

where n_0 is the initial number of particles per unit volume. The time taken to reduce the number of droplets in an emulsion by half can be calculated from the above equation:

$$\tau_{1/2} = \frac{2}{k_B n_0} = \frac{3\eta_1}{4kTn_0} = \left(\frac{\pi\eta_1}{kT}\right)\frac{r^3}{\phi_0} \tag{7.19}$$

For a system where the particles are suspended in water at room temperature, $\tau_{1/2} \approx r^3/\phi_0$, when r is expressed in micrometers. Thus, an oil-in-water emulsion with $\phi = 0.1$ and $r = 1$ μm would have a half-life of about 10 s, which is on the same order as the existence of an emulsion prepared by shaking oil and water together in the absence of a stabilizer or emulsifier. It is also possible to derive an equation to describe the change in the number of dimers, trimers, and other aggregates with time (Evans and Wennerstrom 1994):

$$n_k = n_0 \left(\frac{t}{\tau_{1/2}} \right)^{k-1} \left(1 + \frac{t}{\tau_{1/2}} \right)^{-k-1} \tag{7.20}$$

where n_k is the number of aggregates per unit volume containing k particles. The predicted variation in the total concentration of particles and of the concentration of monomers ($k = 1$), dimers ($k = 2$), and trimers ($k = 3$) with time is shown in Figure 7.8. As would be expected, the total number of particles and the number of monomers decrease progressively with time as flocculation proceeds. The number of dimers, trimers, and other aggregates initially increases with time and then decreases as they interact with other particles and form larger aggregates.

The above equations are only applicable to dilute suspensions containing identical spherical particles suspended in an ideal liquid. Many of the assumptions used in their derivation are not valid for actual food emulsions, which may be concentrated, polydisperse, and have nonideal continuous phases. In addition, the properties of the flocs cannot be assumed to be the same as those of the monomers, and therefore the above theory has to be modified to take into account the dimensions, structure, and hydrodynamic behavior of the flocs (Bremer 1992, Walstra 1996a).

Collisions Due to Gravitational Separation. In polydisperse emulsions, droplet–droplet encounters can occur because of the different creaming (or sedimentation) rates of the differently sized droplets. Large droplets move more quickly than smaller ones and therefore collide with them as they move upward (or downward). The collision frequency for gravitationally induced flocculation is given by (Melik and Fogler 1988, Zhang and Davis 1991):

$$F_G = \pi(v_2 - v_1)(r_1 + r_2)^2 n_1 n_2 \tag{7.21}$$

$$F_G = k_G n_1 n_2 = \frac{g\Delta\rho\phi_1\phi_2}{8\pi\eta_1} \left[\frac{(r_2^2 - r_1^2)(r_1 + r_2)^2}{r_1^3 r_2^3} \right] \tag{7.22}$$

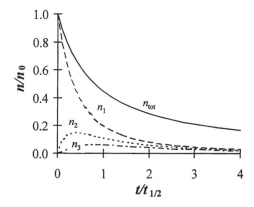

FIGURE 7.8   Dependence of the concentration of the total number of particles (n_T), monomers (n_1), dimers (n_2), and trimers (n_3) on time $t/\tau_{1/2}$. The number of monomers decreases with time, whereas the number of aggregates initially increases and then decreases.

where F_G is the collision frequency due to gravitational separation, v_i is the Stokes creaming velocity of a particle with radius r_i, and $\Delta\rho$ is the density difference between the droplets and the surrounding liquid. This equation indicates that the collision frequency increases as the difference between the creaming velocities of the particles increases. The rate of gravitationally induced flocculation can therefore be retarded by ensuring that the droplet size distribution is not too wide, decreasing the density difference between the oil and aqueous phases, decreasing the droplet concentration, or increasing the viscosity of the continuous phase. Equation 7.22 would have to be modified before it could be applied to systems which do not obey Stokes' law (Section 7.3.1). In addition, it does not take into account the fact that the droplets reach a position at the top or bottom of an emulsion where they cannot move any further and are therefore forced to encounter each other.

Collisions Due to Applied Shear Forces. Food emulsions are often subjected to various kinds of shear flow during their production, storage, and transport. Consequently, it is important to appreciate the effect that shearing has on their stability to flocculation. In a system subjected to Couette flow, the collision frequency is given by (Dickinson 1992, Walstra 1996a):

$$F_S = k_S n^2 = \frac{16}{3} G r^3 n^2 = \left(\frac{3G}{\pi}\right) \frac{\phi^2}{r^3} \tag{7.23}$$

where F_S is the collision frequency due to shear. Thus the frequency of shear-induced collisions can be retarded by decreasing the shear rate, increasing the droplet size, or decreasing the dispersed-phase volume fraction. It should be noted that the collision frequency is independent of the viscosity of the continuous phase.

Relative Importance of Different Collision Mechanisms. In general, each of the above mechanisms may contribute to the droplet collision frequency in an emulsion. In practice, one of the mechanisms usually dominates, depending on the composition and microstructure of the product and the prevailing environmental conditions. To effectively control the collision frequency, it is necessary to establish which mechanism is the most important in the particular system being studied. The ratio of the shear-to-Brownian motion collision frequencies (F_S/F_B) and the gravitational-to-Brownian motion collision frequencies (F_G/F_B) is plotted as a function of shear rate and particle size ratio ($= r_2/r_1$) in Figure 7.9 for a typical emulsion. At low shear rates ($G < 2$ s^{-1}), collisions due to Brownian motion dominate, but at high shear rates those due to mechanical agitation of the system dominate. Gravitationally induced collisions dominate those due to Brownian motion when the particle size ratio exceeds about 5, and thus it is likely to be important in emulsions which have a broad particle size distribution.

7.4.1.2. Collision Efficiency

If every encounter between two droplets led to aggregation, then emulsions would not remain stable long enough to be practically useful. To prevent droplets from flocculating during a collision, it is necessary to have a sufficiently high repulsive energy barrier to stop them from coming too close together (Chapter 3). The height of this energy barrier determines the likelihood that a collision will lead to flocculation (i.e., the collision efficiency). The collision efficiency (E) has a value between 0 (no flocculation) to 1 (every collision leads to flocculation)* and depends on the hydrodynamic and colloidal interactions between the droplets. The flocculation rate therefore depends on the precise nature of the interactions between the

---

* In practice, E can have a value which is slightly higher than 1 because droplet collisions are accelerated when there is a strong attraction between the droplets.

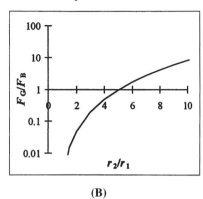

FIGURE 7.9 Relative importance of the different collision mechanisms for typical food emulsions ($\Delta\rho$ = 90 kg m^{-3}, r = 1 μm, ϕ = 0.1). (A) Shear-induced collisions become increasingly important as the shear rate is increased. (B) Gravitationally induced collisions become increasingly important as the ratio of droplet size increases.

emulsion droplets. For collisions induced by Brownian motion (Dukhin and Sjoblom 1996, Walstra 1996b):

$$-\frac{dn_B}{dt} = \frac{4kTn^2 E_B}{3\eta_1} \tag{7.24}$$

$$E_B = \left(2\int_2^\infty \frac{\exp[w(s)\,/\,kT]}{s^2 G(s)}\,ds\right)^{-1} \tag{7.25}$$

where s is the dimensionless center-to-center distance between the droplets ($s = [2r + h]/r$), r is the droplet radius, and h is the surface-to-surface separation. Colloidal interactions are accounted for by the $w(s)$ term and hydrodynamic interactions by the $G(s)$ term (Chapter 3). When there are no colloidal interactions between the droplets, $w(s) = 0$, and no hydrodynamic interactions, $G(s) = 0$, Equation 7.24 becomes equivalent to that derived by Smoluchowski (Equations 7.15 and 7.16). The stability of an emulsion to aggregation is governed primarily by the maximum height of the energy barrier, $w_{max}(s)$, rather than by its width (Friberg 1997). To enhance the stability of an emulsion against flocculation, it is necessary to have an energy barrier which is large enough to prevent the droplets from coming close together. An energy barrier of about 20 kT is usually sufficient to provide long-term stability of emulsions (Friberg 1997). Expressions for the efficiency of shear and gravitationally induced collisions also depend on colloidal and hydrodynamic interactions and have been derived for some simple systems (Zhang and Davis 1991).

7.4.2. Methods of Controlling Flocculation

A knowledge of the physical basis of droplet flocculation facilitates the development of effective strategies for controlling it in food emulsions. These strategies can be conveniently divided into those which influence the collision frequency and those which influence the collision efficiency.

7.4.2.1. Collision Frequency

The rate at which flocculation proceeds can be controlled by manipulating the droplet collision frequency. The most effective means of achieving this depends on the dominant collision mechanism in the emulsion (i.e., Brownian motion, gravity, or mechanical agitation). The rate at which droplets encounter each other in an unstirred emulsion can be reduced by increasing the viscosity of the continuous phase (Equations 7.17 and 7.22). Flocculation may be completely retarded if the continuous phase contains a three-dimensional network of aggregated molecules or particles that prevents the droplets from moving (e.g., a biopolymer gel or a fat crystalline network). The collision frequency increases when an emulsion is subjected to sufficiently high shear rates (Equation 7.23), and therefore it may be important to ensure that a product is protected from mechanical agitation during its storage and transport in order to avoid flocculation. The collision frequency increases as the droplet concentration increases or the droplet size decreases, with the precise nature of this dependence determined by the type of collision mechanism which dominates. The rate of collisions due to gravitational separation depends on the relative velocities of the particles in an emulsion and therefore decreases as the density difference between the droplet and surrounding liquid decreases or as the viscosity of the continuous phase increases (Equation 7.22).

7.4.2.2. Collision Efficiency

The most effective means of controlling the rate and extent of flocculation in an emulsion is to regulate the colloidal interactions between the droplets. Flocculation can be prevented by designing an emulsion in which the repulsive interactions between the droplets are significantly greater than the attractive interactions. A wide variety of different types of colloidal interaction can act between the droplets in an emulsion (Chapter 3). Which of these is important in a given system depends on the type of ingredients present, the microstructure of the emulsion, and the prevailing environmental conditions. To control flocculation in a particular system, it is necessary to identify the most important types of colloidal interaction.

Electrostatic Interactions. Many oil-in-water emulsions used in the food industry are stabilized against flocculation by using electrically charged emulsifiers which generate an electrostatic repulsion between the droplets (e.g., ionic surfactants, proteins, or polysaccharides) (Dickinson 1992, Das and Kinsella 1990). The flocculation stability of electrostatically stabilized emulsions depends mainly on the electrical properties of the emulsifier and the pH and ionic strength of the aqueous phase (Hunter 1986). The number, position, sign, and dissociation constants of the ionizable groups on an emulsifier determine its electrical behavior under different environmental conditions. For each type of food product, it is therefore necessary to select an emulsifier with appropriate electrical characteristics.

Hydrogen ions are potential determining ions for many food emulsifiers (e.g., $COOH \rightarrow COO^- + H^+$ or $NH_2 + H^+ \rightarrow NH_3^+$), and therefore the sign and magnitude of the electrical charge on emulsion droplets are determined principally by the pH of the surrounding solution (Section 3.4.1). The influence of pH on droplet flocculation in a protein-stabilized emulsion is illustrated in Figure 7.10. At pH values sufficiently above or below the isoelectric point (IEP) of the whey proteins (IEP ≈ pH 5), the droplet charge is large enough to prevent flocculation because of the strong electrostatic repulsion between the droplets (Figure 7.10). At pH values near the isoelectric point (IEP ± 1), the net charge on the proteins is relatively low and the electrostatic repulsion between the droplets is no longer sufficiently strong to prevent flocculation. Droplet flocculation leads to a pronounced increase in the viscosity of an emulsion, as well as a decrease in creaming stability, and therefore has important implications for food quality (Demetriades et al. 1997a, Agboola and Dalgleish 1996d).

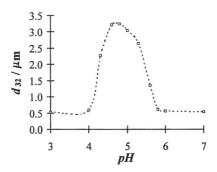

FIGURE 7.10 Influence of pH and ionic strength on the stability of 20% corn-oil-in-water emulsions to flocculation. Extensive flocculation is observed near the isoelectric point of the proteins.

The ionic strength of an aqueous solution depends on the concentration and valency of the ions it contains (Israelachvili 1992). As the ionic strength is increased, the electrostatic repulsion between droplets is progressively screened, until eventually it is no longer strong enough to prevent flocculation (Section 3.11.1). The minimum amount of electrolyte required to cause flocculation is known as the *critical flocculation concentration* or CFC. The CFC decreases as the surface potential of the emulsion droplets decreases and as the valency of the counterions increases. It has been shown that CFC $\propto \Psi_0^4/z^2$ (where Ψ_0 is the surface potential and z is the counterion valency) for droplets with relatively low surface potentials (i.e., $\Psi_0 < 25$ mV) (Hunter 1986). These low surface potentials are found in many food emulsions which are susceptible to flocculation. Under certain conditions, Ψ_0 is inversely proportional to the valency of the counterions, so that CFC $\propto 1/z^6$, which is known as the Schultz–Hardy rule (Hunter 1986). This relationship indicates that a much lower concentration of a multivalent ion than a monovalent ion is required to cause flocculation. The Schultz–Hardy rule can be derived from the DLVO theory by assuming that the CFC occurs when the potential energy barrier, which normally prevents droplets from aggregating, falls to a value of zero due to the addition of salt. Consequently, when two droplets collide with each other, they immediately aggregate into the primary minima. In practice, significant droplet flocculation occurs when the potential energy barrier is slightly higher than zero, and therefore the Schultz–Hardy rule is expected to slightly overestimate the CFC (Hunter 1986).

Some types of ionic species are capable of specifically binding to oppositely charged emulsifier groups and can therefore alter the surface charge. Specifically adsorbed ions may either decrease, increase, or reverse the charge on a droplet depending on their valency and the nature of the specific interactions involved (Section 3.4.2.1). In addition to their influence on surface charge, specifically adsorbed ions are often highly hydrated and therefore increase the short-range hydration repulsion between droplets (Section 3.8).

The presence of multivalent ions in the continuous phase of an electrostatically stabilized emulsion has a strong influence on its flocculation stability (Dickinson et al. 1992; Agboola and Dalgleish 1995, 1996a), first because multivalent ions are much more effective at screening the electrostatic interactions between droplets than monovalent ions and, second, because they are capable of forming electrostatic bridges between two droplets which have the same charge (Section 3.4.4). For example, the addition of calcium ions (Ca^{2+}) to a protein-stabilized emulsion that contains negatively charged droplets can cause droplet flocculation because the ions shield the electrostatic repulsion between the droplets and because they form protein–calcium–protein cross-links between the droplets (Dickinson et al. 1992; Agboola and Dalgleish 1995, 1996a). These multivalent ions may be simple ions (e.g., Ca^{2+}, Mg^{2+}, or

Al^{3+}), small organic ions (amino acids), or charged biopolymers (e.g., proteins or polysaccharides). The influence of multivalent ions on emulsion stability can be reduced by ensuring that they are excluded from the formulation or by adding ingredients which form complexes with them (e.g., EDTA in the case of calcium ions).

The flocculation stability of emulsions that contain electrically charged droplets can be controlled in a variety of ways depending on the system. To prevent flocculation, it is necessary to ensure that the droplets have a sufficiently high surface potential under the prevailing environmental conditions (which often requires that the pH be controlled) and that the electrolyte concentration is below the CFC. In some important types of food, it is necessary to have relatively high concentrations of electrolyte present for nutritional purposes (e.g., mineral-fortified emulsions used in infant, elderly, and athlete formulations). In these systems, the influence of the electrolyte on flocculation stability can be reduced by changing to a nonionic emulsifier or by incorporating ingredients which complex the mineral.

Polymeric Steric Interactions. Many food emulsifiers prevent droplet flocculation through polymeric steric repulsion (Section 3.5). This repulsion must be sufficiently strong and long range to overcome any attractive interactions (Section 3.11.3). Sterically stabilized emulsions are much less sensitive to variations in pH and ionic strength than electrostatically stabilized emulsions (Hunter 1986). Nevertheless, they can become unstable to flocculation under certain conditions. If the composition of the continuous phase or the temperature is altered so that polymer–polymer interactions become more favorable than solvent–solvent/solvent–polymer interactions, then the mixing contribution to the steric interaction becomes attractive and may lead to droplet flocculation. A sterically stabilized emulsion may also become unstable if the thickness of the interfacial membrane is reduced (Section 3.11.3), which could occur if the polymeric segments on the emulsifier were chemically or biochemically cleaved (e.g., by acid or enzyme hydrolysis) or if the continuous phase becomes a poor solvent for the polymer segments. Short-range hydration forces make an important contribution to the flocculation stability of many sterically stabilized emulsions (Israelachvili 1992, Evans and Wennerstrom 1994). In these systems, droplet flocculation may occur when the emulsion is heated, because emulsifier head groups are progressively dehydrated with increasing temperature (Israelachvili 1992, Aveyard et al. 1990).

Biopolymer Bridging. Many types of biopolymer promote flocculation by forming bridges between two or more droplets (Lips et al. 1991). Biopolymers may adsorb either directly to the bare surfaces of the droplets or to the adsorbed emulsifier molecules that form the interfacial membrane (Walstra 1996a). To be able to bind to the droplets, there must be a sufficiently strong attractive interaction between segments of the biopolymer and the droplet surface. The most common types of interaction that operate in food emulsions are hydrophobic and electrostatic (Dickinson 1989, 1992).

When a biopolymer has a number of nonpolar residues along its backbone, some of them may associate with hydrophobic patches on one droplet, while others associate with hydrophobic patches on another droplet. This type of bridging flocculation tends to occur when a biopolymer is used as an emulsifier and there is insufficient present to completely cover the oil–water interface formed during homogenization (Walstra 1996a). Bridging may occur either during the homogenization process or after it is complete (e.g., when a biopolymer is only weakly associated with a droplet, some of its segments can desorb and become attached to a neighboring droplet). This type of bridging flocculation can usually be prevented by ensuring there is a sufficiently high concentration of biopolymer present in the continuous phase prior to homogenization (Dickinson and Euston 1991, Dickinson 1992, Stoll and Buffle 1996).

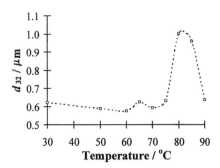

FIGURE 7.11   Influence of temperature on the flocculation stability of 20% corn-oil-in-water emulsions stabilized by whey protein isolate (pH 7, 0 mM NaCl). Flocculation is observed when the emulsions are heated above 70°C because of protein unfolding and exposure of hydrophobic groups.

Bridging flocculation can also occur when a biopolymer in the continuous phase has an electrical charge which is opposite to that of the droplets (Pal 1996). In this case, bridging flocculation can be avoided by ensuring the droplets and biopolymer have similar charges or that either the droplets or biopolymer are uncharged.

Hydrophobic Interactions.   This type of interaction is important in emulsions that contain droplets which have nonpolar regions exposed to the aqueous phase. Their role in influencing the stability of food emulsions has largely been ignored, probably because of the lack of theories to describe them and experimental techniques to quantify them. One of the clearest examples of their importance in food emulsions is the effect of heat on the flocculation stability of oil-in-water emulsions stabilized by globular proteins (Hunt and Dalgleish 1995, Demetriades et al. 1997b). At room temperature, whey protein stabilized emulsions (pH 7, 0 mM NaCl) are stable to flocculation because of the large electrostatic repulsion between the droplets, but when they are heated above 70°C, they become unstable (Figure 7.11). The globular proteins adsorbed to the surface of the droplets unfold above this temperature and expose nonpolar amino acids which were originally located in their interior (Monahan et al. 1996, Dalgleish 1996a). Exposure of these nonpolar amino acids increases the hydrophobic character of the droplet surface and therefore leads to flocculation because of the increased hydrophobic attraction between the droplets (Hunt and Dalgleish 1995, Monahan et al. 1996, Demetriades et al. 1997b).

Hydrophobic interactions are also likely to be important in emulsions in which there is not enough emulsifier present to completely saturate the surfaces of the droplets. This may occur when there is insufficient emulsifier present in an emulsion prior to homogenization or when an emulsion is so diluted that some of the emulsifier desorbs from the droplet surfaces.

Flocculation due to hydrophobic interactions can be avoided by ensuring that there is sufficient emulsifier present to completely cover the droplet surfaces or by selecting an emulsifier which does not undergo detrimental conformational changes at the temperatures used during processing, storage, or handling.

Depletion Interactions.   The presence of nonadsorbing colloidal particles, such as biopolymers or surfactant micelles, in the continuous phase of an emulsion causes an increase in the attractive force between the droplets due to an osmotic effect associated with the exclusion of colloidal particles from a narrow region surrounding each droplet (Section 3.6). This attractive force increases as the concentration of colloidal particles increases, until eventually it may become large enough to overcome the repulsive interactions between the droplets and cause them to flocculate (Aronson 1991; Jenkins and Snowden 1996; Dickinson and Golding

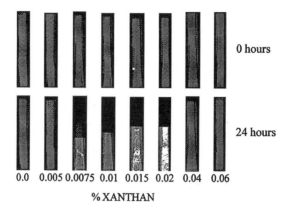

FIGURE 7.12   Influence of xanthan concentration on the stability of 20% corn-oil-in-water emulsions to depletion flocculation. Flocculation is observed above the critical flocculation concentration (i.e., % xanthan > 0.0075 wt%).

1997a,b; Dickinson et al. 1997). This type of droplet aggregation is usually referred to as *depletion flocculation* (Walstra 1996a,b). The lowest concentration required to cause depletion flocculation is referred to as the *critical flocculation concentration* by analogy to the CFC used to characterize the effect of salt on the stability of electrostatically stabilized emulsions. The CFC decreases as the size of the emulsion droplets increases (Section 3.6.3). The flocculation rate initially increases as the concentration of nonadsorbing colloidal particles is increased because of the enhanced attraction between the droplets (i.e., a higher *collision efficiency*). However, once the concentration of colloidal particles exceeds a certain concentration, the flocculation rate may actually decrease because the viscosity of the continuous phase increases so much that the movement of the droplets is severely retarded (i.e., a lower *collision frequency*). This is clearly illustrated in Figure 7.12, which shows the dependence of the height of the creamed layer in a series of corn-oil-in-water emulsions containing different concentrations of xanthan (Basaran et al. 1998). Xanthan is a nonadsorbing biopolymer which is normally used as a stabilizer or thickening agent in food emulsions (BeMiller and Whistler 1996). In the absence of xanthan, the emulsions are stable to creaming over a 24-h period. At xanthan concentrations around 0.01%, there is a net attraction between the droplets which causes them to flocculate and therefore cream rapidly. At higher xanthan concentrations, there is still a strong attraction between the droplets, but they are unable to move because of the large increase in the viscosity of the continuous phase.* Similar observations have been obtained for emulsions to which ionic or nonionic surfactant micelles have been added (Bibette 1991, Aronson 1992, McClements 1994, Jenkins and Snowden 1996). For ionic colloidal particles, such as sodium dodecyl sulfate (SDS) micelles or charged biopolymers, the CFC is expected to be strongly dependent on electrolyte concentration because of its ability to reduce the electrostatic repulsion between colloidal particles and therefore reduce their effective size.

Hydrodynamic Interactions.   The efficiency of the collisions between droplets is also determined by the strength of the hydrodynamic interactions between them (Davis et al. 1989, Dukhin and Sjoblom 1996). As two droplets approach each other, a repulsion arises because

---

* Even when the droplets do flocculate at higher xanthan concentrations, the movement of the flocs themselves will be retarded, and therefore no creaming is observed.

of the resistance associated with the flow of the continuous phase from the thin gap between them. The magnitude of this resistance decreases as the droplet surfaces become more mobile, leading to an increase in the collision efficiency (Section 3.10). On the other hand, the collision efficiency may be reduced when the droplet surfaces are stabilized by small-molecule surfactants because of the Gibbs–Marangoni effect (Walstra 1996a,b).

Influence of Droplet Size. It should be noted that the magnitude of the colloidal interactions between emulsion droplets usually increases with droplet size, which will cause an increase in the height of any energy barriers and in the depth of any energy minima. As a consequence, altering the droplet size may either increase or decrease the stability of an emulsion, depending on the nature of the system.

7.4.3. Structure and Properties of Flocculated Emulsions

The appearance, texture, and stability of emulsions are strongly influenced by the characteristics of any flocs formed (e.g., their number, size, flexibility, and packing) (Walstra 1993a). The structure and properties of flocs are mainly determined by the nature of the colloidal and hydrodynamic interactions between the droplets, but they also depend on the mechanism responsible for the droplet collisions (i.e., Brownian motion, gravity, or mechanical agitation) (Bremer 1992, Evans and Wennerstrom 1994). Valuable insights into the relationship between these parameters and floc characteristics have been obtained from a combination of computer simulations and experimental measurements.

7.4.3.1. *Influence of Colloidal Interactions on Floc Structure*

When the attraction between the droplets is relatively strong compared to the thermal energy, the flocs formed have extremely open structures in which each droplet is only linked to two or three of its neighbors (Figure 7.13). This type of open structure is formed because the droplets "stick" firmly together at the point where they first come into contact and are unable to undergo any subsequent structural rearrangements (Bremer 1992). As a consequence, a droplet which encounters a floc cannot move very far into its interior before becoming attached to another droplet. This type of floc is characterized by a tenuous structure which traps large amounts of continuous phase within it. The volume fraction of particles in such a floc may be as low as 0.13 depending on its size and the strength of the interactions (Dickinson 1992). When the attraction between the droplets is relatively weak compared to the thermal energy, the droplets do not always stick together after a collision and they may be able to roll around each other after sticking together. Thus a droplet which encounters a floc is able to penetrate closer to its center and flocs are able to undergo structural rearrangements, which means that the droplets can pack more closely together (Bijsterbosch et al. 1995). This type of floc is characterized by a more compact structure which traps less of the continuous phase. The volume fraction of the particles in this type of floc may be as high as 0.63, which is close to the value for random packing of monodisperse particles (Dickinson 1992).

Floc with Close   Floc with Open
Packing   Packing

FIGURE 7.13   Schematic diagram of some of the different types of structures that flocs may form.

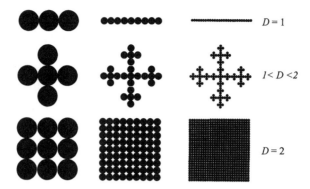

FIGURE 7.14 Examples of two-dimensional structures that can be used to model flocs.

7.4.3.2. Use of Fractal Geometry to Describe Floc Structure

Fractal geometry has proven to be an extremely valuable tool for characterizing the structural organization of droplets in flocs (Bremer 1992, Peleg 1993, Walstra 1996a). Fractal geometry is applicable to systems which exhibit a phenomenon known as self-similarity (i.e., have structures which appear similar when observed at different levels of magnification) (Figure 7.14). There are no truly fractal objects in nature because all real objects have a definite upper and lower size. Nevertheless, many natural objects do show self-similarity over a number of levels of magnification and can therefore be described by fractal geometry (Peleg 1993). Despite their extremely complex structures, these fractal objects can be described by a single parameter (D) known as the fractal dimension.

The concept of a fractal dimension is best illustrated by considering the model two-dimensional fractal flocs shown in Figure 7.14. The floc with the "string-of-beads" type structure has a fractal dimension of one ($D = 1$), because increasing the level of magnification by a factor of X changes the number of particles per floc by a factor of X^1 (i.e., $N \propto X^1$). The floc which contains the array of closely packed particles has a fractal dimension of two ($D = 2$), because increasing the level of magnification by a factor of X changes the number of particles per floc by a factor of X^2 (i.e., $N \propto X^2$). These are the two extreme values of the fractal dimension of a two-dimensional structure. There are many other types of floc which have self-similar structures with intermediate fractal dimensions (Figure 7.14). The number of particles in these flocs is described by the relationship $N \propto X^D$, where D is a noninteger value between 1 and 2. The closer the value is to one, the more tenuous is the floc structure, and the closer it is to two, the more compact. The fractals drawn in Figure 7.14 are characterized by a lower cutoff length (r) which is equal to the radius of the particles and an upper cutoff length (R) which is equal to the radius of the floc. The concept of a fractal dimension therefore only applies on length scales between these limits. The number of particles in a fractal floc is given by the relationship $N = (R/r)^D$.

In nature, flocs are three-dimensional structures, and so D ranges from 1 to 3: the higher the fractal dimension, the more compact the floc structure (Peleg 1993). The volume fraction of droplets in a three-dimensional floc is given by (Bremer 1992):

$$\phi_F = \left(\frac{R}{r}\right)^{D-3} \tag{7.26}$$

This equation indicates that ϕ_F decreases as the size of a floc increases (because $D < 3$) and that the concentration of droplets in a fractal floc decreases as one moves outward from its interior. The fractal dimension of flocs can be determined using a variety of experimental techniques, including rheometry, light scattering, and microscopy (Bremer 1992). Thus it is possible to quantify the influence of different parameters on the structure of the flocs formed (e.g., pH, ionic strength, and temperature).

It should be mentioned that fractal flocs are only kinetically stable structures (Evans and Wennerstrom 1994). When there is a strong attraction between the droplets, the most thermodynamically stable arrangement would be one in which the droplets formed a closely packed structure, as this would maximize the number of favorable attractive interactions. It is because there is a large activation energy associated with the rearrangement of droplets within a floc, which means that it retains its fractal structure.

7.4.3.3. Influence of Floc Structure on Emulsion Properties

The structure and properties of flocs have a pronounced effect on the stability and rheological properties of emulsions. An emulsion that contains flocculated droplets has a higher viscosity than one that contains the same concentration of unflocculated droplets. This is because the effective volume fraction of a floc is greater than the sum of the volume fractions of the individual droplets due to the presence of the continuous phase trapped within it (Dickinson and Stainsby 1982). The viscosity of emulsions usually increases as the floc structure becomes more tenuous for the same reason. Emulsions that contain flocculated droplets exhibit pronounced shear thinning behavior (i.e., the viscosity decreases as the shear rate or shear time increases) (Chapter 8). Shear thinning occurs for two reasons: (1) the flocs are deformed and become aligned with the shear field, which decreases their resistance to flow, and (2) the flocs are disrupted by the shear forces, which decreases their effective volume fraction (Figure 7.15) (Bujannunez and Dickinson 1994, Bower et al. 1997). The ease with which the flocs in an emulsion are deformed and disrupted decreases as the number and strength of the attractive interactions between the droplets increase (Uriev 1994, Pal 1996). Thus one would expect that a higher shear stress would be required to disrupt flocs in emulsions that contain droplets that are strongly bound to each other than one in which they are only weakly bound. Once the shearing forces are removed from an emulsion, the bonds between the droplets may reform. The rate at which this process occurs and the type of structures formed often have an important influence on the quality of food emulsions.

In some cases, shearing an emulsion can actually promote droplet flocculation because the efficiency and frequency of collisions between the droplets are increased (Spicer and Pratsinis 1996). Thus an emulsion which is stable under quiescent conditions may become flocculated when it is subjected to shear forces (Dalgleish 1996a). This type of emulsion initially shows shear thickening behavior because the formation of flocs leads to an increase in viscosity, but

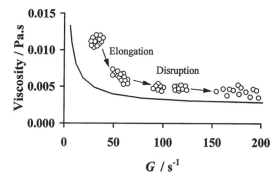

FIGURE 7.15   Schematic diagram of the events that occur during the shearing of a flocculated emulsion and their effect on the emulsion viscosity.

at higher shear rates these flocs may become deformed and disrupted, which leads to a decrease in viscosity (Pal 1996). Emulsions may therefore exhibit complex rheological behavior depending on the nature of the colloidal interactions between the droplets.

At sufficiently high dispersed-phase volume fractions, flocculation leads to the formation of a three-dimensional network of aggregated droplets which extends throughout the emulsion (Bijsterbosch et al. 1995). This type of system is referred to as a particle gel and can be characterized by a yield stress which must be exceeded before the emulsion will flow (Pal 1992, 1996; Pal et al. 1992; Tadros 1994, 1996). The value of the yield stress increases as the dispersed-phase volume fraction increases and as the strength of the attractive forces between the droplets increases (Pal 1996). The minimum dispersed-phase volume fraction required to form a particle gel decreases as the structure of the flocs becomes more tenuous (i.e., $D \to 1$) because this type of structure is able to fill the available space more effectively (Figure 7.6). Thus two emulsions may have exactly the same dispersed-phase volume fractions, yet one could be a low-viscosity liquid and the other a viscoelastic gel, depending on the nature of the colloidal interactions between the droplets.

Flocculation also affects the stability of emulsions to creaming. In dilute emulsions, flocculation increases the creaming rate because the effective size of the particles in the emulsion is increased, which more than compensates for the decrease in $\Delta \rho$ (Equation 7.9). In concentrated emulsions, the droplets are usually prevented from creaming because the formation of a three-dimensional network of aggregated droplets prevents them from moving (Figure 7.6).

7.4.4. Experimental Measurement of Flocculation

A wide variety of experimental techniques have been developed to monitor the extent and rate of flocculation in emulsions. The most direct method is to observe the emulsion through an optical or electron microscope (Chapter 10). The association of droplets with one another can be determined subjectively by eye or more quantitatively by transferring an image of the emulsion to a computer and using image analysis techniques (Jokela et al. 1990, Mikula 1992, Brooker 1995). The major drawback of most microscopic techniques is that the delicate structure of the flocs may be disturbed by the procedures used to prepare the samples for observation. In addition, in concentrated emulsions it is often difficult to tell whether two droplets are flocculated or just in close proximity.

Flocculation can be monitored indirectly by measuring the change in the particle size distribution with time using a particle-sizing instrument (e.g., light scattering, ultrasonic spectrometry, or electrical pulse counting) (Chapter 10). These instruments often provide a fairly qualitative indication of the extent of flocculation in an emulsion, because it is difficult to specify the physical properties of a floc that are needed to mathematically convert the experimental measurements into a true particle size distribution. For example, light-scattering techniques usually assume that the particles in an emulsion are isolated homogeneous spheres, whereas flocs are inhomogeneous and nonspherical aggregates that contain many individual droplets, and therefore the theory used to interpret the experimental data is no longer valid.

In many systems, it is important to determine whether the increase in droplet size is caused by flocculation, coalescence, or Ostwald ripening. The simplest method of achieving this is to alter the emulsion in a way which would be expected to break down any flocs that are present. If there are no flocs present, the particle size will not change after the alteration, but if there are flocs present, there will be a decrease in the particle size. A variety of methods are available for breaking down flocs: (1) altering solvent conditions, such as pH, ionic strength, dielectric constant, or temperature; (2) applying mechanical agitation, such as stirring or sonication; and (3) adding a more surface-active agent, such as a small-molecule surfactant, which displaces the original emulsifier from the droplet interface and which does not cause flocculation itself. The choice of a suitable method depends on the nature of the emulsion and in particular on the type of emulsifier used to stabilize the system. In emulsions where the interfacial membrane consists of proteins that are held together by extensive intermolecular disulfide bonds, it may be necessary to use a combination of mercaptoethanol (to break the disulfide bonds) and a small-molecule surfactant (to displace the proteins).

As with microscopy, the preparation of samples for analysis using particle-sizing instruments often disturbs the structures of the flocs. For example, emulsions often have to be diluted before they can be analyzed, which can cause disruption of the flocs, particularly when there is only a weak attraction between the droplets. In addition, it is important to carry out dilution using a solvent that has similar properties to the continuous phase in which the droplets were originally dispersed (e.g., ionic strength, pH, and temperature); otherwise, the floc structure may be altered. The dispersion of droplets by stirring may also cause some of the flocs to break down, particularly when they are only held together by weak attractive forces. Many of these problems can be overcome using modern particle-sizing techniques based on ultrasonics, nuclear magnetic resonance, or dielectric spectrometry, because they can be used to analyze concentrated emulsions without the need for any sample preparation (Chapter 10).

Flocculation causes an increase in the viscosity of an emulsion and may eventually lead to the formation of a particle gel (Section 7.4.3). As a consequence, flocculation can often be monitored by measuring the change in viscosity or shear modulus of an emulsion (Grover and Bike 1995, Tadros 1996, Pal 1996). An indication of the strength of the attractive forces between flocculated droplets can be obtained from measurements of the viscosity or shear modulus versus shear stress (Pal 1996). As the shear stress is increased, the flocs become deformed and disrupted, so that the viscosity or shear modulus decreases. The stronger the attractive forces between the flocculated droplets, the higher the shear stress needed to disrupt them. Another indirect measurement of the degree of flocculation in an emulsion is to determine the rate at which droplets cream or sediment (Luyten et al. 1993, Basaran et al. 1998). As mentioned earlier, flocculation may either increase or decrease the creaming rate depending on the droplet concentration and the structure of the flocs formed. Experimental methods that can be used to monitor creaming and sedimentation were discussed in Section 7.3.3.

7.5. COALESCENCE

Coalescence is the process whereby two or more liquid droplets merge together to form a single larger droplet (Figure 7.16). It is the principal mechanism by which an emulsion moves toward its most thermodynamically stable state because it involves a decrease in the contact area between the oil and water phases (Section 7.2). Coalescence causes emulsion droplets to cream or sediment more rapidly because of the increase in their size. In oil-in-water

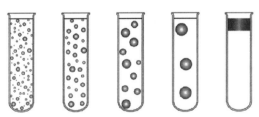

FIGURE 7.16   Droplet coalescence eventually leads to the complete separation of the oil and aqueous phases.

emulsions, coalescence eventually leads to the formation of a layer of oil on top of the material, which is referred to as *oiling off*. In water-in-oil emulsions, it leads to the accumulation of water at the bottom of the material. For these reasons, an understanding of the factors which influence coalescence is important to food manufacturers attempting to create products with extended shelf lives.

7.5.1. Physical Basis of Coalescence

Coalescence is the result of the liquid within two or more emulsion droplets coming into molecular contact (Evans and Wennerstrom 1994, Kabalnov and Wennerstrom 1996, Walstra 1996a,b). Thus this process can only occur when the droplets are close to each other and the interfacial membranes are disrupted. The fact that the droplets must be in close contact means that coalescence is much more dependent on short-range forces, and therefore on the precise molecular details of a system, than either gravitational separation or flocculation. The rate at which coalescence proceeds, and the physical mechanism by which it occurs, is therefore highly dependent on the nature of the emulsifier used to stabilize the system. For this reason, our knowledge of coalescence is much less well developed than that of the other major forms of emulsion instability. Even so, a combination of theoretical and experimental studies has led to a fairly good understanding of the major factors which influence coalescence in a number of important systems. In general, the susceptibility of droplets to coalescence is determined by the nature of the forces that act on and between the droplets (i.e., gravitational, colloidal, hydrodynamic, and mechanical forces) and the resistance of the droplet membrane to rupture. It is useful to distinguish between two different types of situations which can lead to coalescence in food emulsions.

7.5.1.1. Coalescence Induced by Collisions

In this situation, coalescence occurs immediately after two or more droplets encounter each other during a collision. It is the dominant form of coalescence in emulsions in which the droplets are free to move around and collide with each other. The movement of the droplets may be induced by Brownian motion, gravity, or applied mechanical forces (Section 7.4.1.1). The rate at which this type of coalescence occurs is governed by many of the same factors as flocculation (i.e., the collision frequency and efficiency) (Section 7.4.1). However, the collision efficiency is now determined by the probability that the interfacial membranes surrounding the droplets will be ruptured during a collision. In the absence of an emulsifier, the coalescence collision efficiency is high ($E_C \rightarrow 1$), because there is little resistance to the droplets merging together. In the presence of emulsifier, the collision efficiency may be extremely low ($E_C \rightarrow 0$), because of short-range repulsive interactions between the droplets or because of the low probability that the interfacial membrane will be ruptured. For this reason, this type of coalescence is only significant in systems in which there is insufficient emulsifier present to completely saturate the surface of the droplets or during homogenization, where emulsifiers may not have sufficient time to cover the droplet surfaces before a droplet–droplet collision occurs (Chapter 6). It may also be important in emulsions which are subjected to high shear forces, because the impact forces that act on the droplets as the result of a collision may then be sufficient to cause disruption of the interfacial membranes.

7.5.1.2. Coalescence Induced by Prolonged Contact

In this situation, coalescence occurs spontaneously after the droplets have been in contact for a prolonged period. This type of coalescence is therefore important in emulsions which have high droplet concentrations, which contain flocculated droplets, or which contain droplets that have accumulated at the top (or bottom) of the sample due to creaming (or sedimentation). It is no longer convenient to describe the rate at which coalescence occurs in terms of a collision frequency and efficiency, because the droplets are in contact with each other. Instead, it is more useful to characterize this type of coalescence in terms of a *coalescence time,* that is, the length of time the droplets remain in contact before coalescence occurs. In practice, all of the droplets in an emulsion do not usually coalesce after the same time, and therefore it is more appropriate to define an average coalescence time or to stipulate a range of times over which the majority of the droplets coalesce.

One of the most important factors that determine the rate at which this process occurs is the thickness of the layer of continuous phase which normally separates the droplets (Hunter 1989). This thickness is governed mainly by the dependence of the interdroplet pair potential on droplet separation (Chapter 3). Droplets tend to exist at a separation where the interdroplet pair potential is at a minimum, although they may be forced closer together in the presence of external forces, such as gravity or centrifugation. The coalescence stability of an emulsion is usually reduced as the thickness of the layer of continuous phase separating the droplets is decreased (Section 7.5.2).

This type of coalescence is the most important in food emulsions (under quiescent conditions), because in most systems there is usually enough emulsifier present to make the disruption of the interfacial membranes during a collision a highly improbable event.

7.5.1.3. *Coalescence as the Result of "Hole" Formation*

In many emulsions in which coalescence is observed, one would not expect it to occur from an examination of the interdroplet pair potential, because of the extremely high steric repulsion that arises when two droplets closely approach each another (Section 3.5). The interdroplet pair potential, which describes the dependence of the colloidal interactions between the emulsion droplets on separation, assumes that the system is at thermodynamic equilibrium. It therefore predicts that droplets will remain indefinitely in a potential energy minimum. In practice, the stability of an emulsion to coalescence is often governed by nonequilibrium processes associated with the dynamic events that occur on the molecular level (Evans and Wennerstrom 1994, Kabalnov and Wennerstrom 1996).

It is widely accepted that coalescence occurs as the result of the formation of a *hole* which extends across the interfacial membranes surrounding the droplets (Figure 7.17). Once the hole is formed, the liquid in the droplets rapidly flows through it and the droplets merge together to form a single larger droplet. The coalescence rate therefore depends on the likelihood that these holes will form in the interfacial membranes.

Holes may be formed in an interfacial membrane in a variety of different ways, depending on type of emulsifier used and the environmental conditions. They form spontaneously in surfactant monolayers due to thermal fluctuations of their shape (Section 7.5.1.4). They may also be formed as the result of the chemical breakdown of an emulsifier over time (e.g., a protein or polysaccharide may be cleaved into smaller fragments due to enzymic or chemical hydrolysis). These fragments might not be surface active or they may not form a good protective membrane, which leads to the formation of a hole. A hole may also be created when emulsifiers are displaced from the surface of emulsion droplets by more surface-active components which do not form membranes resistant to rupture (e.g., biopolymers may be displaced by high concentrations of alcohol). Which of these mechanisms is important in a particular system depends on the composition and microstructure of the emulsion, as well as environmental conditions such as pH, ionic strength, ingredient interactions, temperature, and mechanical agitation (Section 7.5.3). Quantitative predictions of the rate at which coalescence proceeds in real emulsions are extremely difficult because of the complex nature of the processes involved, the difficulty in developing theories to describe these processes, and the lack of experimental techniques to provide the information required to test these theories. The

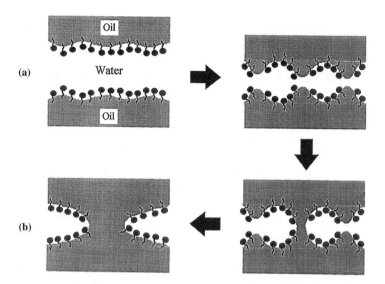

FIGURE 7.17   Coalescence of droplets depends on (a) film thinning and (b) film rupture. (a) As two droplets approach each other, the liquid between them gets increasingly thinner and may continue to thin or may reach some equilibrium value. (b) Two droplets that are in contact may spontaneously merge because of the thermal fluctuation of their membranes, leading to hole formation.

most significant progress in this area has been made in understanding the origins of hole formation and coalescence in emulsions stabilized by small-molecule surfactants (see next section).

7.5.1.4. Coalescence in Surfactant-Stabilized Systems

The susceptibility of emulsions stabilized by small-molecule surfactants to coalescence is governed mainly by the characteristics of the interfacial membrane (i.e., the optimum curvature, interfacial tension, and interfacial rheology) (Section 4.5). Two different physical mechanisms have been proposed for coalescence in surfactant-stabilized emulsions (Evans and Wennerstrom 1994):

1. The surface of an emulsion droplet stabilized by a small-molecule surfactant is highly dynamic and its shape is constantly changing because of its thermal energy. As a consequence of these thermal fluctuations, bare patches which are not protected by surfactant may be temporarily formed. If two of these patches occur simultaneously at a point where the droplets are in contact, then a hole may be formed through which the dispersed phase can flow (Figure 7.17). This mechanism is likely to be important in systems which have very low interfacial tensions and in which the membrane has a low resistance to bending and dilation.

2. The size of the droplets in emulsions is usually orders of magnitude greater than the size of the surfactant molecules, and therefore the interface can be considered to be approximately planar, that is, to have zero curvature ($H = 0$). As a consequence, the surfactant membrane may not be at its optimum curvature (H_0). The formation of a hole across the layer of continuous phase which separates the droplets depends on the development of a highly curved edge (Figure 7.18). If the curvature of the edge is close to the optimum curvature of the surfactant membrane ($H_0 \approx H_{edge}$), then the formation of a hole is energetically favorable. On the other

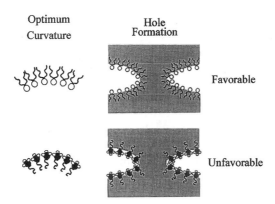

FIGURE 7.18 Coalescence may occur when the optimum curvature of a surfactant is similar to that of the edge of the hole.

hand, if the optimum curvature of the surfactant is opposite to that of the edge, then the formation of a hole is energetically unfavorable. The dependence of hole formation on the optimum curvature of a membrane means that coalescence is related to the molecular geometry of the surfactant molecules (Section 4.5). The relationship between the molecular geometry and coalescence stability of oil-in-water and water-in-oil emulsions is highlighted in Figure 7.18. As with the other coalescence mechanism described above, hole formation depends on thermal fluctuations of the interfacial membrane so that it is able to change its curvature. The coalescence rate will therefore increase as the interfacial tension and rigidity of the membrane decrease.

An additional factor which contributes to the stability of surfactant-stabilized emulsions in which collision-induced coalescence is important is the *Gibbs–Marangoni* effect (Walstra 1993b, 1996a,b). As two droplets approach each other, the liquid in the continuous phase is forced out of the narrow gap which separates them. As the liquid is squeezed out, it drags some of the surfactant molecules along the droplet surface, which leads to the formation of a region where the surfactant concentration on the surfaces of the two emulsion droplets is lowered (Figure 7.19). This causes a surface tension gradient at the interface, which is energetically unfavorable. The surfactant molecules therefore have a tendency to flow toward the region of low surfactant concentration and high interfacial tension, dragging some of the liquid in the surrounding continuous phase along with them. This motion of the continuous

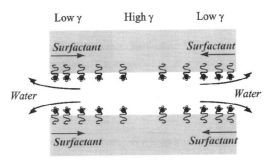

FIGURE 7.19 Gibbs–Marangoni effect.

phase is in the opposite direction of the outward flow that occurs when it is squeezed from between the droplets, and therefore it opposes the movement of the droplets toward each other and therefore increases their stability to coalescence.

7.5.2. Methods of Controlling Coalescence

The rate at which coalescence occurs is strongly dependent on the structure and dynamics of the interfacial membranes which surround the droplets. As a consequence, the most appropriate method of controlling coalescence is highly dependent on the type of emulsifier used to stabilize the system, as well as the prevailing environmental conditions. Even so, it is possible to give some general advice about the most effective methods of avoiding coalescence. These methods can be conveniently divided into two categories: those which prevent extended close contact and those which prevent membrane disruption.

7.5.2.1. Prevention of Extended Close Contact

The coalescence rate can be decreased by reducing the length of time that droplets are in close contact. The droplet contact time can be reduced in a number of different ways, including decreasing their collision frequency, ensuring that they do not flocculate, and preventing them from forming a concentrated layer at the top (or bottom) of an emulsion due to creaming (or sedimentation).

Even when the droplets are in contact with each other for extended periods (e.g., when they are in a floc or in a creamed layer), they can still be prevented from coalescing by ensuring that they do not get too close. The probability of coalescence occurring increases as the thickness of the layer of continuous phase separating them decreases, because the thermal fluctuations in the membrane shape may then become large enough to form a hole which extends from one droplet to another. The thickness of this layer is determined principally by the magnitude and range of the various attractive and repulsive forces that act between the droplets (Chapter 3). The coalescence stability can therefore be enhanced by ensuring there is a sufficiently large repulsive interaction to prevent the droplets from coming into close contact. This can be achieved in a number of ways, including varying the emulsifier type, pH, ionic strength, or temperature.

7.5.2.2. Prevention of Membrane Disruption

The formation of a hole in an interfacial membrane depends on the fluctuations in its shape caused by thermal energy or applied mechanical forces (Evans and Wennerstrom 1994). The magnitude of these fluctuations is governed by the interfacial tension and rheology of the membrane (Israelachvili 1992, Evans and Wennerstrom 1994). The lower the interfacial tension, the greater the magnitude of the fluctuations and therefore the greater the likelihood a hole will form. The higher the viscoelasticity, the greater the short-range steric repulsion between the droplets and the greater the damping of the membrane fluctuations. Consequently, coalescence becomes less likely as the interfacial tension or the viscoelasticity of the membrane increases (Dickinson 1992). The thickness of the droplet membranes is also likely to influence the coalescence stability of emulsions. Droplets with thicker membranes would be expected to provide the greatest stability because they are less likely to be ruptured and they provide a greater steric repulsion between the droplets. For small-molecule surfactants, it is important to select an emulsifier that has an optimum curvature which does not favor coalescence (Section 7.5.1.4).

The likelihood of a hole forming in a given region within a membrane increases as its area increases. Consequently, the larger the contact area between two droplets, the greater the coalescence rate. The coalescence rate therefore increases as the size of the droplets in an emulsion increases, or when the droplets become flattened against one another. Droplet flattening occurs in highly concentrated emulsions, in creamed layers, or when emulsions are subjected to centrifugal forces (Dickinson and Stainsby 1982, Walstra 1996a). Large droplets are more prone to flattening than smaller droplets because the interfacial forces which tend to keep them in a spherical shape are lower (Chapter 6). Large droplets are also more prone to collision-induced coalescence because the impact forces generated during a collision are greater and the magnitude of the attractive forces between the droplets is larger. On the other hand, increasing the size of the droplets decreases the frequency of the encounters between droplets, which may be the dominant effect in emulsions where the rate-limiting step is the collision frequency.

7.5.3.   Influence of Emulsifier Type and Environmental Conditions

Protein emulsifiers have been found to be extremely effective at providing protection against coalescence (Dickinson 1992). The main reason for this is that proteins are capable of producing emulsions with small droplet sizes, they provide strong repulsive forces between droplets (due to a combination of electrostatic and steric interactions), the interfacial tension is relatively high, and they form membranes which are highly viscoelastic. Partially hydro-lyzed proteins are less effective at preventing coalescence because they tend to form thinner, less viscoelastic interfacial layers that are easier to rupture.

Coalescence of emulsions stabilized by small-molecule surfactants is largely governed by their ability to keep droplets apart, rather than the resistance of the droplet membrane to rupture. Nonionic surfactants, such as the Tweens, do this by having polymeric hydrophilic head groups that provide a large steric overlap and hydration repulsion (McClements and Dungan 1997). Ionic surfactants, such as SDS and fatty acids, achieve this mainly through electrostatic repulsion (Section 3.4). Nevertheless, this electrostatic repulsion is only appre-ciable when the aqueous phase has a low ionic strength. At high ionic strengths, the electro-static repulsion is screened by the counterions, so that the droplets come close together, and are prone to coalescence, because the membrane is easily disrupted due to its low interfacial viscosity and interfacial tension.

Food manufacturers often need to create emulsions with extended shelf lives, and so it is important for them to understand the influence of various types of processing and storage conditions on droplet coalescence. Under quiescent conditions, emulsions stabilized by milk proteins are fairly stable to coalescence, provided the droplets are completely liquid (Das and Kinsella 1990). However, droplet coalescence may occur when the droplets are subjected to shear forces or brought into close contact for extended periods (e.g., in a creamed layer or concentrated emulsion) (Dickinson and Williams 1994). The presence of small amounts of low-molecular-weight surfactants in the aqueous phase greatly enhances the tendency of emulsion droplets to coalesce during shearing (Chen et al. 1993, Dickinson et al. 1993b, Lips et al. 1993). Emulsion instability probably occurs because small-molecule surfactants are adsorbed to the interface and increase the mobility of the proteins, therefore increasing the ease with which the interfacial layer is ruptured (Dickinson et al. 1993b). The stability of emulsions to shear forces therefore depends on the structure and properties of the adsorbed interfacial layer. Further study is needed to increase our understanding of the relationship between the physicochemical properties of emulsions during shear and the properties of the adsorbed protein.

When an oil-in-water emulsion is frozen, only part of the water is initially crystallized and the oil droplets are forced into the remaining liquid region (Sherman 1968, Berger 1976, Dickinson and Stainsby 1982). The ionic strength of this region is increased significantly because of the concentration of salts and other components. The combination of forcing the droplets into a more confined space and altering the solvent conditions is often sufficient to disrupt the droplet membranes and promote coalescence once an emulsion is thawed (Berger 1976). In addition, the oil droplets may crystallize during the freezing process, which can lead to emulsion instability through partial coalescence (Section 7.6). The development of freeze–thaw stable emulsions is a major challenge to food scientists.

Coalescence may also be promoted when an emulsion is dried into a powder, because drying may disrupt the integrity of the interfacial layer surrounding the droplets (Young et al. 1993a), which leads to coalescence once the emulsion is reconstituted. The coalescence stability can often be improved by adding relatively high concentrations of protein or carbohydrates to the system prior to drying (Young et al. 1993a,b). These molecules form a thick interfacial membrane around the droplets which is less prone to disruption during the dehydration process.

The centrifugation of an emulsion may also lead to extensive coalescence because the droplets are forced together into a compact droplet-rich layer with sufficient force to flatten the droplets and disrupt the membranes (Dickinson and Stainsby 1982).

The coalescence of droplets in an emulsion may also be influenced by various chemical or biochemical changes that occur over time. Lipid oxidation leads to the development of surface-active reaction products which may be capable of displacing emulsifier molecules from the droplet surface and thereby promoting coalescence (Coupland and McClements 1996). Extensive enzymatic hydrolysis of proteins or polysaccharides may cause an interfacial layer to be disrupted, again promoting droplet coalescence. An understanding of the various chemical and biochemical factors that determine coalescence is therefore essential in creating emulsions with extended shelf lives or that can be broken down under specific conditions.

7.5.4. Experimental Measurement of Droplet Coalescence

Experimental characterization of coalescence can be carried out using a variety of experimental techniques, many of which are similar to those used to monitor flocculation (Section 7.4.3). The most direct approach is to observe droplet coalescence using an optical microscope (Mikula 1992). An emulsion is placed on a microscope slide and the change in the droplet size distribution is measured as a function of time, by counting the individual droplets manually or by using a computer with image-processing software. It is possible to observe individual coalescence events using a sufficiently rapid camera, but these events are often so improbable in food emulsions that they are difficult to follow directly (Dickinson 1992, Walstra 1996a).

An alternative microscopic method involves the observation of the coalescence of single emulsion droplets at a planar oil–water interface (Dickinson et al. 1988). An oil droplet is released from a capillary tube into an aqueous phase and moves to the oil–water interface due to gravity (Figure 7.20). The time taken for the droplet to merge with the interface after it has arrived there is measured. The results from this type of experiment demonstrate that there are two stages to droplet coalescence: (1) a lag phase corresponding to film thinning, where the droplet remains at the interface but no coalescence occurs, and (2) a coalescence phase where the membrane spontaneously ruptures and the droplets merge with the bulk liquid. Droplets exhibit a spectrum of coalescence times because membrane rupture is a chance process. Consequently, there is an approximately exponential decrease in the number of uncoalesced droplets remaining at the interface with time after the lag phase. The major disadvantages of this technique are that only droplets above about 1 μm can be observed and coalescence often occurs so slowly that it is impractical to monitor it continuously using a microscope. To detect coalescence over a reasonably short period, it is necessary to have relatively low concentrations of emulsifier at the surfaces of the droplets, which is unrealistic because the droplets in food emulsions are nearly always saturated with emulsifier.

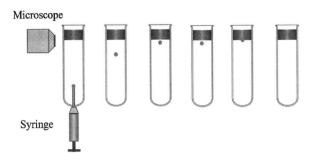

FIGURE 7.20 Microscopic technique for monitoring coalescence.

Droplet coalescence can also be monitored by measuring the time dependence of the droplet size distribution using an instrumental particle-sizing technique, such as light scattering, electrical pulse counting, or ultrasonics (Chapter 10). These techniques are fully automated and provide a measurement of the size of a large number of droplets in only a few minutes. Nevertheless, it is important to establish whether the increase in droplet size is due to coalescence, flocculation, or Ostwald ripening. Coalescence can be distinguished from flocculation by measuring the droplet size distribution in an emulsion, then changing the environmental conditions so that any flocs are broken down, and remeasuring the droplet size distribution (Section 7.4.3). If no flocs are present, the average droplet size remains constant, but if there are flocs present, it decreases. Coalescence is more difficult to distinguish from Ostwald ripening because they both involve an increase in the average size of the individual droplets with time.

As mentioned earlier, studies of coalescence can take a considerable time to complete because of the very slow rate of the coalescence process. Coalescence studies can be accelerated by applying a centrifugal force to an emulsion so that the droplets are forced together: the more resistant the membrane is to disruption, the higher the centrifugation force it can tolerate or the longer the time it will last at a particular speed before membrane disruption is observed (Sherman 1995). Alternatively, coalescence can be accelerated by subjecting the emulsions to high shear forces and measuring the shear rate at which coalescence is first observed or the length of time that the emulsion must be sheared at a constant shear rate before coalescence is observed (Dickinson and Williams 1994). Nevertheless, these accelerated coalescence tests may not always give a good indication of the long-term stability of an emulsion. For example, chemical or biochemical changes may occur in an emulsion which is stored for a long period that eventually lead to coalescence, but they may not be detected in an accelerated coalescence test. Alternatively, there may a critical force that is required to cause membrane rupture which is exceeded in a centrifuge or shearing device, but which would never be exceeded under normal storage conditions. As a consequence, these accelerated coalescence tests must be used with caution.

7.6. PARTIAL COALESCENCE

Partial coalescence occurs when two or more partly crystalline oil droplets come into contact and form an irregularly shaped aggregate (Figure 7.21). The aggregate partly retains the shape of the droplets from which it was formed because the fat crystal network within the droplets prevents them from completely merging together (Mulder and Walstra 1974, Boode 1992, Dickinson and McClements 1995, Walstra 1996a). Partial coalescence is particularly important in dairy products, because milk fat globules are partly crystalline over a fairly wide range of temperatures (Mulder and Walstra 1974, Walstra and van Beresteyn 1975). The applica-

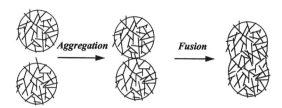

FIGURE 7.21 Partial coalescence occurs between two droplets that are partly crystalline when a crystal from one droplet penetrates into the liquid portion of another droplet. With time, the droplets merge together.

tion of shear forces or temperature cycling to cream containing partly crystalline milk fat globules can cause partial coalescence, which leads to a marked increase in viscosity ("thickening") and subsequent phase separation (Van Boekel and Walstra 1981, Boode et al. 1991, Boode 1992). Partial coalescence is an essential process in the production of ice cream, whipped toppings, butter, and margarine (Dickinson and Stainsby 1982; Goff et al. 1987; Barford and Krog 1987; Barford et al. 1987, 1991; Moran 1994). Oil-in-water emulsions are cooled to a temperature where the droplets are partly crystalline and a shear force is applied which leads to droplet aggregation via partial coalescence (Mulder and Walstra 1974). In butter and margarine, aggregation results in phase inversion (Moran 1994), whereas in ice cream and whipped cream, the aggregated fat droplets form a network that surrounds the air cells and provides the necessary mechanical strength required to produce good stability and texture (Barford et al. 1987, Goff 1993).

7.6.1. Physical Basis of Partial Coalescence

Partial coalescence is initiated when a solid fat crystal from one droplet penetrates into the liquid oil portion of another droplet (Boode 1992; Boode and Walstra 1993a,b; Boode et al. 1993; Walstra 1996a). Normally, the crystal would be surrounded by the aqueous phase, but when it penetrates into another droplet, it is surrounded by liquid oil. This causes the droplets to remain aggregated because it is energetically more favorable for a fat crystal to be surrounded by oil molecules than by water molecules (i.e., the fat crystal is wetted better by liquid oil than by water). Over time, the droplets merge more closely together because this reduces the surface area of oil exposed to water (Figure 7.21).

Partial coalescence may occur immediately after two droplets come into contact with each other, or it may occur after the droplets have been in contact for an extended period (Boode 1992). It is affected by many of the same factors which influence normal coalescence, including contact time, collision frequency, droplet separation, colloidal and hydrodynamic interactions, interfacial tension, and membrane viscoelasticity (Section 7.5.1). Nevertheless, there are also a number of additional factors which are unique to partial coalescence, the most important being the fact that the oil phase is crystalline (Boode and Walstra 1993a,b; Walstra 1996a).

The *solid fat content* (ϕ_{SFC}) is the percentage of fat that is crystalline at a particular temperature, varying from 0% for a completely liquid oil to 100% for a completely solid fat. Partial coalescence only occurs in emulsions that contain partly crystalline droplets, because a solid fat crystal from one droplet must penetrate into the liquid oil region of another droplet (Boode and Walstra 1993a,b). If the droplets were completely liquid, they would undergo normal coalescence, and if they were completely solid, they would undergo flocculation rather than partial coalescence because the droplets are not able to merge together. Increasing the ϕ_{SFC} from 0% causes an initial increase in the partial coalescence rate until a maximum

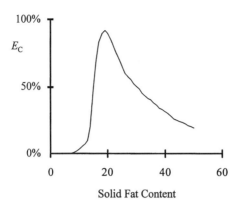

FIGURE 7.22   Schematic diagram of the influence of SFC on the rate of partial coalescence in emulsions.

value is reached, after which the partial coalescence rate decreases (Boode and Walstra 1993a,b). Consequently, there is a certain ϕ_{SFC} at which the maximum rate of partial coalescence occurs (Figure 7.22). The solid fat content at which this maximum rate occurs depends on the morphology and location of the crystals within the droplets, as well as the magnitude of the applied shear field (Boode 1992, Boode and Walstra 1993a,b).

The dimensions and location of the fat crystals within an emulsion droplet play an important role in determining its susceptibility to partial coalescence (Campbell 1991, Darling 1982). The further a fat crystal protrudes into the aqueous phase, the more likely it is to penetrate another droplet and therefore cause partial coalescence (Darling 1982, Walstra 1996a). The different types of partially crystalline fat globule commonly observed in milk are represented in Figure 7.23 (Walstra 1967). Milk fat usually crystallizes as small platelets that appear needle-shaped when observed by polarized light microscopy (Walstra 1967, Boode 1992). These crystals may be evenly distributed within the interior of the droplet (type N), located exclusively at the oil–water interface (type L), or a combination of the two (type M). The type of crystals formed depends on whether the nucleation is homogeneous, surface heterogenous, or volume heterogeneous, as well as the cooling conditions (Dickinson and McClements 1995).

It is possible to alter the distribution of fat crystals within an emulsion droplet after they have been formed by carefully cycling the temperature (Boode et al. 1991, Boode 1992). When milk is cooled rapidly, there is a tendency for the fat crystals within the droplets to be fairly evenly distributed (type N). Thermodynamically, it is actually more favorable for these crystals to be located at the oil–water interface rather than in the interior of the droplets; however, their movement to the interface is restricted because they are trapped within the fat crystal network. If the emulsion is heated to a temperature where most (but not all) of the crystals melt, the crystal network is broken down, and the remaining crystals move to the

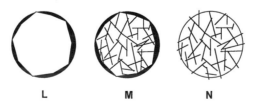

FIGURE 7.23   Typical distributions of fat crystals found in milk fat globules.

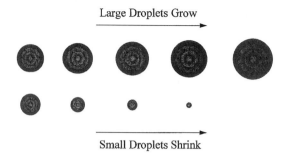

FIGURE 7.24 Ostwald ripening involves the growth of large droplets at the expense of smaller ones.

interface unhindered by the crystal network. When the emulsion is cooled, these crystals act as nucleation sites and crystal growth is restricted to the droplet surface. Thus L-type droplets are formed with large crystals located at the oil–water interface. This phenomenon is believed to be responsible for the increase in the rate of partial coalescence which occurs during temperature cycling of oil-in-water emulsions (Boode et al. 1991). The crystals at the interface in L-type droplets are larger and protrude farther into the aqueous phase and are therefore more likely to cause partial coalescence (Boode 1992).

In other types of food emulsions, there may be different crystal structures within the droplets than those shown in Figure 7.24. The crystal structure formed will depend on factors such as the chemical structure of the fat, the cooling rate, temperature cycling, the application of shear forces, droplet size distribution, the type of emulsifier used to stabilize the emulsion droplets, and the presence of any impurities that can poison or catalyze crystal growth. Further work is needed to elucidate the relationship between the morphology and location of crystals within emulsion droplets and their propensity to undergo partial coalescence.

7.6.2. Methods of Controlling Partial Coalescence

7.6.2.1. Prevention of Extended Close Contact

Partial coalescence is more likely to occur in emulsions that contain droplets which remain in contact for extended periods (i.e., flocculated emulsions, concentrated emulsions, and creamed layers). For example, experiments have shown that partial coalescence occurs more rapidly when oil droplets are located in a creamed layer than when they are freely suspended in the aqueous phase (Boode 1992). This is because the contact time is greater and the interdroplet separation is reduced. In emulsions that contain freely suspended droplets, the rate of partial coalescence is proportional to the frequency of encounters between emulsion droplets. Thus anything which increases the collision frequency will increase the rate of partial coalescence (provided it does not also reduce the collision efficiency). The precise nature of these effects depends on whether the droplet collisions are induced by Brownian motion, shear, or gravity (Van Boekel and Walstra 1981, Boode et al. 1991). For example, many emulsions are stable to partial coalescence at rest, but are prone to "thickening" when sheared.

Partial coalescence can only occur when droplets get so close together that a fat crystal from one droplet protrudes into another droplet. Thus any droplet–droplet interaction which prevents droplets from coming into close contact should decrease the rate of partial coalescence (e.g., electrostatic, polymeric steric overlap, hydration, and hydrodynamic repulsion). On the other hand, any droplet–droplet interaction which causes the droplets to come into close contact should increase the rate of partial coalescence (e.g., van der Waals, hydrophobic, depletion, mechanical, centrifugational, and gravitational forces).

The size of the droplets in an emulsion affects partial coalescence in a number of ways. The efficiency of partial coalescence can increase with droplet size because there is a greater probability of there being a crystal present in the contact zone between the droplets (Boode 1992). On the other hand, increasing the droplet size decreases their collision frequency, which can lead to a decrease in partial coalescence with increasing size (McClements et al. 1994). The influence of droplet size is therefore quite complex and depends on the rate-limiting step in the partial coalescence process.

7.6.2.2. Prevention of Membrane Disruption

Emulsifiers that form thicker and more viscoelastic films at the oil–water interface are more resistant to penetration by fat crystals and so provide greater stability to partial coalescence (Van Boekel and Walstra 1981, Boode et al. 1991, Boode 1992). Consequently, the rate of partial coalescence is less for droplets stabilized by proteins than for those stabilized by small-molecule surfactants (Dickinson and McClements 1995). The chemical structure of the emulsifier molecules at the interface of the emulsion droplets has an important impact on the stability of a number of foods. The presence of phospholipids in dairy emulsions has been observed to decrease their stability to thickening under the influence of shear forces, which has been attributed to the displacement of protein molecules from the oil–water interface by the more surface-active phospholipids, leading to the formation of an emulsifier film which is more susceptible to penetration by fat crystals (Boode 1992). When an ice cream mix or dairy whipped topping is aged in the presence of small-molecule surfactants prior to cooling and shearing, the resulting product has improved texture and better stability (Goff et al. 1987; Barford and Krog 1987; Barford et al. 1987, 1991). This is because emulsifiers displace the milk proteins from the oil–water interface and so enhance the tendency for partial coalescence to occur during subsequent cooling and shearing. For these reasons, small-molecule surfactants are often added to ice creams and dairy toppings to improve their physical characteristics. Van Boekel and Walstra (1981) calculated that only about one in a million encounters between droplets leads to partial coalescence. This suggests that the rate-limiting step of partial coalescence is the penetration of the emulsifier film by a crystal and subsequent nucleation, rather than the collision frequency.

7.6.2.3. Control of Crystal Concentration, Structure, and Location

The degree of partial coalescence in an emulsion can be regulated by controlling the concentration, structure, and location of the fat crystals within the droplets. The most effective method of preventing partial coalescence is to ensure that all the droplets are either completely liquid or completely solid. This can be achieved by carefully controlling the temperature of the emulsion so that it is above or below some critical level. If it is not possible to alter the temperature, then the solid fat content could be controlled by selecting a fat with a different melting profile (Moran 1994). The susceptibility of an emulsion to partial coalescence also depends on the morphology and the precise location of the crystals within the emulsion droplets. Thus two emulsions with the same solid fat content may have very different stabilities because of differences in crystal structure. The structure of the crystals can be controlled by varying the cooling rate at which they are formed or by selecting an appropriate fat source.

7.6.3. Experimental Measurement of Partial Coalescence

A variety of experimental techniques have been used to monitor the susceptibility of food emulsions to partial coalescence (Dickinson and McClements 1995). The physical state of fats is monitored using experimental techniques that utilize differences in the physicochemical properties of the solid and liquid phases (e.g., density, compressibility, birefringence, molecular mobility, or packing) or changes associated with the solid–liquid phase transition (e.g., absorption or release of heat). The physicochemical properties which are of most interest to food scientists are (1) the final melting point of the fat; (2) the variation of the solid fat content with temperature; (3) the morphology, interactions, and location of the crystals within the droplets; (4) the packing of the fat molecules within the crystals; and (5) the influence of droplet crystallization on the overall stability and rheology of the emulsion.

The variation of the solid fat content with temperature can be measured using a variety of techniques, including density measurements, differential scanning calorimetry, nuclear magnetic resonance, ultrasonic velocity measurements, and electron spin resonance (Dickinson and McClements 1995). The technique used in a particular experiment depends on the equipment available, the information required, and the nature of the sample being tested. The position of fat crystals relative to an oil–water interface depends on the relative magnitude of the oil–water, oil–crystal, and crystal–water interfacial tensions and is characterized by the contact angle (Darling 1982, Campbell 1991). Darling (1982) has described a technique which can be used to measure the contact angle between liquid oil, solid fat, and aqueous phases.

Two fats can have exactly the same solid fat content but very different physical characteristics because of differences in the crystal habit and spatial distribution of the crystals. The location of crystals within a system and their crystal habit can be studied by polarized light microscopy (Boode 1992, Kellens et al. 1992) or electron microscopy (Soderberg et al. 1989), depending on the size of the crystals. The packing of the molecules in the crystals can be determined by techniques which utilize the scattering or adsorption of radiation. X-ray diffraction and small-angle neutron scattering have been used to determine the long and short spacings of the molecules in fat crystals (Hernqvist 1990, Cebula et al. 1992). Infrared and Raman spectroscopy have been used to obtain information about molecular packing via its effect on the vibration of certain chemical groups in fat molecules (Chapman 1965). Each polymorphic form has a unique spectra which can be used to identify it. The polymorphic form of fat crystals can also be identified by measuring the temperature at which phase transitions occur and the amount of heat absorbed/released using differential scanning calorimetry (McClements et al. 1993a,b).

Partial coalescence leads to an increase in the size of the particles in an emulsion, which can be followed by microscopic, light-scattering, electrical pulse counting, or ultrasonic techniques (Chapter 10). Ultimately, a food scientist is interested in the influence of droplet crystallization on the bulk physicochemical properties of a food emulsion, such as its appearance, stability, and texture. Partial coalescence causes an increase in emulsion viscosity and may eventually lead to the formation of a three-dimensional network of aggregated droplets, so that it can be monitored by measuring the increase in viscosity or shear modulus, as a function of either time or temperature (Boode 1992, Boode and Walstra 1993a,b). The stability of emulsion droplets to creaming or sedimentation is also influenced by partial coalescence, which can be followed by the techniques described in Section 7.3.3.

7.7. OSTWALD RIPENING

Ostwald ripening is the process whereby large droplets grow at the expense of smaller ones because of mass transport of dispersed phase from one droplet to another through the intervening continuous phase (Figure 7.24) (Kabalnov and Shchukin 1992, Taylor 1995). It is negligible in most food emulsions because the mutual solubilities of triacylglycerols and water are so low that the mass transport rate is insignificant (Dickinson and Stainsby 1982). Nevertheless, it may be important in oil-in-water emulsions which contain more water-soluble lipids (e.g., flavor oils or when the aqueous phase contains alcohol (e.g., cream liqueurs). In this type of system, the food manufacturer may have to consider methods for retarding the rate of Ostwald ripening.

7.7.1. Physical Basis of Ostwald Ripening

Ostwald ripening occurs because the solubility of the material in a spherical droplet increases as the size of the droplet decreases (Dickinson 1992):

$$S(r) = S(\infty) \exp\left(\frac{2\gamma V_m}{RTr}\right) \qquad (7.27)$$

Here, V_m is the molar volume of the solute, γ is the interfacial tension, $S(\infty)$ is the solubility of the solute in the continuous phase for a droplet with infinite curvature (a planar interface), and $S(r)$ is the solubility of the solute when contained in a spherical droplet of radius r. The increase in solubility with decreasing droplet size means that there is a higher concentration of solubilized material around a small droplet than around a larger one. The solubilized molecules therefore move from the smaller droplets to the larger droplets because of this concentration gradient. Once steady state has been achieved, the rate of Ostwald ripening is given by (Kabalnov and Shchukin 1992):

$$\frac{d\langle r\rangle^3}{dt} = \frac{8\gamma V_m S(\infty) D}{9RT} \qquad (7.28)$$

where D is the translation diffusion coefficient of the solute through the continuous phase. This equation indicates that the change in droplet size with time becomes more rapid as the equilibrium solubility of the molecules in the continuous phase increases.

Equation 7.28 assumes that the emulsion is dilute, whereas most food emulsions are concentrated, and so the growth or shrinkage of a droplet cannot be considered to be independent of its neighbors. Thus the growth rate must be modified by a correction factor that takes into account the spatial distribution of the droplets (Kabalnov and Shchukin 1992). Equation 7.28 also assumes that the rate-limiting step is the diffusion of the solute molecules through the continuous phase. Most food emulsions contain droplets that are surrounded by interfacial membranes which may retard the diffusion of solute molecules, and in this case the above equation must be modified (Kabalnov and Shchukin 1992):

$$\frac{d\langle r\rangle^3}{dt} = \frac{3}{4\pi}\left(\frac{S_m - S_c}{R_m + R_c}\right) \qquad (7.29)$$

where R_m and R_c are the diffusion resistances of the membrane and the continuous phase, respectively:

$$R_m = \frac{1}{4\pi r D_m} \qquad R_c = \frac{\delta C_{m,\infty}}{4\pi r^2 D_c C_{c,\infty}} \qquad (7.30)$$

Here, δ is the thickness of the droplet membrane, $C_{i,\infty}$ is the solubility of the solute in the specified phase, and the subscripts m and c refer to the properties of the membrane and the continuous phase, respectively. When the diffusion of the solute molecules through the interfacial membrane is limiting, the growth rate of the droplet size is proportional to r^2 rather than r^3 (Kabalnov and Shchukin 1992).

7.7.2. Methods of Controlling Ostwald Ripening

There are a number of ways in which the rate of Ostwald ripening in emulsions can be controlled.

7.7.2.1. Droplet Size Distribution

The rate of Ostwald ripening increases as the average size of the droplets in an emulsion decreases because the solubility of the dispersed phase increases with decreasing radius. The initial rate also increases as the width of the particle size distribution increases (Kabalnov and Shchukin 1992). Ostwald ripening can therefore be retarded by ensuring that an emulsion has a narrow droplet size distribution and that the droplets are fairly big. Nevertheless, there may be other problems associated with having relatively large droplets in an emulsion, such as accelerated creaming, flocculation, or coalescence.

7.7.2.2. Solubility

The greater the equilibrium solubility of the dispersed phase in the continuous phase, the faster the rate of Ostwald ripening (Equation 7.28). Ostwald ripening is therefore extremely slow in oil-in-water emulsions that contain lipids which are sparingly soluble in water, such as long-chain triacylglycerols, but increases as the lipid molecules become smaller and more polar (Dickinson and Stainsby 1988). Certain substances are capable of increasing the water solubility of lipids in water and are therefore able to enhance the rate of Ostwald ripening (e.g., alcohols or surfactant micelles) (McClements et al. 1993b, Weiss et al. 1996, Coupland et al. 1997). Ostwald ripening could therefore be retarded by excluding these substances from the emulsion or by using lipids with a low water solubility.

7.7.2.3. Interfacial Membrane

The rate of Ostwald ripening increases as the interfacial tension increases (Equation 7.28). Consequently, it is possible to retard its progress by using an emulsifier which is highly effective at reducing the interfacial tension (Kabalnov et al. 1995). The mass transport of molecules from one droplet to another depends on the rate at which the molecules diffuse across the interfacial membrane (Kabalnov and Shchukin 1992). It may therefore be possible to retard Ostwald ripening by decreasing the diffusion coefficient of the dispersed phase in the membrane or by increasing the thickness of the membrane. Little work has been carried out in this area; however, it may prove to be an extremely useful means of controlling the stability of some food emulsions.

7.7.2.4. Droplet Composition

The rate of Ostwald ripening is particularly sensitive to the composition of emulsion droplets which contain chemical substances with different solubilities in the continuous phase (Kabalnov and Shchukin 1992, Dickinson and McClements 1995, Arlauskas and Weers 1996). Consider an oil-in-water emulsion that contains droplets comprised of two different types of organic molecules: M_{low} has a low water solubility and M_{high} has a high water solubility. The diffusion of M_{high} molecules from the small to the large droplets occurs more rapidly than the M_{low} molecules. Consequently, there is a greater percentage of M_{high} in the larger droplets than in the smaller droplets. Differences in the composition of emulsion droplets are thermodynamically unfavorable because of the entropy of mixing: it is entropically more favorable to have the two oils distributed evenly throughout all of the droplets rather than concentrated in particular droplets. Consequently, there is a thermodynamic driving force that operates in opposition to the Ostwald ripening effect. After a certain time, the driving force for Ostwald ripening (differences in droplet size) is exactly compensated for by the driving force for the entropy of mixing (differences in droplet composition), and so the size and composition of the droplets remain constant (Kabalnov and Shchukin 1992). It is therefore possible to control the rate of Ostwald ripening in oil-in-water emulsions by using an oil phase which contains a mixture of lipids with different water solubilities.

7.7.3. Experimental Measurement of Ostwald Ripening

Methods of monitoring Ostwald ripening are fairly similar to those used to monitor droplet coalescence (i.e., techniques that measure changes in droplet size distribution with time) (Section 7.5.4). If the droplets are sufficiently large (>1 μm), then optical microscopy can be used (Kabalnov and Shchukin 1992); otherwise, instrumental particle-sizing techniques, such as light scattering, electrical conductivity, or ultrasonics, can be used. Nevertheless, it is often difficult to directly distinguish between coalescence and Ostwald ripening using these particle-sizing techniques because both instability mechanisms lead to an increase in the average size of the droplets over time.

7.8. PHASE INVERSION

Phase inversion is the process whereby a system changes from an oil-in-water emulsion to a water-in-oil emulsion or vice versa (Figure 7.25). Phase inversion is an essential step in the manufacture of a number of important food products, including butter and margarine (Mulder and Walstra 1974, Dickinson and Stainsby 1982, Moran 1994). In other foods, phase inversion is undesirable because it has an adverse effect on their appearance, texture, stability, and taste. In these products, a food manufacturer wants to avoid the occurrence of phase inversion.

7.8.1. Physical Basis of Phase Inversion

Phase inversion is usually triggered by some alteration in the composition or environmental conditions of an emulsion (e.g., dispersed-phase volume fraction, emulsifier type, emulsifier concentration, solvent conditions, temperature, or mechanical agitation) (Shinoda and Friberg 1986, Dickinson 1992, Campbell et al. 1996). Only certain types of emulsion are capable of

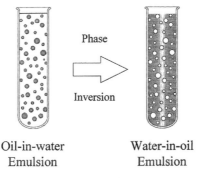

Phase

Inversion

Oil-in-water
Emulsion

Water-in-oil
Emulsion

FIGURE 7.25   Phase inversion involves the conversion of an oil-in-water emulsion to a water-in-oil emulsion or vice versa.

undergoing phase inversion, rather than being completely broken down into their component phases. These emulsions are capable of existing in a kinetically stable state after the phase inversion has taken place. It is usually necessary to agitate an emulsion during the phase inversion process; otherwise it will separate into its component phases.

The physicochemical basis of phase inversion is believed to be extremely complex, involving aspects of flocculation, coalescence, partial coalescence, and emulsion formation. At the point where phase inversion occurs, which is often referred to as the "balance point," the system may contain regions of oil-in-water emulsion, water-in-oil emulsion, multiple emulsion, and bicontinuous phases. Phase inversion in food emulsions can be conveniently divided into two different categories according to its origin.

7.8.1.1.   Surfactant-Induced Phase Inversion

This type of phase inversion occurs in emulsions which are stabilized by small-molecule surfactants and is caused by changes in the molecular geometry of the surfactant molecules (Shinoda and Friberg 1986, Salager 1988, Evans and Wennerstrom 1994). An emulsion stabilized by a nonionic surfactant undergoes a transition from an oil-in-water emulsion to a bicontinuous system to a water-in-oil emulsion on heating because of the progressive dehydration of the head groups (Lehnert et al. 1994). This process is characterized by a phase inversion temperature, which is governed by the molecular geometry of the surfactant molecules (Davis 1994b). An oil-in-water emulsion stabilized by an ionic surfactant exhibits a similar kind of behavior when the concentration of electrolyte in the aqueous phase is increased (Salager 1988, Binks 1993). Increasing the ionic strength causes the system to undergo a transition from an oil-in-water emulsion to a bicontinuous system to a water-in-oil emulsion, because the electrical charge on the surfactant head groups is progressively screened by the counterions (Section 4.5). Surfactant-induced phase inversions are reversible because when the emulsion is cooled or the ionic strength is decreased, the water-in-oil emulsion will revert back to an oil-in-water emulsion. Nevertheless, it is usually necessary to continuously agitate the system during this process, or else it will separate into the individual oil and water phases. Even though the phase inversion is reversible, there is often an effect which is similar to that of supercooling (i.e., the temperature at which the phase inversion occurs on heating is different from that on cooling) (Dickinson 1992, Vaessen and Stein 1995). This is because there is an activation energy which must be overcome before a system can be transformed from one state to another. The principal factor determining reversible phase inversion is the molecular geometry of the surfactant molecules and its dependence on environmental conditions.

7.8.1.2.   Fat Crystallization–Induced Phase Inversion

When an oil-in-water emulsion that contains completely liquid droplets is cooled to a temperature where the droplets are partly crystalline and then sheared, it undergoes a phase inversion to a water-in-oil emulsion (Mulder and Walstra 1974, Dickinson and Stainsby 1982). The principal cause of this type of phase inversion is partial coalescence of the droplets, which leads to the formation of a continuous fat crystal network which traps water droplets within it. This is one of the principal manufacturing steps in the production of margarine and butter (Dickinson and Stainsby 1982). When the emulsion is heated to a temperature where the fat crystals melt, the emulsion breaks down because the water droplets are released and sediment to the bottom of the sample, where they coalesce with other water droplets. This is clearly seen when one melts margarine or butter and then cools it back to the original temperature: the system before and after heating is very different. This type of phase inversion depends mainly on the crystallization of the fat and on the resistance of the droplets to partial coalescence (see Section 7.6.2).

7.8.2. Methods of Controlling Phase Inversion

The propensity for phase inversion to occur in an emulsion can be controlled in a number of ways.

7.8.2.1. Dispersed Volume Fraction

When the dispersed-phase volume fraction of an emulsion is increased while all the other experimental variables are kept constant (e.g., emulsifier type, emulsifier concentration, temperature, shearing rate), a critical volume fraction (ϕ_c) is reached where the system either undergoes a phase inversion or completely breaks down so that the excess dispersed phase forms a layer on top of the emulsion. There is usually a range of volume fractions over which an emulsion can exist as either a water-in-oil emulsion or an oil-in-water emulsion (Dickinson 1992). Within this region, the emulsion can be converted from one state to another by altering some external property, such as the temperature or shear rate. From geometrical packing considerations, it has been estimated that this range extends from $1 - \phi_{cp} < \phi < \phi_{cp}$, where ϕ_{cp} refers to the volume fraction when the droplets are packed closely together without being distorted. In practice, factors other than simple geometric considerations will influence this range, including the fact that the droplets can become deformed and the chemical structure of the emulsifier used. Phase inversion can therefore be prevented by ensuring that the droplet concentration is kept below ϕ_{cp}.

7.8.2.2. Emulsifier Type and Concentration

The most important factor determining the susceptibility of an emulsion to surfactant-induced phase inversion is the molecular geometry of the surfactant used to stabilize the droplets (Evans and Wennerstrom 1994). Surfactant-stabilized emulsions undergo a phase inversion when some change in the environmental conditions causes the optimum curvature of the surfactant monolayer to tend toward zero (e.g., temperature, ionic strength, or the presence of a co-surfactant) (Section 4.5). These types of surfactants usually have an intermediate HLB number or are electrically charged. Alternatively, mixtures of two different types of surfactants can be used, one to stabilize the oil-in-water emulsion and the other the water-in-oil emulsion. The point at which phase inversion occurs is then sensitive to the overall concentration and ratio of the surfactants used (Dickinson 1992). Emulsions stabilized by proteins do not exhibit this type of phase inversion because proteins are incapable of stabilizing water-in-oil emulsions.

The emulsifier type and concentration are also extremely important in determining the stability of emulsions to fat crystallization–induced phase inversion. Emulsifiers that form thick viscoelastic membranes are more likely to protect an emulsion from this type of phase inversion because they retard partial coalescence (Section 7.6). It is therefore extremely important to select an emulsifier which exhibits the appropriate behavior over the experimental conditions that a food emulsion experiences during its lifetime.

7.8.2.3. Mechanical Agitation

It is often necessary to subject an emulsion to a high shearing force to induce phase inversion in the region where it can possibly exist as either an oil-in-water or water-in-oil emulsion. The higher the shearing force, the more likely phase inversion is to occur.

7.8.2.4. Temperature

Increasing or decreasing the temperature of emulsions is one of the most important means of inducing phase inversion. The mechanism by which this process occurs depends on whether the phase inversion is induced by surfactant changes or crystallization. Cooling an oil-in-water emulsion to a temperature where the oil partly crystallizes and shearing causes fat crystallization–induced phase inversion. On the other hand, heating an oil-in-water emulsion stabilized by a surfactant may cause surfactant-stabilized phase inversion above the phase inversion temperature.

Fat crystallization–induced phase inversion is the most important type in the food industry because it is an essential step in the manufacture of butter and margarine (Dickinson and Stainsby 1982). Surfactant-induced phase inversion may be important in emulsions stabilized by nonionic surfactants that must be heated to high temperatures. In these systems, it is important for the food manufacturer to ensure that the phase inversion temperature of the emulsifier is above the highest temperature that the emulsion will experience during processing, storage, and handling.

7.8.3. Experimental Measurement of Phase Inversion

Phase inversion can be monitored using a variety of experimental techniques (Lehnert et al. 1994). When an emulsion changes from the oil-in-water to the water-in-oil emulsion type, or vice versa, there is usually a significant change in emulsion viscosity. This is because the viscosity of an emulsion is governed principally by the viscosity of the continuous phase (which is different for oil and water) and the dispersed-phase volume fraction (which may also be altered) (Chapter 8). An oil-in-water emulsion has an aqueous continuous phase and so its electrical conductivity is much greater than that of a water-in-oil emulsion (Lehnert et al. 1994, Keikens 1997). Thus there is a dramatic reduction in the electrical conductivity of an oil-in-water emulsion when phase inversion occurs. Information about the process of phase inversion may also be obtained by monitoring the emulsion under a microscope or measuring the size of the droplets using a particle-sizing technique. Measurements of the temperature dependence of the interfacial tension between the oil and water phases can also be used to predict the likelihood that phase inversion will occur in an emulsion (Shinoda and Friberg 1986, Lehnert et al. 1994).

7.9. CHEMICAL AND BIOCHEMICAL STABILITY

The majority of this chapter has been concerned with the physical instability of food emulsions, rather than with their chemical instability. This is largely because emulsion scientists have historically focused mainly on the physical aspects of food emulsions. Nevertheless, there are many types of chemical or biochemical reactions which can have adverse effects on the quality of food emulsions (Fennema 1996a). For this reason, there has been a growing interest in the influence of various chemical and biochemical reactions on the stability of food emulsions in recent years.

One of the most common forms of instability in foods which contain fats is lipid oxidation (Nawar 1996). Lipid oxidation leads to the development of undesirable "off-flavors" (rancidity) and potentially toxic reaction products. In addition, it may also promote the physical instability of some emulsions (Coupland and McClements 1996). For example, many of the reaction products generated during lipid oxidation are surface active and may therefore be able to interact with the interfacial membrane surrounding the droplets in such a way as to lead to droplet coalescence. The importance of lipid oxidation in food emulsions has led to a considerable amount of research being carried out in this area over the past few years (Frankel 1991; Roozen et al. 1994a,b; Coupland and McClements 1996; Coupland et al. 1996; Huang et al. 1994, 1996a,b,c; Frankel et al. 1994; Frankel 1996; Mei et al. 1998). The main emphasis of this work is to develop effective strategies for retarding lipid oxidation in emulsions by incorporating antioxidants, controlling storage conditions, or engineering droplet interfacial properties.

There is also an increasing interest in the influence of biochemical reactions on the properties of food emulsions (Dalgleish 1996a). A number of studies have recently been carried out to determine the influence of certain enzymes on the stability and physicochemical properties of food emulsions (Agboola and Dalgleish 1996b,c). These studies have shown that the properties of food emulsions can be altered appreciably when the adsorbed proteins are cleaved by enzyme hydrolysis. A number of studies have also shown that globular proteins become denatured after they have been adsorbed to the surface of an emulsion droplet and that they remain in this state after they are desorbed (Corredig and Dalgleish 1995, de Roos and Walstra 1996). This has important implications for the action of many enzymes in food emulsions. The activity of an enzyme may be completely lost when it is adsorbed to the surface of the droplets in an emulsion, and therefore any biochemical reactions catalyzed by it will cease (de Roos and Walstra 1996).

Given their obvious importance for food quality, it seems likely that there will be an increasing emphasis on the influence of biochemical and chemical reactions on the stability of food emulsions in the future.

8 Emulsion Rheology

8.1. INTRODUCTION

The application of a force to a material causes it to become deformed and/or to flow (Whorlow 1992, Macosko 1994). The extent of the deformation and flow depends on the physicochemical properties of the material. *Rheology* is the science which is concerned with the relationship between applied forces and the deformation and flow of matter (Macosko 1994). Most rheological tests involve the application of a force to a material and a measurement of the resulting flow or change in shape (Whorlow 1992).

A knowledge of the rheological properties of emulsions is important to food scientists for a number of reasons (Sherman 1970, Dickinson and Stainsby 1982, Race 1991, Shoemaker et al. 1992, Rao et al. 1995, Rao 1995). Many of the sensory attributes of food emulsions are directly related to their rheological properties (e.g., creaminess, thickness, smoothness, spreadability, pourabilty, flowability, brittleness, and hardness). A food manufacturer must therefore be able to design and produce a product which has the rheological properties expected by the consumer. The shelf life of many food emulsions depends on the rheological characteristics of the component phases (e.g., the creaming of oil droplets depends on the viscosity of the aqueous phase) (Section 7.3). Information about the rheology of food products is used by food engineers to design processing operations which depend on the way that a food behaves when it flows through a pipe, is stirred, or is packed into containers. Rheological measurements are also used by food scientists as an analytical tool to provide fundamental insights about the structural organization and interactions of the components within emulsions (e.g., measurements of viscosity versus shear rate can be used to provide information about the strength of the colloidal interactions between droplets) (Hunter 1993, Tadros 1994). In this chapter, the basic principles of rheology, the rheological characteristics of food emulsions, and the instruments available for carrying out rheological measurements are covered.

Food emulsions are compositionally and structurally complex materials which can exhibit a wide range of different rheological behaviors, ranging from low-viscosity fluids (such as milk and fruit juice beverages) to fairly hard solids (such as refrigerated margarine or butter). Food scientists aim to develop theories which can be used to describe and predict the rheological behavior of food emulsions and experimental techniques to characterize these properties. Despite the diversity of food emulsions, it is possible to characterize many of their rheological properties in terms of a few simple models: the *ideal solid,* the *ideal liquid*, and the *ideal plastic* (Sherman 1970, Tung and Paulson 1995). More complex systems can then be described by combining two or more of these simple models. In the following sections, the concepts of the ideal solid, ideal liquid, and ideal plastic are introduced, as well as some of the deviations from these models that are commonly observed in food emulsions.

8.2. RHEOLOGICAL PROPERTIES OF MATERIALS

8.2.1.   Solids

In our everyday lives, we come across solid materials which exhibit quite different rheological characteristics. Some may be soft, others hard; some may be brittle, others rubbery; some may break easily, others may not. Despite this range of different behaviors, it is possible to characterize the rheological properties of many solid foods in terms of a few simple concepts.

8.2.1.1.   Ideal Elastic Solids

An ideal elastic solid is often referred to as a *Hookean* solid, after Robert Hooke, the scientist who first described this type of behavior (Whorlow 1992, Macosko 1994, Rao et al. 1995). Hooke observed experimentally that there is a linear relationship between the deformation of a solid material and the magnitude of the force applied to it, provided the deformation is not too large (Figure 8.1). He also observed that when the force was removed from the material, it returned back to its original length. In general, Hooke found that the force per unit area (or *stress*) was proportional to the relative deformation (or *strain*). Hooke's law can therefore be summarized by the following statement:

$$\text{stress } (\tau) = \text{constant } (E) \times \text{strain } (\gamma) \tag{8.1}$$

A stress can be applied to a material in a number of different ways, including simple shear, simple compression, and bulk compression (Figure 8.2). Equation 8.1 is applicable to each of these situations, but the values of the stress, strain, and constant used depend on the nature of the deformation (Table 8.1).

The equations given in Table 8.1 assume that the material is homogeneous and isotropic (i.e., its properties are the same in all directions). To characterize the rheological constants of an ideal elastic solid, it is therefore necessary to measure the change in its dimensions when a force of known magnitude is applied.

The elastic behavior of a solid is related to the intermolecular forces which hold the molecules together. When a stress is applied to a material, the bonds between the molecules are compressed or expanded, and therefore they store energy. When the stress is removed, the bonds give up this energy and the material returns to its original shape. The elastic modulus

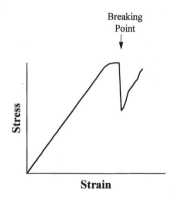

FIGURE 8.1   At small deformations, there is a linear relationship between the applied stress and the resultant strain for an ideal elastic solid. At higher deformations, the stress is no longer linearly related to strain and the material will eventually break.

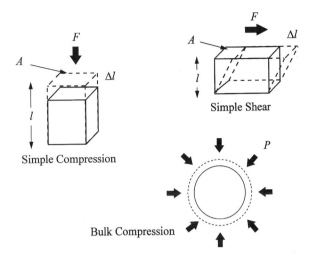

FIGURE 8.2   An elastic solid can be deformed in a number of different ways.

of an ideal elastic solid is therefore related to the strength of the interactions between the molecules within it.

8.2.1.2.   Nonideal Elastic Solids

Hooke's law is only strictly applicable to elastic materials at low strains, and therefore most fundamental rheological studies of solid foods are carried out using very small material deformations (<1%). Nevertheless, the rheological behavior of foods at large deformations is often more relevant to their actual use (e.g., spreadability, slicing, or mastication) (van Vliet 1995). For this reason, it is also important to characterize the rheological behavior of solids at large deformations. At strains just above the region where Hooke's law is obeyed, the stress is no longer proportional to the strain, and therefore an *apparent modulus* is defined, which is equal to the stress/strain at a particular value of the strain. It is therefore necessary to stipulate the strain (or stress) at which an apparent modulus of a material is measured. Even though the material does not strictly obey Hooke's law, it still returns to its original shape once the force is removed. Above a certain deformation, however, a solid may not return to its original shape after the applied stress is removed, because it either breaks or flows. A

TABLE 8.1
Rheological Parameters for Different Types of Deformation of Elastic Solids

Deformation	Stress	Strain	Elastic modulus
Simple shear	$\tau = \dfrac{F}{A}$	$\gamma = \dfrac{\Delta l}{l} = \tan \phi$	$G = \dfrac{F}{A \tan \phi}$
Simple compression	$\tau = \dfrac{F}{A}$	$\gamma = \dfrac{\Delta l}{l}$	$Y = \dfrac{Fl}{A\Delta l}$
Bulk compression	$\tau = \dfrac{F}{A} = P$	$\gamma = \dfrac{\Delta V}{V}$	$K = \dfrac{PV}{\Delta V}$

Note: G is the shear modulus, Y is Young's modulus, K is the bulk modulus, P is the pressure, ϕ is the deformation angle, and the other symbols are as defined in Figure 8.2.

material that breaks at low strains is referred to as being *brittle*, whereas a material that flows is referred to as being *plastic* or *viscoelastic* (see later). The stress at which a material breaks is referred to as the *breaking stress*, whereas the strain at which it breaks is referred to as the *breaking strain*. A material usually ruptures or flows when the forces holding the atoms or molecules in the material together are exceeded. This often begins at regions where the bonds holding the material together are relatively weak (e.g., a crack). A knowledge of the breaking stress or breaking strain required to disrupt a material is often a useful indication of its ability to be broken up during mastication, cut with a knife, or its sensitivity to rupture during storage and transport.

8.2.2. Liquids

Liquid food emulsions also exhibit a wide range of rheological properties. Some have low viscosities and flow easily, like milk, while others are very viscous, like mayonnaise. Even so, it is possible to characterize their rheological properties using a few simple concepts.

8.2.2.1. Ideal Liquids

The ideal liquid is often referred to as a *Newtonian* liquid, after Isaac Newton, the scientist who first described its behavior (Whorlow 1992, Macosko 1994, Rao 1995). When a shear stress is applied to an ideal liquid, it continues to flow as long as the stress is applied. Once the stress is removed, there is no elastic recovery of the material (i.e., it does not return to its original shape).

   The viscosity of a liquid is a measure of its resistance to flow: the higher the viscosity, the greater the resistance (Macosko 1994). The concept of viscosity can be understood by considering a liquid which is contained between two parallel plates (Figure 8.3). The bottom plate is at rest, while the top plate moves in the *x*-direction with a constant velocity (*v*). It is assumed that the liquid between the plates consists of a series of infinitesimally thin layers. The liquid layers in direct contact with the bottom and top plates are assumed to "stick" to them, so that they have velocities of 0 and *v*, respectively. The intervening liquid layers slide over each other with velocities that range between 0 and *v*, the actual value being given by $dy(dv/dy)$, where dy is the distance from the bottom plate and dv/dy is the velocity gradient between the plates. The shear stress applied to the fluid is equal to the shear force divided by the area over which it acts ($\tau = F/A$). The rate of strain is given by the change in displacement of the layers per unit time: $d\gamma/dt$ (or $\dot{\gamma}$) $= dv/dy$. For an ideal liquid, the shear stress is proportional to the rate of strain (Figure 8.4):

$$\tau = \eta\dot{\gamma} \tag{8.2}$$

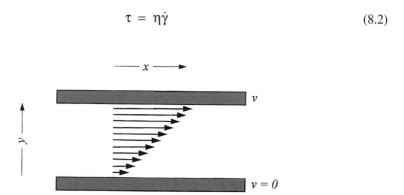

FIGURE 8.3   The viscosity of a liquid is related to the friction between the liquid layers as they slide across each other: the greater the friction, the higher the viscosity.

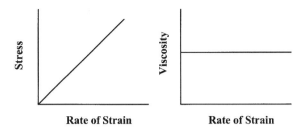

FIGURE 8.4   Stress is proportional to *rate* of strain for an ideal liquid.

where the constant of proportionality (η) is called the *viscosity*. The viscosity arises from the friction between the liquid layers as they slide past one another (Macosko 1994). The lower the viscosity of a liquid, the less resistance between the liquid layers, and therefore the smaller the force required to cause the top plate to move with a given velocity, or the faster the top plate moves when a given force is applied. The ideal viscous fluid differs from the ideal elastic solid because the shear stress is proportional to the *rate* of strain (Figure 8.3), rather than the strain (Figure 8.1).

The units of shear stress (τ) are N m^{-2} (or Pa), and those of shear rate ($\dot{\gamma}$) are s^{-1}; thus the viscosity (η) has units of N s m^{-2} (or Pa s) in the SI system. Viscosity can also be expressed in the older cgs units of poise, where 1 Pa s = 10 P. Thus the viscosity of water can be quoted as 1 mPa s, 0.001 Pa s, 0.01 P, or 1 cP, depending on the units used.

Ideally, a Newtonian liquid should be incompressible (its volume does not change when a force is applied to it), isotropic (its properties are the same in all directions), and structureless (it is homogeneous). Although many liquid foods do not strictly meet these criteria, their rheological behavior can still be described excellently by Equation 8.2 (e.g., milk). Nevertheless, there are many others that exhibit nonideal liquid behavior and so their properties cannot be described by Equation 8.2.

The type of flow depicted in Figure 8.3 occurs at low shear rates and is known as *laminar flow,* because the liquid travels in a well-defined laminar pattern. At higher shear rates, eddies form in the liquid and the flow pattern is much more complex. This type of flow is referred to as *turbulent,* and it is much more difficult to mathematically relate the shear stress to the rate of strain under these conditions. For this reason, instruments that measure the viscosity of liquids are designed to avoid turbulent flow.

8.2.2.2.   Nonideal Liquids

Nonideality may manifest itself in a number of different ways; for example, the viscosity of a liquid may depend on the *shear rate* and/or the *time* over which the shear stress is applied, or the fluid may exhibit some elastic as well as viscous properties (Macosko 1994, Tung and Paulson 1995). Plastic and viscoelastic materials, which have some elastic characteristics, are considered in later sections.

Shear-Rate-Dependent Nonideal Liquids.   In an ideal liquid, the viscosity is independent of shear rate and the length of time the liquid is sheared (i.e., the ratio of the shear stress to the shear rate does not depend on shear rate or time) (Figure 8.5). In practice, many food emulsions have viscosities which depend on the shear rate and the length of time the emulsion is sheared (Dickinson 1992). In this section, emulsions in which the viscosity depends on shear rate but is independent of the shearing time are examined (Dickinson 1992). In the following section, emulsions in which the viscosity depends on both the shear rate and shearing time are examined.

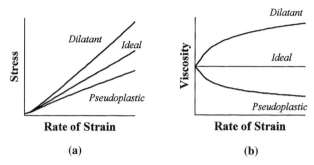

FIGURE 8.5 Comparison of the viscosity of ideal and nonideal liquids. (a) Shear stress versus shear rate. (b) Viscosity versus shear rate.

The viscosity of an emulsion may either increase or decrease as the shear rate is increased, rather than staying constant, as for a Newtonian liquid (Figure 8.5). In these systems, the viscosity at a particular shear rate is referred to as the *apparent* viscosity. The dependence of the apparent viscosity on shear rate means that it is crucial to stipulate the shear rate used to carry out the measurements when reporting data. The choice of shear rate when measuring the apparent viscosity of a nonideal liquid is a particularly important consideration when carrying out rheological measurements in a laboratory which are supposed to mimic some process which occurs in a food naturally (e.g., flow through a pipe, creaming of an emulsion droplet, or mastication). The test in the laboratory should use a shear rate that is as close as possible to that which the food experiences in practice.

The two most common types of shear-rate-dependent nonideal liquids are:

1. ***Pseudoplastic fluids.*** Pseudoplastic flow is the most common type of nonideal behavior exhibited by food emulsions. It manifests itself as a *decrease* in the apparent viscosity of a fluid as the shear rate is increased and is therefore often referred to as *shear thinning* (Figure 8.5). Pseudoplasticity may occur for a variety of reasons in food emulsions (e.g., the spatial distribution of the particles may be altered by the shear field, nonspherical particles may become aligned with the flow field, solvent molecules bound to the particles may be removed, or flocs may be deformed and disrupted) (Hunter 1993, Mewis and Macosko 1994).

2. ***Dilatant fluids.*** Dilatant behavior is much less common than pseudoplastic behavior. It manifests itself as an increase in the apparent viscosity as the shear rate is increased and is therefore often referred to as *shear thickening* (Figure 8.5). Dilantancy is often observed in concentrated emulsions or suspensions where the particles are packed tightly together (Hunter 1989). At intermediate shear rates, the particles form two-dimensional "sheets" which slide over each other relatively easily, but at higher shear rates, these sheets are disrupted and so the viscosity increases (Pal 1996). Shear thickening may also occur when the particles in an emulsion become flocculated because of an increased collision frequency (Section 7.4.1.1); therefore, this process usually leads to time-dependent behavior and so will be considered in the following section.

Liquids that exhibit pseudoplastic behavior often have a viscosity versus shear rate profile similar to that shown in Figure 8.6. The viscosity decreases from a constant value at low shear rates (η_0) to another constant value at high shear rates (η_∞). A number of mathematical equations have been developed to describe the rheological behavior of shear-rate-dependent nonideal liquids. The major difference is the range of shear rates over which they are

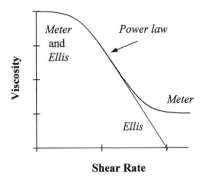

FIGURE 8.6   Typical viscosity profile for a pseudoplastic material. The viscosity decreases from a constant value (η_0) at low shear rates to another constant value (η_∞) at high shear rates. (Adapted from Hunter 1993.)

applicable. If measurements are carried out across the whole shear rate range (which often extends many orders of magnitude), then the viscosity can often be described by the Meter equation (Hunter 1989):

$$\eta = \eta_\infty + \frac{\eta_0 - \eta_\infty}{1 + (\tau / \tau_i)^n} \tag{8.3}$$

where τ_i is the shear stress where the viscosity is midway between the low and high shear rate limits and n is the power index. The rheological properties of this type of system can therefore be characterized by four parameters: η_0, η_∞, τ_i, and n.

   If measurements are only carried out at shear rates that are sufficiently less than the high shear rate plateau (Figure 8.6), then the rheology can be described by the *Ellis* equation (Hunter 1993):

$$\eta = \frac{\eta_0}{1 + (\tau / \tau_i)^n} \tag{8.4}$$

   If measurements are carried out at intermediate shear rates (i.e., above the low shear plateau and below the high shear plateau), the rheology can often be described by a simple power-law model (Hunter 1993):

$$\tau = A(d\gamma / dt)^B \qquad \text{or} \qquad \eta = A(d\gamma / dt)^{B-1} \tag{8.5}$$

   The constants A and B are usually referred to as the *consistency index* and the *power index*, respectively (Dickinson 1992). For an ideal liquid, $B = 1$; for an emulsion which exhibits shear thinning, $B < 1$; and for an emulsion which exhibits shear thickening, $B > 1$. Equations 8.5 are easy to use since they only contain two unknown parameters, which can simply be obtained from a plot of log (τ) versus log ($d\gamma/dt$). Nevertheless, these equations should only be used after it has been proven experimentally that the relationship between log (τ) and log ($d\gamma/dt$) is linear over the shear rates used.

Time-Dependent Nonideal Liquids.   The apparent viscosity of the fluids described in the previous section depended on the shear rate, but not on the length of time that the shear was applied. There are many food emulsions whose apparent viscosity either increases or decreases with time during the application of shear. In some cases, this change is reversible and the fluid will recover its original rheological characteristics if it is allowed to stand at rest for a sufficiently long period. In other cases, the change brought about by shearing the sample is irreversible, and the sample will not recover its original characteristics.

An appreciation of the time dependency of the flow properties of food emulsions is of great practical importance in the food industry. The duration of pumping or mixing operations, for instance, must be carefully controlled so that the food sample has an apparent viscosity which is suitable for the next processing operation. If a food is mixed or pumped for too long, it may become too thick or too runny and thus lose its desirable rheological properties.

The dependence of the rheology of a liquid on time is often associated with some kind of relaxation process (Hunter 1993, Mewis and Macosko 1994). When an external force is applied to a system that is initially at equilibrium, the material takes a certain length of time to reach the new equilibrium conditions, which is characterized by a relaxation time (τ_R). When the measurement time is of the same order as the relaxation time, it is possible to observe changes in the properties of the system with time. Thus the rheological properties of an emulsion depend on the time scale of the experiment. Time-dependent nonideal fluids are classified in two different categories:

1. *Thixotropic behavior.* A thixotropic fluid is one in which the apparent viscosity *decreases* with time when the fluid is subjected to a constant shear rate (Figure 8.7). Emulsions which exhibit this type of behavior often contain particles (droplets, crystals, or biopolymers) which are aggregated by weak forces. Shearing of the material causes the aggregated particles to be progressively deformed and disrupted, which decreases the resistance to flow and therefore causes a reduction in viscosity over time. If the relaxation time associated with the disruption of the flocs is shorter than the measurement time, then the viscosity will be observed to tend to a constant final value. This value may correspond to the point where the rate of structure disruption is equal to the rate of structure reformation or where there is no more structure to be broken down. In pseudoplastic liquids, the breakdown of the aggregated particles occurs so rapidly that the system almost immediately attains its new equilibrium position, and so it appears as though the viscosity is independent of time.

2. *Rheopectic.* In some food emulsions, the apparent viscosity of the fluid *increases* with time when it is subjected to a constant shear rate (Figure 8.7). One of the most common reasons for this type of behavior is that shearing increases both the frequency and efficiency of collisions between droplets, which leads to enhanced aggregation (Section 7.4.1) and consequently an increase in apparent viscosity over time.

In some fluids, the time-dependent rheological properties are irreversible (i.e., once the shear force is removed, the system may not fully regain its initial rheological properties). Liquids that experience this type of permanent change are called *rheodestructive.* This type of behavior might occur when flocs are disrupted by an intense shear stress and are unable to reform when the shear stress is removed. Otherwise, the structure and rheological properties of a material may return to their original values. In this case, the recovery time is often an important characteristic of the material.

The rheological properties of time-dependent nonideal liquids can be characterized by measuring the change in their viscosity over time. From these measurements, one can obtain a relaxation time for the structural rearrangements that occur in the emulsion. Nevertheless,

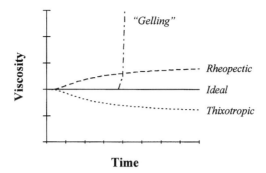

FIGURE 8.7 Comparison of the viscosity of ideal and time-dependent nonideal liquids. The viscosity may increase or decrease to a constant value with time. Alternatively, it may increase steeply due to network formation.

this type of experiment is often inconvenient if one also wants to obtain information about the dependence of the viscosity on shear rate. One would have to establish the relaxation time for the structural rearrangements at each shear rate, and then ensure that the samples were sheared for a time that was long enough for them to reach their steady-state rheology. If the relaxation time of a sample is relatively long, this type of measurement would be time consuming and laborious to carry out. Instead, it is often more convenient to measure the viscosity of a fluid when the shear rate is increased from zero to a certain value and then decreased back to zero again (Figure 8.8). When there is a significant structural relaxation in a system, the upward curve is different from the downward curve and one obtains a *hysteresis loop*. The area within the loop depends on the degree of relaxation that occurs and the rate at which the shear rate is altered. The slower the shear rate is altered, the more time the system has to reach its equilibrium value, and therefore the smaller the area within the hysteresis loop. By carrying out measurements as a function of the rate at which the shear rate is increased, it is possible to obtain information about the relaxation time.

8.2.3. Plastics

A number of food emulsions exhibit rheological behavior known as *plasticity* (e.g., margarine, butter, and certain spreads) (Sherman 1968a,c, 1970; Tung and Paulson 1995). A plastic material has elastic properties below a certain applied stress, known as the *yield stress*, but flows like a fluid when this stress is exceeded.

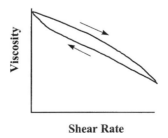

FIGURE 8.8 Hysteresis curve for liquids whose viscosity depends on the length of time they are sheared.

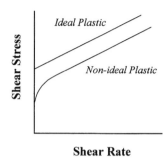

FIGURE 8.9   Rheological behavior of ideal and nonideal plastics.

8.2.3.1. Ideal Plastics

The ideal plastic material is referred to as a *Bingham plastic*, after the scientist who first proposed this type of rheological behavior (Sherman 1970). Two equations are needed to describe the rheological behavior of a Bingham plastic, one below the yield stress and one above it:

$$\tau = G\gamma \qquad (\text{for } \tau < \tau_0) \tag{8.6}$$

$$\tau - \tau_0 = \eta\dot{\gamma} \qquad (\text{for } \tau \geq \tau_0) \tag{8.7}$$

where G is the shear modulus, η is the viscosity, and τ_0 is the yield stress. The rheological properties of an ideal plastic are shown in Figure 8.9.

Foods that exhibit plastic behavior usually consist of a network of aggregated molecules or particles dispersed in a liquid matrix (Clark 1987, Edwards et al. 1987, Tung and Paulson 1995). For example, margarine and butter consist of a network of tiny fat crystals dispersed in a liquid oil phase (Moran 1994). Below a certain applied stress, there is a small deformation of the sample, but the weak bonds between the crystals are not disrupted. When the critical yield stress is exceeded, the weak bonds are broken and the crystals slide past one another, leading to flow of the sample. Once the force is removed, the flow stops. A similar type of behavior can be observed in emulsions containing three-dimensional networks of aggregated droplets.

8.2.3.2. Nonideal Plastics

Above the yield stress, the fluid flow may exhibit non-Newtonian behavior similar to that described earlier for liquids (e.g., pseudoplastic, dilatant, thixotropic, or rheopectic). The material may also exhibit nonideal elastic behavior below the yield stress (e.g., the yield point may not be sharply defined; instead, the stress may increase dramatically, but not instantaneously, as the shear rate is increased) (Figure 8.9). This would occur if the material did not all begin to flow at a particular stress, but there was a gradual breakdown of the network structure over a range of stresses (Sherman 1968a,c).

8.2.4. Viscoelastic Materials

Many food emulsions are not pure liquids or pure solids, but have rheological properties that are partly viscous and partly elastic (Sherman 1968a,c, 1970; Dickinson 1992). Plastic materials exhibit elastic behavior below a certain value of the applied stress and viscous behavior above this value. In contrast, viscoelastic materials exhibit both viscous and elastic behavior simultaneously. In an ideal elastic solid, all the mechanical energy applied to the material is stored in the deformed bonds and is returned to mechanical energy once the force is removed (i.e., there is no loss of mechanical energy). On the other hand, in an ideal liquid, all of the mechanical energy applied to the material is dissipated due to friction (i.e., the mechanical energy is converted to heat). In a viscoelastic material, part of the energy is stored as mechanical energy within the material, and part of the energy is dissipated. For this reason, when a force is applied to a *viscoelastic* material, it does not instantaneously adopt its new dimensions, nor does it instantaneously return to its undeformed state when the force is removed (as an ideal elastic material would). In addition, the material may even remain permanently deformed once the force is removed. The rheological properties of a viscoelastic material are characterized by a complex elastic modulus (E^*) which is comprised of an elastic and a viscous contribution:

$$E^* = E' + iE'' \tag{8.8}$$

Here, E' is known as the *storage* modulus and E'' as the *loss* modulus.

Two types of experimental tests are commonly used to characterize the rheological properties of viscoelastic materials: one based on transient measurements and the other on dynamic measurements (Whorlow 1992). Both types of tests can be carried out by the application of simple shear, simple compression, or bulk compression to the material being analyzed. Simple shear tests are the most commonly used to analyze food emulsions, and therefore only these will be considered here. Nevertheless, the same basic principles are also relevant to compression tests.

8.2.4.1. *Transient Tests*

In a transient experiment, a constant stress is applied to a material and the resulting strain is measured as a function of time or vice versa.

Creep. In a creep experiment, a constant stress is applied to a material and the change in its dimensions with time are monitored, which results in a strain versus time curve (Sherman 1968c, 1970). The data are usually expressed in terms of a parameter called the *compliance* (J), which is equal to the ratio of the strain to the applied stress (and is therefore the reciprocal of the modulus). The compliance is proportional to the strain, but it is a better parameter to use to characterize the rheological properties of the material because it takes into account the magnitude of the applied stress. The time dependence of the compliance of a material can also be measured when the stress is removed, which is referred to as a *creep recovery* experiment. A typical compliance versus time curve for a viscoelastic material is shown in Figure 8.10 (Sherman 1968a). This curve can be divided into three regions:

1. A region of instantaneous elastic deformation in which the bonds between the particles are stretched elastically. In this region, the material acts like an elastic solid with a compliance J_0 given by the ratio of the strain to the applied stress.
2. A region of retarded elastic compliance in which some bonds are breaking and some are reforming. In this region, the material has viscoelastic properties and its compliance is given by $J_R = J_M [1 - \exp(-t/\tau_M)]$, where J_M and τ_M are the mean compliance and retardation time.
3. A region of Newtonian compliance (J_N) when the bonds are disrupted and do not reform so that the material only flows: $J_N = t/\eta_N$.

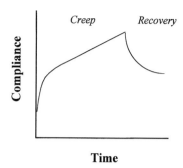

FIGURE 8.10   A typical creep versus time curve for a viscoelastic material, such as ice cream.

The total creep compliance of the system is therefore given by:

$$J(t) = J_0 + J_R(t) + J_N(t) = J_0 + J_M[1 - \exp(-t / \tau_M)] + t / \eta_N \qquad (8.9)$$

This type of material is usually referred to as a *viscoelastic liquid,* because it continues to flow for as long as the stress is applied. Some materials exhibit a different type of behavior and are referred to as *viscoelastic solids.* When a constant stress is applied to a viscoelastic solid, the creep compliance increases up to a finite equilibrium value (J_E) at long times rather than continuously increasing. When the force is removed, the compliance returns to zero, unlike a *viscoelastic liquid,* which does not return to its initial shape.

Stress Relaxation.   Instead of applying a constant force and measuring the change in the strain with time, it is also possible to apply a constant strain and measure the change in the stress acting on the material with time. This type of experiment is referred to as a *stress relaxation.* The same type of information can be obtained from creep and stress relaxation experiments, and the method used largely depends on the type of rheological instrument available.

8.2.4.2. Dynamic Tests

In a dynamic experiment, a sinusoidal stress is applied to a material and the resulting sinusoidal strain is measured or vice versa (Tung and Paulson 1995, Liu and Masliyah 1996). In this section, only the case where a stress is applied to the sample and the resultant strain is measured is considered. The applied stress is characterized by its maximum amplitude (τ_0) and its angular frequency (ω). The resulting strain has the same frequency as the applied stress, but its phase is different because of relaxation mechanisms associated with the material (Whorlow 1992). Information about the viscoelastic properties of the material can therefore be obtained by measuring the maximum amplitude (γ_0) and phase shift (δ) of the strain (Figure 8.11). The amplitude of the applied stress used in this type of test is usually so small that the material is in the *linear viscoelastic region* (i.e., the stress is proportional to the strain), and the properties of the material are not affected by the experiment (van Vliet 1995, Liu and Masliyah 1996).

If the applied stress varies sinusoidally with time, then (Whorlow 1992):

$$\tau = \tau_0 \cos(\omega t) \qquad (8.10)$$

and the resulting harmonic strain is

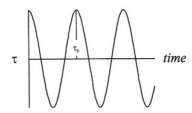

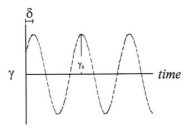

FIGURE 8.11 The rheological properties of a viscoelastic material can be determined by measuring the relationship between an applied sinusoidal stress and the resultant sinusoidal strain.

$$\gamma = \gamma_0 \cos(\omega t - \delta) \tag{8.11}$$

The compliance of the material is therefore given by:

$$J(t) = \frac{\gamma}{\tau_0} = \frac{\gamma_0}{\tau_0} (\cos \delta \cos \omega t + \sin \delta \sin \omega t) \tag{8.12}$$

or

$$J(t) = J' \cos \omega t + J'' \sin \omega t \tag{8.13}$$

where J' $(= \gamma_0 \cos \delta / \tau_0)$ is known as the *storage compliance*, which is the in-phase component of the compliance, and J'' $(= \gamma_0 \sin \delta / \tau_0)$ is known as the *loss compliance*, which is the 90° out-of-phase component of the compliance. The in-phase component of the compliance is determined by the elastic properties of the material, whereas the 90° out-of-phase component is determined by the viscous properties. This is because the stress is proportional to the strain ($\tau \propto \gamma$) for elastic materials, whereas it is proportional to the *rate* of strain ($\tau \propto d\gamma/dt$) for viscous materials (Macosko 1994).

The dynamic rheological properties of a material can therefore be characterized by measuring the frequency dependence of the applied stress and the resulting strain and then plotting a graph of J' and J'' versus frequency. Alternatively, the data are often presented in terms of the magnitude of the complex compliance ($J* = J' - iJ''$) and the phase angle:

$$|J*| = \sqrt{J'^2 + J''^2} \tag{8.14}$$

$$\delta = \tan^{-1}\left(\frac{J''}{J'}\right) \tag{8.15}$$

of a material provides a useful insight into its viscoelastic properties: δ
elastic solid, δ = 90° for a perfectly viscous fluid, and 0 < δ < 90° for
ial. The more elastic a material (at a particular frequency), the smaller the
lower the amount of energy dissipated per cycle.
onvenient to express the rheological properties of a viscoelastic material
ulus rather than its compliance (Whorlow 1992). The complex, storage,
and loss moduli of a material can be calculated from the measured compliances using the
following relationships:

$$G^* = G' + iG'' \qquad G' = \frac{J'}{J'^2 + J''^2} \qquad G'' = \frac{J''}{J'^2 + J''^2} \qquad (8.16)$$

8.3. MEASUREMENT OF RHEOLOGICAL PROPERTIES

Food emulsions can exhibit a wide range of different types of rheological behavior, including
liquid, solid, plastic, and viscoelastic (Dickinson and Stainsby 1982, Dickinson 1992). Consequently, a variety of instrumental methods have been developed to characterize their
rheological properties. Instruments vary according to the type of deformation they apply to
the sample (shear, compression, elongation, or some combination), the property they measure, their cost, their sophistication, and their ease of operation (Whorlow 1992).

In many industrial applications, it is necessary to have instruments that make measurements which are rapid, low cost, simple to carry out, and reproducible, rather than give
absolute fundamental data (Sherman 1970, Rao 1995). Thus simple empirical instruments are
often used in quality assurance laboratories, rather than the more sophisticated and expensive
instruments used in research and development. The information obtained from these empirical instruments is often difficult to relate to the fundamental rheological constants of a
material because the applied stresses and strains are not easily measured or defined. Rather
than a simple elongation, shear, or compression, different types of forces may be applied
simultaneously. For example, when a blade cuts through a meat product, both shear and
compression forces are applied together, and the sample is deformed beyond the limit where
Hooke's law is applicable. To compare data from different laboratories, it is necessary to
carefully follow standardized test procedures. These procedures may define experimental
parameters such as the sample size and preparation procedure, the magnitude of the force or
deformation, the design of the device used, the speed of the probe, the length of time the force
is applied, and the measurement temperature.

For food scientists involved in research and development, it is usually necessary to use
instruments which provide information about the fundamental rheological constants of
the material being tested. These instruments are designed to apply well-defined stresses
and strains to a material in a controlled manner so that stress–strain relationships can be
measured and interpreted using available mathematical theories. Rheological properties
determined using these techniques can be compared with measurements made by other
workers or in other laboratories. In addition, measured rheological properties can be
compared with predictions made using various mathematical theories which have been
developed to relate the structure and composition of materials to their fundamental rheological properties.

It is convenient to categorize rheological instruments according to whether they utilize
simple compression (or elongation) or shear forces.*

---

* At present, few instruments utilize bulk compression to analyze the rheological properties of food emulsions.

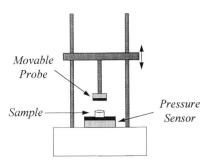

Movable Probe

Sample

Pressure Sensor

FIGURE 8.12 Universal Testing Machine for measuring the rheological properties of a material by a compression or elongation test.

8.3.1. Simple Compression and Elongation

This type of test is most frequently carried out on solid or semisolid foods that are capable of supporting their own weight (e.g., butter, margarine, and frozen ice cream) (Bourne 1982, Rao et al. 1995). Measurements are often carried out using instruments referred to as *Universal Testing Machines*. The solid sample to be analyzed is placed between a fixed plate and a moving probe (Figure 8.12). The probe can have many different designs depending on the type of information required, including a flat plate, a blade, a cylindrical spike, and even a set of teeth!

The probe can be moved vertically, either upward or downward, at a controlled speed. Either the probe or the plate contains a pressure sensor which measures the force exerted on the sample when it is deformed by the probe. The instrument also records the distance that the probe moves through the sample. The stress and strain experienced by a material can therefore be calculated from a knowledge of its dimensions and the force and deformation recorded by the instrument.

Some of the common tests carried out using Universal Testing Machines are:

1.   *Stress–strain curve.*   The stress on a sample is measured as a function of strain as it is compressed at a fixed rate (Figure 8.1). The resulting stress–strain curve is used to characterize the rheological properties of the material being tested. The slope of stress versus strain at small deformations is often a straight line, with a gradient equal to the elastic modulus (Table 8.1). At intermediate deformations, the stress may no longer be proportional to the strain and some flow may occur, so that when the stress is removed the sample does not return to its original shape. At larger deformations, the sample may rupture and the breaking stress, strain, and modulus can be determined. The operator must decide the distance and speed at which the probe will move through the sample. For viscoelastic materials, the shape of the upward and downward curves may be different and depends on the speed at which the probe moves. This type of test is used commonly to test solid samples and gels, such as margarine, butters, spreads, and desserts.

2.   *Repeated deformation.*   The sample to be analyzed is placed between the plate and the probe, and then the probe is lowered and raised a number of times at a fixed speed so that the sample experiences a number of compression cycles (Rao et al. 1995). An ideal elastic solid would show the same stress–strain curve for each cycle. However, the properties of many materials are altered by compression (e.g., due to rupture or flow), and therefore successive compression cycles give different stress–strain curves. This type of test is often used to give some indication of the

processes that occur when a food is chewed in the mouth (i.e., the breakdown of food structure).

3. ***Transient experiments.*** The sample is placed between the plate and the probe and then compressed to a known deformation, and the relaxation of the stress with time is measured (stress relaxation). Alternatively, a constant stress could be applied to the sample, and the variation of the strain is measured over time (creep). This type of experiment is particularly useful for characterizing the rheological properties of viscoelastic food emulsions (see Section 8.2.4).

By using different fixtures, the same type of instrument can be used to carry out elongation experiments. A sample is clamped at both ends, and then the upper clamp is moved upward at a controlled speed and the force required to elongate the sample is measured by the pressure sensor as a function of sample deformation. Again, the elastic modulus and breaking strength of the material can be determined by analyzing the resulting stress–strain relationship. Universal Testing Machines can also be adapted to perform various other types of experiments, such as bending or slicing.

Recently, a number of more sophisticated instruments, based on dynamic rheological measurements, have been developed to characterize the rheological properties of solids, plastics, and viscoelastic materials (Harwalker and Ma 1990, Wunderlich 1990, Whorlow 1992). In addition to carrying out the standard compression measurements mentioned above, they can also be used to carry out dynamic compression measurements on viscoelastic materials. The sample to be analyzed is placed between a plate and a probe, and an oscillatory shear stress of known amplitude and frequency is applied to it. The amplitude and phase of the resulting strain are measured and converted into a storage and loss modulus using suitable equations (Section 8.2.4.2). The amplitude of the applied stress must be small enough to be in the linear viscoelastic region of the material. These instruments are relatively expensive to purchase and therefore only tend to be used by research laboratories in large food companies, government institutions, and universities. Nevertheless, they are extremely powerful tools for carrying out fundamental studies of food emulsions. The rheological properties of a sample can be measured as a function of time or temperature, and thus processes such as gelation, aggregation, crystallization, melting, and glass transitions can be monitored. The measurement frequency can also be varied, which provides valuable information about relaxation processes within a sample.

Some complications can arise when carrying out simple compression experiments. There may be friction between the compressing plates and the sample, which can lead to the generation of shear as well as compressional forces (Whorlow 1992). For this reason, it is often necessary to lubricate the sample with oil to reduce the effects of friction. In addition, the cross-sectional area of the sample may change during the course of the experiment, which would have to be taken into account when converting the measured forces into stresses. Finally, for viscoelastic materials, some stress relaxation may occur during the compression or expansion, and so the results depend on the rate of sample deformation.

8.3.2. Shear Measurements

Instruments which utilize shear measurements are used to characterize the rheological properties of liquids, viscoelastic materials, plastics, and solids (Whorlow 1992, Rao 1995). The type of instrument and test method used in a particular situation depend on physicochemical characteristics of the sample being analyzed, as well as the kind of information required. Some instruments can be used to characterize the rheological properties of both solids and liquids, whereas others can only be used for either solids or liquids. Certain types of viscometers are capable of measuring the viscosity of fluids over a wide range of shear rates and can therefore be used to analyze both ideal and nonideal liquids, whereas the shear rate cannot be controlled in others and so they are only suitable for analyzing ideal liquids. A number of instruments can be used to characterize the rheological behavior of viscoelastic materials using both transient and dynamic tests, whereas others can only use either one or the other type of test. To make accurate and reliable measurements, it is important to select the most appropriate instrument and test method and to be aware of possible sources of experimental error.

8.3.2.1. Capillary Viscometers

The simplest and most commonly used capillary viscometer is called the *Ostwald viscometer* (Hunter 1986, Whorlow 1992). This device consists of a glass U-tube into which the sample to be analyzed is poured. The whole arrangement is placed in a thermostated water bath to reach the measurement temperature (Figure 8.13). The viscosity of the liquid is measured by sucking it into one arm of the tube using a slight vacuum and then measuring the time it takes to flow back through a capillary of fixed radius and length. The time it takes to travel through the capillary is related to the viscosity by the following equation:

$$t = C \frac{\eta}{\rho} \tag{8.17}$$

where ρ is the density of the fluid, t is the measured flow time, and C is a constant which depends on the precise size and dimensions of the U-tube. The higher the viscosity of the fluid, the longer it takes to flow through the tube. The simplest method for determining the viscosity of a liquid is to measure its flow time and compare it with that of a liquid of known viscosity, such as distilled water:

$$\eta_s = \left(\frac{t_s}{t_0} \frac{\rho_s}{\rho_0} \right) \eta_0 \tag{8.18}$$

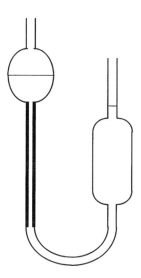

FIGURE 8.13   Capillary viscometer used to measure the viscosity of Newtonian liquids.

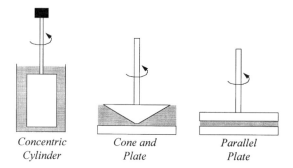

Concentric   Cone and   Parallel
Cylinder   Plate   Plate

FIGURE 8.14   Different types of measurement cells commonly used with dynamic shear rheometers and viscometers.

where the subscripts s and 0 refer to the sample being analyzed and the reference fluid, respectively. This type of viscometer is used principally to measure the viscosity of Newtonian liquids. It is unsuitable for analyzing non-Newtonian liquids because the sample does not experience a uniform and controllable shear rate (Hunter 1986). U-tubes with capillaries of various diameters are available to analyze liquids with different viscosities: the larger the diameter, the higher the viscosity of the sample which can be analyzed.

8.3.2.2.   Mechanical Viscometers and Dynamic Rheometers

A number of mechanical rheological instruments have been designed to measure the shear properties of liquids, viscoelastic materials, plastics, and solids (Whorlow 1992, Macosko 1994). These instruments are usually computer controlled and can carry out sophisticated test procedures as a function of time, temperature, shear rate, or frequency (Figure 8.14). Basically, the sample to be analyzed is placed in a thermostated measurement cell, where it is subjected to a controlled shear stress (or strain). The resulting strain (or stress) is measured by the instrument, and so the rheological properties of the sample can be determined from the stress–strain relationship. The type of rheological test carried out depends on whether the sample is liquid, solid, or viscoelastic. The instruments can be divided into two different types: *constant stress* instruments, which apply a constant torque to the sample and measure the resultant *strain* or *rate of strain*, and *constant strain* instruments, which apply a constant *strain* or *rate of strain* and measure the torque generated in the sample. For convenience, only constant stress instruments are discussed here, although both types are commonly used in the food industry.

A number of different types of measurement cells can be used to contain the sample during an experiment (Pal et al. 1992):

1. **Concentric cylinder.**   The sample is placed in the narrow gap between two concentric cylinders. The inner cylinder is driven at a constant *torque* (angular force) and the resultant *strain* (angular deflection) or *rate of strain* (speed at which the cylinder rotates) is measured, depending on whether one is analyzing a predominantly solid or liquid sample.* For a solid, the angular deflection of the inner cylinder from its rest position is an indication of its elasticity: the larger the deflection, the smaller the shear modulus. For a liquid, the speed at which the inner cylinder rotates is governed by the viscosity of the fluid between the plates: the

---

* In some instruments, the outer cylinder rotates and the torque on the inner cylinder is measured.

faster it spins at a given torque, the lower the viscosity of the liquid being analyzed. The torque can be varied in a controlled manner so that the (apparent) elastic modulus or viscosity can be measured as a function of shear stress. This instrument can be used for measuring the viscosity of non-Newtonian liquids, the viscoelasticity of semisolids, and the elasticity of solids.

2. *Parallel plate.* In this type of measurement cell, the sample is placed between two parallel plates. The lower plate is stationary, while the upper one can rotate. A constant *torque* is applied to the upper plate, and the resultant *strain* or *rate of strain* is measured, depending on whether one is analyzing a predominantly solid or liquid sample. The main problem with this type of experimental arrangement is that the shear strain varies across the sample: the shear strain in the middle of the sample is less than that at the edges. The parallel plate arrangement is therefore only suitable for analyzing samples which have rheological properties that are independent of shear rate, and it is therefore unsuitable for analyzing nonideal liquids or solids.

3. *Cone and plate.* This is essentially the same design as the parallel plate instrument, except that the upper plate is replaced by a cone. The cone has a slight angle which is designed to ensure that a more uniform shear stress acts across the sample. The cone and plate arrangement can therefore be used to analyze nonideal materials.

Any of these arrangements can be used to carry out simple viscosity measurements on fluids, by measuring the variation of shear stress with shear rate. However, some of the more sophisticated ones can also be used to carry out transient and dynamic rheological tests. Typically, the rheological properties of samples are measured as a function of time or temperature.

A number of possible sources of experimental error are associated with rheological measurements carried out using shear viscometers and rheometers (Sherman 1970, Hunter 1989, Pal et al. 1992). First, the gap between the cylinders or plates should be at least 20 times greater than the diameter of the droplets, so that the emulsion appears as a homogeneous material within the device (Pal et al. 1992). On the other hand, the gap must be narrow enough to ensure a fairly uniform shear stress across the whole of the sample. Second, a phenomenon known as *wall slip* may occur within a viscometer or rheometer, which can cause serious errors in the measurements if not properly taken into account (Sherman 1970). It is normally assumed that the liquid in direct contact with the surfaces of the cylinders (or plates) moves with them at the same velocity (Hunter 1986). This assumption is usually valid for simple liquids because the small molecules are caught within the surface irregularities on the cylinder and are therefore dragged along with it. For an emulsion, this assumption may not hold because the droplets or flocs are greater in size than the surface irregularities. Under these circumstances, a phase separation occurs at the cylinder surface where a thin layer of continuous phase acts as a lubricant and *slip* occurs. Wall slip effects can be taken into account by roughening the surfaces of the cylinders or by using a range of different gap widths (Hunter 1986, Pal et al. 1992). Third, the rheological properties of many samples depend on their previous thermal and shear history, and so this must be carefully controlled in order to obtain reproducible measurements. For example, the viscosity of many foods decreases substantially upon shearing due to disruption of an aggregated network of particles or molecules, and the recovery of the initial viscosity takes a certain length of time to achieve after the shear stress is removed. Fourth, many emulsions may be susceptible to creaming or sedimentation during the course of an experiment, which should be avoided if accurate rheological measurements are required.

8.3.3.   Empirical Techniques

Many of the techniques mentioned above are unsuitable for application in the food industry because the instrumentation is too expensive, it requires skilled operators, or measurements take too long to carry out (Sherman 1970, Rao et al. 1995). For these reasons, a large
number of empirical techniques have been developed by food scientists that provide simple
and rapid determinations of the rheological properties of a sample. Many of these empirical techniques have become widely accepted for analyzing specific food types. Typical
examples include penetrometers to measure the hardness of butters, margarines, and spreads
(Sherman 1970); devices for measuring the time it takes for liquids to flow through a
funnel (Liu and Masliyah 1996); or devices that measure the time it takes for a spherical
ball to fall through a sample contained within a glass tube (Becher 1957). It is difficult
to analyze the data from these devices using fundamental rheological concepts because it
is difficult to define the stresses and strains involved. Nevertheless, these devices are
extremely useful when rapid empirical information is more important than fundamental
understanding.

8.4.   RHEOLOGICAL PROPERTIES OF EMULSIONS

Food emulsions exhibit a wide range of different rheological properties, ranging from low-
viscosity liquids to fairly rigid solids. The rheological behavior of a particular food depends
on the type and concentration of ingredients it contains, as well as the processing and storage
conditions it has experienced. In this section, the relationship between the rheological properties of emulsions and their composition and microstructure is discussed. We begin by
considering the rheology of dilute suspensions of noninteracting rigid spheres, because the
theory describing the properties of this type of system is well established (Hiemenz 1986,
Hunter 1986, Mewis and Macosko 1994, Tadros 1994). Nevertheless, many food emulsions
are concentrated and contain nonrigid, nonspherical, and/or interacting droplets (Dickinson
1992). The theoretical understanding of these types of systems is less well developed,
although some progress has been made, which will be reviewed.

8.4.1.   Dilute Suspensions of Rigid Spherical Particles

The viscosity of a liquid increases upon the addition of rigid spherical particles because the
particles disturb the normal flow of the fluid, causing greater energy dissipation due to
friction (Hunter 1986, Mewis and Macosko 1994). Einstein derived an equation to relate the
viscosity of a suspension of rigid spheres to its composition:

$$\eta = \eta_0(1 + 2.5\phi) \tag{8.19}$$

where η_0 is the viscosity of the liquid surrounding the droplets and ϕ is the dispersed-phase
volume fraction. This equation assumes that the liquid is Newtonian, the particles are rigid
and spherical, that there are no particle–particle interactions, that there is no slip at the
particle–fluid interface, and that Brownian motion effects are unimportant. The Einstein
equation predicts that the viscosity of a dilute suspension of spherical particles increases
linearly with particle volume fraction and is independent of particle size and shear rate. The
Einstein equation gives excellent agreement with experimental measurements for suspensions
that conform to the above criteria, often up to particle concentrations of about 5%.

   It is convenient to define a parameter known as the *intrinsic viscosity* of a suspension, [η]
(Dickinson and Stainsby 1982):

$$[\eta] = \frac{\eta / \eta_0 - 1}{\phi} \tag{8.20}$$

so that

$$\eta = \eta_0 (1 + [\eta]\phi) \tag{8.21}$$

For rigid spherical particles, the intrinsic viscosity tends to 2.5 as the volume fraction tends to zero. For nonspherical particles or for particles that swell due to the adsorption of solvent, the intrinsic viscosity is larger than 2.5 (Hiemenz 1986), whereas it may be smaller for fluid particles (Sherman 1968a,c; Dickinson and Stainsby 1982). For these systems, measurements of $[\eta]$ can provide valuable information about their shape or degree of solvation.

8.4.2. Dilute Suspensions of Fluid Spherical Particles

Food emulsions usually contain fluid, rather than solid, particles. In the presence of a flow field, the liquid within a droplet is caused to circulate because it is dragged along by the liquid (continuous phase) that flows past the droplet (Sherman 1968a,c; Dickinson and Stainsby 1982). Consequently, the difference in velocity between the materials on either side of the droplet surface is less than for a solid particle, which means that less energy is lost due to friction and therefore the viscosity of the suspension is lower. The greater the viscosity of the fluid within a droplet, the more it acts like a rigid sphere, and therefore the higher the viscosity of the suspension.

The viscosity of a suspension of noninteracting spherical droplets is given by (Tadros 1994):

$$\eta = \eta_0 \left[1 + \left(\frac{\eta_0 + 2.5\eta_{drop}}{\eta_0 + \eta_{drop}} \right) \phi \right] \tag{8.22}$$

where η_{drop} is the viscosity of the liquid in the droplets. For droplets containing relatively high-viscosity liquids ($\eta_{drop}/\eta_0 \gg 1$), the intrinsic viscosity tends to 2.5, and therefore this equation tends to that derived by Einstein (Equation 8.19). For droplets that contain relatively low-viscosity fluids ($\eta_{drop}/\eta_0 \ll 1$), such as air bubbles, the intrinsic viscosity tends to unity, and so the suspension viscosity is given by $\eta = \eta_0(1 + \phi)$. One would therefore expect the viscosity of oil-in-water or water-in-oil emulsions to be somewhere between these two extremes. In practice, the droplets in most food emulsions are coated by a layer of emulsifier molecules that forms a viscoelastic membrane. This membrane retards the transmittance of the tangential stress from the continuous phase into the droplet and therefore hinders the flow of the fluid within the droplet (Pal et al. 1992, Tadros 1994). For this reason, most food emulsions contain droplets which act like rigid spheres, and so their viscosities at low concentrations can be described by the Einstein equation.

At sufficiently high flow rates, the hydrodynamic forces can become so large that they overcome the interfacial forces holding the droplets together and cause the droplets to become deformed and eventually disrupted (Chapter 6). The shear rates required to cause droplet disruption are usually so high that the flow profile is turbulent rather than laminar, and so it is not possible to make viscosity measurements. Nevertheless, a knowledge of the viscosity of fluids at high shear rates is important for engineers who design mixers and homogenizers.

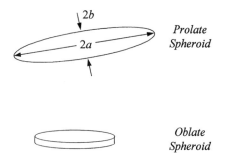

FIGURE 8.15   Examples of oblate and prolate spheroids.

8.4.3.   Dilute Suspensions of Rigid Nonspherical Particles

Many of the particles in food emulsions may have nonspherical shapes (e.g., flocculated droplets, partially crystalline droplets, fat crystals, ice crystals, or biopolymer molecules) (Dickinson 1992). Consequently, it is important to appreciate the effects of particle shape on suspension viscosity. The shape of many particles can be approximated as prolate spheroids (rod-like) or oblate spheroids (disk-like). A spheroid is characterized by its axis ratio $r_p = a/b$, where a is the major axis and b is the minor axis (Figure 8.15). For a sphere, $a = b$; for a prolate spheroid, $a > b$; and for an oblate spheroid, $a < b$. The flow profile of a fluid around a nonspherical particle causes a greater degree of energy dissipation than that around a spherical particle, which leads to an increase in viscosity (Hunter 1986, Hiemenz 1986, Mewis and Macosko 1994). The magnitude of this effect depends on the rotation and orientation of the spherical particle. For example, the viscosity of a rod-like particle is much lower when it is aligned parallel to the fluid flow, rather than perpendicular, because the parallel orientation offers less resistance to flow.

The orientation of a spheroid particle in a flow field is governed by a balance between the hydrodynamic forces that act upon it and its rotational Brownian motion (Mewis and Macosko 1994). The hydrodynamic forces favor the alignment of the particle along the direction of the flow field, because this reduces the energy dissipation. On the other hand, the alignment of the particles is opposed by their *rotational* Brownian motion, which favors the complete randomization of their orientations. The relative importance of the hydrodynamic and Brownian forces is expressed in terms of a dimensionless number, known as the Peclet number (Pe). For simple shear flow (Mewis and Macosko 1994):

$$\text{Pe} = \frac{\dot{\gamma}}{D_R} \tag{8.23}$$

where $\dot{\gamma}$ is the shear rate and D_R is the *rotational* Brownian diffusion coefficient, which depends on particle shape:

$$D_R = \frac{kT}{8\pi\eta r^3} \qquad \text{for rigid spheres} \tag{8.24}$$

$$D_R = \frac{3kT}{32\pi\eta b^3} \qquad \text{for circular disks} \tag{8.25}$$

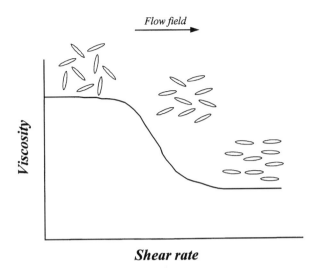

FIGURE 8.16   At low shear rates, the particles rotate freely in all directions, but as the shear rate increases, they become more and more aligned with the flow field. This causes a reduction in the viscosity with increasing shear rate (i.e., pseudoplasticity).

$$D_R = \frac{3kT}{8\pi\eta r^3} (\ln 2r_p - 0.5) \qquad \text{for long thin rods} \qquad (8.26)$$

When the Peclet number is much less than unity (Pe << 1), the rotational Brownian motion dominates, and the particles tend to rotate freely in the liquid. This type of behavior is observed when the particles are small, the shear rate is low, and/or the viscosity of the surrounding fluid is low. When the Peclet number is much greater than unity (Pe >> 1), the hydrodynamic forces dominate, and the particles become aligned with the flow field (Figure 8.16). This type of behavior is observed when the particles are large, the shear rate is high, and/or the viscosity of the surrounding liquid is high.

The viscosity of a suspension of nonspherical particles therefore depends on the shear rate. At low shear rates (i.e., Pe << 1), the viscosity has a constant high value. As the shear rate is increased, the hydrodynamic forces become more important, and so the particles become oriented with the flow field, which causes a reduction in the viscosity. At high shear rates (i.e., Pe >> 1), the hydrodynamic forces dominate and the particles remain aligned with the shear field, and therefore the viscosity has a constant low value (Figure 8.16). Thus suspensions of nonspherical particles exhibit shear thinning behavior. The shear rate at which the viscosity starts to decrease depends on the size and shape of the particles, as well as the viscosity of the surrounding liquid. Mathematical formulae similar to Equation 8.3 have been developed to calculate the influence of shear rate on the viscosity of suspensions of nonspherical particles, but these usually have to be solved numerically. Nevertheless, explicit expressions are available for systems that contain very small or very large particles.

8.4.4.   Dilute Suspensions of Flocculated Particles

When the attractive forces between the droplets dominate the repulsive forces, and are sufficiently greater than the thermal energy of the system, then droplets can aggregate into a primary or secondary minimum (Chapter 3). The rheological properties of many food emulsions are dominated by the fact that the droplets are flocculated, and so it is important to understand the factors which determine the rheological characteristics of these systems. It is often convenient to categorize systems as being either strongly flocculated ($w_{attractive} > 20$ kT) or weakly flocculated ($1\ kT < w_{attractive} < 20\ kT$), depending on the strength of the attraction between the droplets (Liu and Masliyah 1996).

A dilute suspension of flocculated droplets consists of flocs which are so far apart that they do not interact with each other through colloidal or hydrodynamic forces. This type of suspension has a higher viscosity than a suspension that contains the same concentration of isolated particles because the particles in the flocs trap some of the continuous phase and therefore have a higher *effective* volume fraction than the actual volume fraction (Liu and Masliyah 1996). In addition, the flocs may rotate in solution because of their rotational Brownian motion, sweeping out an additional amount of the continuous phase and thus increasing their *effective* volume fraction even more.

Suspensions of flocculated particles tend to exhibit pronounced shear thinning behavior (Figure 8.17). At low shear rates, the hydrodynamic forces are not large enough to disrupt the bonds holding the particles together, and so the flocs act like particles with a fixed size and shape, resulting in a constant viscosity. As the shear rate is increased, the hydrodynamic forces become large enough to cause flocs to become deformed and eventually disrupted. The deformation of the flocs results in their becoming elongated and aligned with the shear field, which results in a reduction in the viscosity. The disruption of the flocs decreases their effective volume fraction and therefore also contributes to a decrease in the suspension viscosity. The viscosity reaches a constant value at high shear rates, either because all of the flocs are completely disrupted so that only individual droplets remain or because the number of flocculated droplets remains constant since the rate of floc formation is equal to that of floc disruption (Campanella et al. 1995).

Depending on the nature of the interdroplet pair potential (Chapter 3), it is also possible to observe shear thickening due to particle flocculation under the influence of the shear field (de Vries 1963). Some emulsions contain droplets which are not flocculated under quiescent conditions because there is a sufficiently high energy barrier to prevent the droplets from falling into a primary minimum. However, when a shear stress is applied to the emulsions, the frequency of collisions and the impact force between the droplets increase, which can

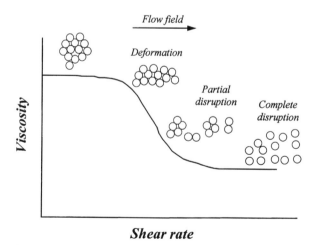

FIGURE 8.17   An emulsion that contains flocculated droplets exhibits shear thinning behavior because the flocs are deformed and disrupted in the shear field.

cause the droplets to gain sufficient energy to "jump" over the energy barrier and become flocculated, therefore leading to shear thickening.

Quite complicated behavior can therefore be observed in some emulsions (Pal et al. 1992, Liu and Masliyah 1996). For example, an emulsion that contains droplets which are weakly flocculated in a secondary minimum exhibits shear thinning at fairly low shear rates, but shows shear thickening when the shear rate exceeds some critical level where the droplets have sufficient energy to "jump" over the energy barrier and fall into the primary minimum. The value of this critical shear rate increases as the height of the energy barrier increases. A knowledge of the interdroplet pair potential is therefore extremely useful for understanding and predicting the rheological behavior of food emulsions.

The size, shape, and structure of flocs largely determine the rheological behavior of dilute suspensions of flocculated particles (Dickinson and Stainsby 1982, Liu and Masliyah 1996). Flocs formed by the aggregation of emulsion droplets often have structures that are fractal (Chapter 7). The effective volume fraction (ϕ_{eff}) of a fractal floc is related to the size of the floc and the fractal dimension by the following expression (Bremer 1992):

$$\phi_{eff} = \phi\left(\frac{R}{r}\right)^{3-D} \tag{8.27}$$

where r is the droplet radius and R is the floc radius. The viscosity of a dilute emulsion containing fractal flocs can therefore be established by substituting this expression into the Einstein equation:*

$$\eta = \eta_0(1 + [\eta]\phi_{eff}) = \eta_0[1 + [\eta]\phi(R/r)^{3-D}] \tag{8.28}$$

If it is assumed that the flocs are approximately spherical, then $[\eta] = 2.5$. This equation offers a useful insight into the relationship between the rheology and microstructure of flocculated emulsions.

Flocs with fairly open structures (i.e., lower D) have higher viscosities than those with compact structures because they have higher effective volume fractions (Equation 8.28). As mentioned earlier, the viscosity decreases with increasing shear rate partly because of disruption of the flocs (i.e., a decrease in R). The shear stress at which the viscosity decreases depends on the magnitude of the forces holding the droplets together within a floc. The greater the strength of the forces, the larger the shear rate required to deform and disrupt the flocs. Thus the dependence of the viscosity of an emulsion on shear stress can be used to provide valuable information about the strength of the bonds holding the droplets together (Sherman 1970, Dickinson and Stainsby 1982, Hunter 1989).

8.4.5. Concentrated Suspensions of Nonflocculated Particles in the Absence of Colloidal Interactions

When the concentration of particles in a suspension exceeds a few percent, the particles begin to interact with each other through a combination of hydrodynamic and colloidal interactions, and this alters the viscosity of the system (Hunter 1986, Mewis and Macosko 1994, Tadros 1994). In this section, we examine the viscosity of concentrated suspensions in the absence

---

* Alternatively, the expression for the effective volume fraction can be placed in the Dougherty–Krieger equation developed for more concentrated emulsions (Barnes 1994).

of long-range colloidal interactions between the particles (i.e., it is assumed that the particles act like hard spheres). The more complicated situation of suspensions in which long-range colloidal interactions are important is treated in the following section. Hydrodynamic interactions are the result of the relative motion of neighboring particles and are important in all types of nondilute suspensions.

At low concentrations, hydrodynamic interactions are mainly between pairs of particles, but as the concentration increases, three or more particles may be involved. As the concentration increases, the measured viscosity becomes larger than that predicted by the Einstein equation because these additional hydrodynamic interactions lead to a greater degree of energy dissipation. The Einstein equation can be extended to account for the effects of these interactions by including additional volume-fraction terms (Pal et al. 1992):

$$\eta = \eta_0 (1 + a\phi + b\phi^2 + c\phi^3 + \cdots) \tag{8.29}$$

The value of the constants a, b, c, etc. can be determined either experimentally or theoretically. For a suspension of rigid spherical particles, the value of a is 2.5, so that Equation 8.29 tends to the Einstein equation at low-volume fractions. A rigorous theoretical treatment of the interactions between pairs of droplets has established that $b = 6.2$ for rigid spherical particles (Hunter 1986). It is extremely difficult to theoretically calculate the value of higher order terms because of the complexity of the mathematical treatment of interactions among three or more particles. In addition, each successive constant only extends the applicability of the equation to a slightly higher volume fraction. For this reason, it has proven to be more convenient to adopt a semiempirical approach to the development of equations which describe the viscosity of concentrated suspensions. One of the most widely used equations was derived by Dougherty and Krieger and is applicable across the whole volume-fraction range (Hunter 1986, Mewis and Macosko 1994):

$$\frac{\eta}{\eta_0} = \left(1 - \frac{\phi}{\phi_c}\right)^{-[\eta]\phi_c} \tag{8.30}$$

where ϕ_c is an adjustable parameter which is related to the volume fraction at which the spheres become closely packed and $[\eta]$ is the intrinsic viscosity. Typically, the value of ϕ_c is between about 0.6 and 0.7 for spheres which do not interact via long-range colloidal interactions, but it may be considerably lower for suspensions in which there are strong long-range attractive or repulsive interactions between the droplets (see following sections). The intrinsic viscosity is 2.5 for spherical particles, but may be much larger for nonspherical, swollen, or aggregated particles (Hiemenz 1986).

The viscosity of concentrated suspensions often exhibits shear thinning behavior due to Brownian motion effects (Pal et al. 1992, Mewis and Macosko 1994, Liu and Masliyah 1996). It has already been mentioned that shear thinning occurs when the shear stress is large enough to overcome the rotational Brownian motion of nonspherical particles. Shear thinning can also occur because of the translational Brownian motion of particles. At low shear stresses, the particles have a three-dimensional isotropic and random distribution because of their Brownian motion (Hunter 1993). As the shear stress increases, the particles become more ordered along the flow lines to form "strings" or "layers" of particles that offer less resistance to the fluid flow and therefore cause a decrease in the suspension viscosity.

The decrease in viscosity with increasing shear stress can be described by the following equation:

$$\eta = \eta_\infty + \frac{\eta_0 - \eta_\infty}{1 + (\tau / \tau_i)} \qquad (8.31)$$

where τ_i is a critical shear stress that is related to the size of the droplets ($\tau_i = kT/\beta r^3$), and β is a dimensionless constant with a value of about 0.431 (Hunter 1989). The value of τ_i is a characteristic of a particular system which describes the relative importance of the *translational* Brownian motion and hydrodynamic shear forces. When $\tau \ll \tau_i$, Brownian motion dominates and the particles have a random distribution, but when $\tau \gg \tau_i$, the shear forces dominate and the particles become organized into "strings" or "layers" along the lines of the shear field, which causes less energy dissipation. This equation indicates that the viscosity decreases from a constant value at low shear stresses (η_0) to another constant value at high shear stresses (η_∞). The viscosity can decrease by as much as 30% from its low shear rate value, with the actual amount depending on the dispersed-phase volume fraction (Hunter 1989). The shear rate at which the viscosity starts to decrease from its η_0 value is highly dependent on the particle size. For large particles, τ_i is often so low that it is not possible to observe any shear thinning behavior, but for smaller particles, shear thinning behavior may be observed at the shear rates typically used in a rheological experiment.

The Dougherty–Krieger equation can still be used to describe the dependence of the suspension viscosity on dispersed-phase volume fraction, but the value of ϕ_c used in the equation is shear rate dependent. This is because droplets can pack more efficiently at higher shear rates, and therefore ϕ_c increases with shear rate (Hunter 1989).

8.4.6.   Suspensions of Nonflocculated Particles with Repulsive Interactions

Most food emulsions contain droplets which have various types of colloidal interaction acting between them (e.g., van der Waals, electrostatic, steric, hydrophobic, depletion, etc.) (Chapter 3). The precise nature of these interactions has a dramatic influence on the rheology of particulate suspensions. For example, two emulsions with the same droplet concentration could have rheological properties ranging from a low-viscosity Newtonian liquid to a highly viscoelastic material, depending on the nature of the colloidal interactions. In this section, suspensions that contain isolated spherical particles with repulsive interactions are considered. In the following section, the rheological properties of flocculated emulsions are considered.

The major types of droplet repulsion in most food emulsions are due to electrostatic and steric interactions. These repulsive interactions prevent the droplets from coming into close contact when they collide with each other and therefore increase the effective volume fraction of the droplets (Tadros 1994, Mewis and Macosko 1994):

$$\phi_{eff} = \phi\left(1 + \frac{\delta}{r}\right)^3 \qquad (8.32)$$

where δ is equal to half the distance of closest separation between the two droplets.

For steric stabilization, δ is approximately equal to the thickness of the adsorbed layer. For electrostatically stabilized systems, it is related to the Debye length (κ^{-1}) and can be described by the following equation at low shear stresses (Mewis and Macosko 1994):

$$\delta = \kappa^{-1} \ln\left(\frac{\alpha}{\ln[\alpha /(\ln \alpha)]}\right) \tag{8.33}$$

where $\alpha = 4\pi \, \varepsilon_0\varepsilon_R\Psi_0r^2\kappa \, \exp(2r\kappa)/kT$, ε_0 is the dielectric permittivity of a vacuum, ε_R is the relative dielectric permittivity of the continuous phase, Ψ_0 is the electrical potential at the droplet surface, r is the radius, κ^{-1} is the Debye length, k is Boltzmann's constant, and T is the absolute temperature. For electrically charged oil droplets, the distance of closest contact therefore decreases as the surface charge decreases or as the ionic strength of the aqueous phase increases.

It is convenient to categorize droplets with repulsive interactions as being either "hard" particles or "soft" particles (Liu and Masliyah 1996). A hard particle is incompressible, and so its effective size is independent of shear rate or droplet concentration. On the other hand, a soft particle is compressible, and so its effective size may be reduced at high shear rates or droplet concentrations. Sterically stabilized droplets with dense interfacial layers are usually considered to act like hard particles because the layer is relatively incompressible, whereas electrostatically stabilized droplets or sterically stabilized droplets with open interfacial layers are usually considered to act like soft particles because the layer is compressible.

The viscosity of an emulsion that contains droplets with repulsive interactions can be related to the dispersed-phase volume fraction by replacing the value of ϕ in the Dougherty–Krieger equation (Equation 8.33) with the effective volume fraction (ϕ_{eff}). A plot of viscosity versus ϕ_{eff} then falls on the same curve as for an emulsion that contains droplets with no long-range colloidal interactions. In emulsions that contain "soft" particles, it is also necessary to replace the value of ϕ_c with $\phi_{c,eff}$, to take into account that the particles may be compressed at higher volume fractions and can therefore pack more efficiently. As a consequence, the viscosity of an emulsion that contains soft particles is lower than one that contains hard particles at the same *effective* volume fraction.

The influence of repulsive interactions on the rheology of emulsions depends on the magnitude of δ relative to the size of the particles. For relatively large particles (i.e., $\delta \ll r$), this effect is negligible, but for small droplets or droplets with thick layers around them (i.e., $\delta \approx r$), this effect can significantly increase the viscosity of a suspension (Tadros 1994).

The rheological properties of electrostatically stabilized systems that contain small emulsion droplets are particularly sensitive to pH and salt concentration. The viscosity would decrease initially with increasing salt concentration because screening of the charges decreases δ. Above a certain salt concentration, the interactions between the droplets would become attractive, rather than repulsive, and therefore flocculation will occur, causing an increase in emulsion viscosity with salt. The rheological properties of electrostatically stabilized emulsions are therefore particularly sensitive to the pH, salt concentration, and type of ions present.

8.4.7. Concentrated Suspensions of Flocculated Particles

In concentrated emulsions, the flocs are close enough together to interact with each other, through hydrodynamic interactions, colloidal interactions, or entanglement. The viscosity of flocculated emulsions can be described by the Dougherty–Krieger equation by assuming that the flocs are "soft" particles with the *actual* droplet volume fraction replaced by the *effective* droplet volume fraction given by Equation 8.27. At a given actual droplet volume fraction, the emulsion viscosity increases as the size of the flocs increases or the packing of the flocs becomes more open (lower D).

At sufficiently high droplet concentrations, flocculation may lead to the formation of a three-dimensional network of aggregated droplets (Sherman 1970, Goodwin and Ottewill 1991, Pal 1996). The more open the structure of the droplets within the flocs, the lower the value of the actual droplet volume fraction where the network is formed. Network formation causes a suspension of particles to exhibit plastic and/or viscoelastic characteristics (Pal 1996). The network of aggregated droplets acts like a solid at low shear stresses because the applied forces are not sufficient to overcome the forces holding the droplets together. Once a critical shear stress is exceeded, the bonds between the droplets are disrupted and so the droplets can flow past one another. If some of the bonds are capable of reforming during the shearing process, then the emulsion will exhibit viscoelastic behavior (Sherman 1968a). At higher shear stresses, the rate of bond disruption greatly exceeds that of bond formation and the emulsion acts like a liquid. Consequently, a suspension that contains a three-dimensional network of aggregated particles often has a yield stress, below which it acts like an elastic solid and above which it acts like a liquid. Above the yield stress, the suspension often exhibits strong shear thinning behavior as more and more flocs are deformed and disrupted. The magnitude of the yield stress depends on the strength of the attractive forces holding the particles together: the greater the attractive forces, the greater the yield stress (Pal 1996). The rheology of the system is also sensitive to the structural organization of the droplets (e.g., whether they are loosely or densely packed and the number of bonds per droplet) (Bremer 1992, Pal 1996).

8.4.8. Emulsions with Semisolid Continuous Phases

A number of food emulsions consist of droplets dispersed in a continuous phase which is either partly crystalline or gelled (Sherman 1970, Dickinson and Stainsby 1982, Dickinson 1992, Moran 1994). Butter and margarine consist of water droplets suspended in a liquid oil phase which contains a three-dimensional network of aggregated fat crystals. Many meat products, desserts, and sauces consist of oil droplets suspended in an aqueous phase of aggregated biopolymer molecules. The rheological properties of these systems are usually dominated by the properties of the continuous phase, and therefore their properties can be described using theories developed for networks of aggregated fat crystals or biopolymer molecules (Clark 1987). Nevertheless, in some systems the presence of the emulsion droplets does play a significant role in determining the overall rheological behavior (see below).

It is important that spreadable products, such as butter, margarine, and low-fat spreads, retain their shape when they are removed from the refrigerator, but spread easily when a knife is applied (Sherman 1970, Moran 1994). These products must therefore be designed so that they exhibit plastic properties (i.e., they have a yield stress below which they are elastic and above which they are viscous). The plastic behavior of this type of product is usually attributed to the presence of a three-dimensional network of aggregated fat crystals. Low shear stresses are not sufficiently large to disrupt the bonds which hold the aggregated crystals together, and so the product exhibits solid-like behavior. Above the yield stress, the applied shear stress is sufficiently large to cause the bonds to be disrupted, so that the fat crystals flow over each other and the product exhibits viscous-like behavior. After the stress is removed, the bonds between the fat crystals reform over time, and therefore the product regains its elastic behavior. The creation of a product with the desired rheological characteristics involves careful selection and blending of various food oils, as well as control of the cooling and shearing conditions used during the manufacture of the product. To the author's knowledge, little systematic research has been carried out to establish the influence of the characteristics of the water droplets on the rheological properties of these products (e.g., particle size distribution, emulsifier type, and dispersed-phase volume fraction).

A great deal of research has recently been carried out to determine the influence of oil droplets on the rheology of filled gels (Jost et al. 1986; Aguilera and Kessler 1989; Xiong et al. 1991; Xiong and Kinsella 1991; Yost and Kinsella 1993; McClements et al. 1993c; Dickinson and Hong 1995a,b, 1996; Dickinson and Yamamoto 1996). These filled gels are created by heating oil-in-water emulsions which contain a significant amount of whey protein in the continuous phase above a temperature where the proteins unfold and form a three-dimensional network of aggregated molecules. The oil droplets can act as either structure *promoters* or structure *breakers,* depending on the nature of their interaction with the gel network. When the droplets are stabilized by dairy proteins, the attractive interactions between the adsorbed proteins and those in the network reinforce the network and increase the gel strength. Conversely, when the droplets are stabilized by small-molecule surfactants (which do not interact strongly with the protein network), the presence of the droplets tends to weaken the network and decrease the gel strength. Quite complex rheological behavior can be observed in emulsions that contain mixtures of proteins and small-molecule surfactants (Dickinson and Hong 1995a,b; Dickinson and Yamamoto 1996). The shear modulus of filled gels that contain protein-stabilized oil droplets increases dramatically when a small amount of surfactant is added to the system, but decreases at higher surfactant values. The incorporation of surfactants into filled gels may therefore prove to be an effective means of controlling their rheological properties.

The influence of the emulsion droplets also depends on their size relative to the pore size of the gel network (McClements et al. 1993c, Yost and Kinsella 1993). If the droplets are larger than the pore size, they tend to disrupt the network and decrease the gel strength, but if they are smaller than the pore size, they are easily accommodated into the network without disrupting it.

8.5. MAJOR FACTORS THAT DETERMINE EMULSION RHEOLOGY

8.5.1. Dispersed-Phase Volume Fraction

The viscosity of an emulsion increases with dispersed-phase volume fraction. At low droplet concentrations, this increase is linearly dependent on volume fraction (Equation 8.19), but it becomes steeper at higher concentrations (Equation 8.30). Above a critical dispersed-phase volume fraction (ϕ_c), the droplets are packed so closely together that they cannot easily flow past each other, and so the emulsion has gel-like properties. The precise nature of the dependence of the viscosity on volume fraction is mainly determined by the nature of the colloidal interactions between the droplets.

8.5.2. Rheology of Component Phases

The viscosity of an emulsion is directly proportional to the viscosity of the continuous phase (Equations 8.19 and 8.30), and so any alteration in the rheological properties of the continuous phase has a corresponding influence on the rheology of the whole emulsion. For this reason, the presence of a thickening agent in the aqueous phase of an oil-in-water emulsion (Pettitt et al. 1995, Pal 1996) or the presence of a fat crystal network in the oil phase of a water-in-oil emulsion (Moran 1994) largely determines the overall rheological properties of the system.

The rheology of the dispersed phase has only a minor influence on the rheology of food emulsions because the droplets are covered with a fairly viscoelastic membrane, which means they have properties similar to rigid spheres (Tadros 1994, Walstra 1996a).

8.5.3. Droplet Size

The influence of the droplet size and the droplet size distribution on the rheology of an emulsion depends on the dispersed-phase volume fraction and the nature of the colloidal interactions. The viscosity of dilute emulsions is independent of the droplet size when there are no long-range attractive or repulsive colloidal interactions between the droplets (Pal et al. 1992, Pal 1996). When there is a relatively long-range repulsion between the droplets (i.e., $\delta \sim r$), their *effective* volume fraction is much greater than their *actual* volume fraction, and so there is a large increase in the viscosity of the emulsion (Tadros 1994, Pal 1996). The droplet size also influences the degree of droplet flocculation in an emulsion (Chapters 3 and 7), which has an impact on the emulsion rheology. For example, the greater the extent of droplet flocculation, or the more open the structure of the flocs formed, the larger the emulsion viscosity.

The mean droplet size and degree of polydispersity have a particularly significant influence on the rheology of concentrated emulsions (Liu and Masliyah 1996, Pal 1996). In emulsions that contain nonflocculated droplets, the maximum packing factor (ϕ_c) depends on the polydispersity. Droplets are able to pack more efficiently when they are polydisperse, and therefore the viscosity of a concentrated polydisperse emulsion is less than that of a monodisperse emulsion with the same droplet volume fraction. Emulsions that contain flocculated droplets are able to form a three-dimensional gel network at lower volume fractions when the droplet size decreases.

The droplet size also alters the rheology of emulsions due to its influence on the relative importance of rotational and translation Brownian motion effects compared to the shear stress (Mewis and Macosko 1994).

8.5.4. Colloidal Interactions

The nature of the colloidal interactions between the droplets in an emulsion is one of the most important factors determining its rheological behavior. When the interactions are long range and repulsive, the effective volume fraction of the dispersed phase may be significantly greater than its actual volume fraction, $\phi_{eff} = \phi(1 + \delta/r)^3$, and so the emulsion viscosity increases (Section 8.4.6). When the interactions between the droplets are sufficiently attractive, the effective volume fraction of the dispersed phase is increased due to droplet flocculation, which results in an increase in emulsion viscosity (Section 8.4.4). The rheological properties of an emulsion therefore depend on the relative magnitude of the attractive (mainly van der Waals, hydrophobic, and depletion interactions) and repulsive (mainly electrostatic, steric, and thermal fluctuation interactions) interactions between the droplets (Chapter 3). A food scientist can therefore control the rheological properties of food products by manipulating the colloidal interactions between the droplets. Increases in the viscosity of oil-in-water emulsions due to droplet flocculation have been induced by adding biopolymers to increase the depletion attraction (Dickinson and Golding 1997a,b), by adding biopolymers to cause bridging flocculation (Dickinson and Golding 1997a), by altering the pH or ionic strength to reduce electrostatic repulsion (Hunt and Dalgleish 1994, Demetriades et al. 1997a), and by heating protein-stabilized emulsions to increase hydrophobic attraction (Demetriades et al. 1997b).

8.5.5. Particle Charge

The charge on an emulsion droplet can influence the rheological properties of an emulsion in a number of ways. First, the charge determines whether the droplets are aggregated or unaggregated and the distance of closest approach (Section 8.5.4). Second, the droplet charge influences the rheology due to the *primary electroviscous effect* (Pal 1996). As a

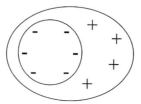

FIGURE 8.18   The primary electroviscous effect increases the viscosity of an emulsion.

charged droplet moves through a fluid, the cloud of counterions surrounding it becomes distorted (Figure 8.18). This causes an attraction between the charge on the droplet and that associated with the cloud of counterions that lags slightly behind it. This attraction opposes the movement of the droplets and therefore increases the emulsion viscosity because more energy is needed to cause the droplets to move at the same rate as uncharged droplets.

9 Appearance and Flavor

9.1. INTRODUCTION

The modern consumer is faced with a huge variety of different types of food products from which to choose, and within each category there are a number of different brand names. The initial choice of a particular product is governed by many factors, including its cost, quality, packaging, ease of preparation, and nutritional value. Once a consumer has made a decision to purchase a certain brand name, the manufacturer wants to ensure that he or she is satisfied with the product and will purchase it again. It is therefore important to make certain that the product has desirable and reproducible quality attributes (i.e., appearance, taste, texture, and shelf life). In previous chapters, we considered emulsion rheology, which is related to texture (Chapter 8), and emulsion stability, which is related to shelf life (Chapter 7). In this chapter, focus is on the factors which influence the flavor and appearance of food emulsions.

9.2. EMULSION FLAVOR

The term "flavor" refers to those volatile components in foods which are sensed by receptors in the nose (aroma) and those nonvolatile components which are sensed by receptors in the tongue and the inside of the mouth (taste) (Thomson 1986). In addition, certain components in foods may also contribute to flavor because of their influence on the perceived texture (mouthfeel) (Kokini 1987). The flavor of a food is therefore a combination of aroma, taste, and mouthfeel, with the former usually the most important (Taylor and Linforth 1996). Flavor perception is an extremely complicated process which depends on a combination of physicochemical, biological, and psychological phenomena (Thomson 1986, Bell 1996). Before a food is placed in the mouth, its flavor is perceived principally through those volatile components which are inhaled directly into the nasal cavity. After the food is placed in the mouth, the flavor is determined by nonvolatile molecules which leave the food and are sensed by receptors on the tongue and the inside of the mouth, as well as by those volatile molecules which are drawn into the nasal cavity through the pharynx at the back of the mouth (Thomson 1986). The interactions between flavor molecules and human receptors which lead to the perceived flavor of a food are extremely complicated and are still fairly poorly understood (Thomson 1986). In addition, expectations and eating habits vary from individual to individual, so that the same food may be perceived as tasting differently by two separate individuals or by a single individual at different times. This section focuses on the physicochemical aspects of flavor partitioning and release in foods, because these are the most relevant topics to emulsion science. Although the physiological and psychological aspects of flavor are extremely important, they are beyond the scope of this book.

It is widely recognized that the perceived flavor of a food is not simply determined by the type and concentration of the flavor molecules present (Taylor and Linforth 1996), but by a number of other physicochemical factors as well, including:

1. The environment (matrix) in which the flavor molecules are located (e.g., the lipid, aqueous, or interfacial regions)
2. The physical state of the environment (e.g., gas, liquid or solid)
3. The structural organization of the components (e.g., emulsion versus nonemulsion)
4. The chemical state of the flavor molecules (e.g., degree of ionization or self-association)
5. The physical and chemical interaction of flavors with other molecules (e.g., proteins, carbohydrates, surfactants, or minerals)
6. The rate at which flavor molecules move from one environment to another

Given the large number of factors which contribute to food flavor, it is extremely difficult to accurately predict the flavor of a final product from a knowledge of its physicochemical characteristics. For this reason, the formulation of food flavors is often the result of art and craft, rather than the application of fundamental scientific principles. Even so, a more rigorous scientific approach to this topic would have great benefits for the food industry because it would enable manufacturers to design foods in a more systematic and cost-effective manner. In this section, some of the more fundamental aspects of the science of emulsion flavor are reviewed.

9.2.1. Flavor Partitioning

The perception of a flavor depends on the precise location of the flavor molecules within an emulsion. Most flavors are perceived more intensely when they are present in the aqueous phase rather than the oil phase (McNulty 1987, Kinsella 1989). The aroma is determined by the presence of volatile molecules in the vapor phase above an emulsion (Overbosch et al. 1991, Taylor and Linforth 1996). Certain flavor molecules may associate with the interfacial region, which alters their concentration in the vapor and aqueous phases (Wedzicha 1988). It is therefore important to establish the factors which determine the partitioning of flavor molecules within an emulsion. An emulsion system can be conveniently divided into four phases between which the flavor molecules distribute themselves: the interior of the droplets, the continuous phase, the oil–water interfacial region, and the vapor phase above the emulsion.* The relative concentration of the flavor molecules in each of these regions depends on their molecular structure and the properties of each of the phases (Baker 1987, Bakker 1995). In this section, we start by examining flavor partitioning in some simple model systems and then move on to some more complicated and realistic model systems. It should be stressed that most of the physical principles of flavor partitioning in emulsions are also applicable to the partitioning of other types of food ingredients, such as antioxidants, colors, preservatives or vitamins (Wedzicha 1988, Coupland and McClements 1996, Huang et al. 1997).

9.2.1.1. Partitioning in Homogenous Liquids

The simplest situation to consider is the partitioning of a flavor between a homogeneous liquid and the vapor above it (Figure 9.1). At thermodynamic equilibrium, the flavor distributes itself between the liquid and vapor according to the equilibrium partition coefficient (Wedzicha 1988):

---

* Flavors could also be present at the air–emulsion interface, but the interfacial area of this region is usually so small that the amount of flavor involved is negligible.

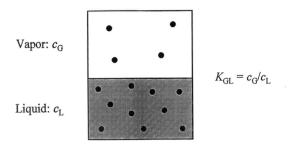

FIGURE 9.1 Flavor molecules distribute themselves between a bulk liquid and its vapor phase according to their equilibrium partition coefficient ($K_{GL} = c_G/c_L$).

$$K_{GL} = \frac{a_G}{a_L} \tag{9.1}$$

where a_G and a_L are the activity coefficients of the flavor in the gas and liquid phases, respectively. The concentration of flavors in foods is usually very low, and so the activity coefficients can be replaced by concentrations, since interactions between flavor molecules are insignificant (Wedzicha 1988, Overbosch et al. 1991):

$$K_{GL} = \frac{c_G}{c_L} \tag{9.2}$$

where c_G and c_L are the concentrations of the flavor in the gas and liquid phases, respectively. It is often more convenient to represent the partitioning of a flavor as the mass fraction of the total amount of flavor in the system which is in the vapor phase:

$$\Phi_m = \frac{1}{1 + V_L / (V_G K_{GL})} \tag{9.3}$$

where V_G and V_L are the volumes of the gas and liquid phases, respectively.

The magnitude of the partition coefficient depends on the relative strength of the interactions between a flavor molecule and its surroundings in the gas and liquid phases (Tinoco et al. 1985, Israelachvili 1992):

$$K_{GL} = \exp\left(-\frac{\Delta G_{GL}}{RT}\right) \tag{9.4}$$

where ΔG_{GL} is the difference in free energy per mole of the solute in the gas and liquid phases.

The free energy term depends on the change in molecular interactions and configurational entropy which occurs when a flavor molecule moves from the vapor phase into the liquid phase. The configurational entropy favors the random distribution of the flavor molecules throughout the whole volume of the system, rather than their confinement to just the liquid phase, and therefore it favors volatilization. The change in energy associated with the molecular interactions is determined by the formation of solvent–flavor bonds ($\approx z w_{SF}$) and the disruption of solvent–solvent bonds ($\approx \frac{1}{2} z w_{SS}$), which occurs when a flavor molecule moves

TABLE 9.1
Equilibrium Partition Coefficients of Some Common Flavors in Oil and Water at 25°C

Compound	Air–water (K_{GW})	Air–oil (K_{GO})
Aldehydes		
Acetaldehyde	2.7×10^{-3}	
Propanal	3.0×10^{-3}	
Butanal	4.7×10^{-3}	2.3×10^{-3}
Pentanal	6.0×10^{-3}	1.0×10^{-3}
Hexanal	8.7×10^{-3}	0.35×10^{-3}
Heptanal	11×10^{-3}	0.10×10^{-3}
Octanal	21×10^{-3}	0.04×10^{-3}
Nonanal	30×10^{-3}	
Ketones		
Butan-2-one	1.9×10^{-3}	1.9×10^{-3}
Pent-2-one	2.6×10^{-3}	
Heptan-2-one	5.9×10^{-3}	1.03×10^{-3}
Oct-2-one	7.7×10^{-3}	
Nonan-2-one	15×10^{-3}	
Undecan-2-one	26×10^{-3}	
Alcohols		
Methanol	0.18×10^{-3}	
Ethanol	0.21×10^{-3}	10×10^{-3}
Propanol	0.28×10^{-3}	
Butanol	0.35×10^{-3}	0.57×10^{-3}
Pentanol	0.53×10^{-3}	
Hexanol	0.63×10^{-3}	0.36×10^{-3}
Heptanol	0.77×10^{-3}	
Octanol	0.99×10^{-3}	0.09×10^{-3}

Data compiled from Buttery et al. (1969, 1971, 1973) and Overbosch et al. (1991).

from the vapor into the solvent.* Here, z is the coordination number of the flavor molecules, and w_{SS} and w_{SF} are the solvent–solvent and solvent–flavor interaction energies (Chapter 2). The overall change therefore depends on the relative strength of both the solvent–solvent and solvent–flavor interactions: $\Delta w \approx z w_{SF} - \frac{1}{2} z w_{SS}$.

The molecular characteristics of flavor molecules largely determine their partition coefficient in different solvents (Table 9.1). When a nonpolar flavor molecule is dispersed in a nonpolar solvent, or when a polar flavor molecule is dispersed in a polar solvent, there is a decrease in volatility with increasing molecular weight (Buttery et al. 1973). This is because the number of favorable attractive interactions between the flavor molecules and the solvent increases as the size of the flavor molecules increases. On the other hand, when a nonpolar flavor molecule is dispersed in a polar solvent, there is a decrease in volatility with increasing molecular weight (Buttery et al. 1969, 1971; Bomben et al. 1973; Franzen and Kinsella 1975). Nonpolar flavors are less volatile in nonpolar solvents than in polar solvents (Buttery et al. 1973, Bakker 1995), because a number of relatively strong hydrogen bonds have to be

---

* This simple analysis assumes that the sizes of the solvent and flavor molecules are approximately equal.

replaced by relatively weak van der Waals bonds when a nonpolar molecule is introduced into a polar solvent, which is energetically unfavorable because of the hydrophobic effect. In addition, polar flavors are less volatile in polar solvents than in nonpolar solvents, because the hydrogen bonds which form in polar solvents are more strongly attractive than the van der Waals bonds which form in nonpolar solvents.

9.2.1.2. Influence of Flavor Ionization

A number of water-soluble flavors have chemical groups which are capable of undergoing proton association–dissociation as a result of changes in pH (e.g., $-COOH \rightarrow -COO^- + H^+$ or $-NH_3^+ \rightarrow -NH_2 + H^+$). The volatility and flavor characteristics of the different ionic forms of a molecule are different because of changes in their molecular interactions with the solvent (Baldwin et al. 1973, Wedzicha 1988, Guyot et al. 1996). For example, the ionized form of a flavor is less volatile in an aqueous solution than the nonionized form because of strong ion–dipole interactions between it and the surrounding water molecules. It is therefore important to take into account the effect of ionization on the partitioning of flavor molecules. The concentration of a specific ionic form at a certain pH can be determined using the Henderson–Hasselbach equation: $pH = pK_a - \log(c_{Acid}/c_{Base})$, where $pK_a = -\log(K_a)$ and K_a is the dissociation constant of the acidic group (Atkins 1994).

The volatility of the ionized form of a flavor is usually much lower than that of the nonionized form, and so the concentration of the flavor in the vapor phase is determined principally by the amount of nonionized flavor ($c_{L,N}$) present in the liquid phase. The partition coefficient is therefore given by:

$$K_{GL} = \frac{c_G}{c_{L,N}} = \frac{c_G}{c_L(1 + 10^{pH-pK_a})^{-1}} \tag{9.5}$$

In practice, it is more convenient to define an *effective* partition coefficient, which is equal to the concentration of the flavor in the vapor phase relative to the *total* amount of flavor in the liquid phase ($c_L = c_{L,I} + c_{L,N}$), where $c_{L,I}$ is the concentration of the ionized form of the flavor:

$$K_{GL}^e = \frac{c_G}{c_L} = \frac{K_{GL}}{(1 + 10^{pH-pK_a})} \tag{9.6}$$

When the pH of the aqueous solution is well below the pK_a value of the acid group, the flavor molecule is almost exclusively in the nonionized form ($K_{GL}^e = K_{GL}$), and so the flavor in the vapor phase is at its most intense. As the pH is raised toward the pK_a value of the acid group, the fraction of flavor molecules in the nonionized form decreases ($K_{GL}^e < K_{GL}$), and so the flavor in the vapor phase becomes less intense.

It should be stressed that the ionization of a flavor molecule may also influence the partition coefficient because it alters its interactions with other charged molecules within the aqueous phase. For example, there may be attractive electrostatic interactions between an ionized flavor molecule and an oppositely charged biopolymer, which leads to flavor binding and therefore a reduction of its concentration in the vapor phase (Section 9.2.1.3).

9.2.1.3. Influence of Flavor Binding

Many proteins and carbohydrates are capable of binding flavor molecules and therefore altering their distribution within an emulsion (Franzen and Kinsella 1975, Bakker 1995,

O'Neill 1996, Hansen and Booker 1996, Hau et al. 1996). Flavor binding can cause a significant alteration in the perceived flavor of a food. This alteration is often detrimental to food quality because it changes the characteristic flavor profile, but it can also be beneficial when the bound molecules are off-flavors. A flavor chemist must therefore take binding effects into account when formulating the flavor of a particular product.

The equilibrium partition coefficient of a flavor in the presence of a biopolymer is given by:

$$K_{GL} = \frac{c_G}{c_{L,F}} \tag{9.7}$$

where $c_{L,F}$ is the concentration of free (unbound) flavor present in the liquid. The concentration of free flavor depends on the nature of the binding between the flavor and biopolymer (Overbosch et al. 1991). Binding may be the result of covalent bond formation or physical interactions, such as electrostatic, hydrophobic, van der Waals, or hydrogen bonds (Chapter 2). It may take place at specific sites on the surface of a biopolymer molecule or nonspecifically at any location on the surface. It may be reversible or irreversible.

The extent of flavor binding to a biopolymer can be conveniently characterized by a *binding coefficient*:

$$K^* = \frac{c_{L,B}}{c_{L,F}} \tag{9.8}$$

where $c_{L,B}$ is the concentration of bound flavor in the liquid phase ($c_L = c_{L,F} + c_{L,B}$). The stronger the binding between the flavor and a biopolymer, the greater the value of K^*.

It is convenient to define an *effective* partition coefficient, which relates the concentration of flavor in the gas phase to the total amount of flavor in the liquid phase (Overbosch et al. 1991):

$$K_{GL}^e = \frac{c_G}{c_L} = \frac{K_{GL}}{1 + K^*} \tag{9.9}$$

This equation indicates that the concentration of flavor in the vapor phase is not influenced by the biopolymer when the binding coefficient is small ($K^* \ll 1$), but that there is a large reduction in the concentration in the vapor phase when the binding is strong ($K^* > 1$).

Binding constants are frequently measured experimentally by equilibrium dialysis. A biopolymer solution is placed inside a semipermeable dialysis bag, which is then suspended in a solution of the flavor molecules (Figure 9.2). The large biopolymer molecules are restricted to the inside of the dialysis bag, while the small flavor molecules can move through it. After the system has reached equilibrium, the amount of flavor bound to the biopolymer molecules is determined by measuring the concentration of flavor in the bag above that which would be expected in the absence of biopolymer. An alternative technique which is also widely used to measure flavor binding is head space analysis (Section 9.2.3). In this technique, the partition coefficient is determined by measuring the equilibrium concentration of volatiles in the head space above a solution. By measuring the partition coefficient at different biopolymer concentrations, it is possible to determine the binding constant.

Flavor molecules, such as alkanes, aldehydes, and ketones, have been shown to be capable of specifically binding to various different types of protein (e.g., casein, β-lactoglobulin,

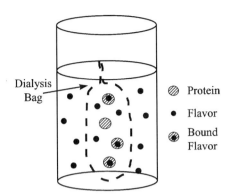

FIGURE 9.2 The principles of equilibrium dialysis. A biopolymer solution is placed in a dialysis bag which is then suspended in a flavor solution. The amount of flavor bound to the biopolymer molecules is determined by analyzing the flavor concentration in the cell.

bovine serum albumin, and soy protein) (Langourieux and Crouzet 1995, O'Neill 1996, Hansen and Booker 1996, Boudaud and Dumont 1996). These proteins have little flavor themselves, but can cause significant changes in the flavor profile of an emulsion by binding either desirable or undesirable flavors. The extent of flavor binding depends on the molecular structure of the flavor and protein molecules. Nonpolar flavors are believed to bind to nonpolar patches on the surfaces of proteins through hydrophobic attraction. An increase in binding of flavors to β-lactoglobulin has been observed in the presence of urea or on heating, because these treatments cause partial unfolding of the proteins, which increases their surface hydrophobicity (O'Neill 1996). Flavors may also bind to polysaccharides, but in this case the molecular interactions involved are more likely to be van der Waals, electrostatic, or hydrogen bonds (Hau et al. 1996). In some systems, there may be irreversible binding due to chemical reactions between biopolymer and flavor molecules.

9.2.1.4. Influence of Surfactant Micelles

Surfactants are normally used to physically stabilize emulsion droplets against aggregation by providing a protective membrane around the droplet (Chapter 7). Nevertheless, there is often enough free surfactant present in an aqueous phase to form surfactant micelles (Section 4.5). These surfactant micelles are capable of solubilizing nonpolar molecules in their hydrophobic interior, which increases the affinity of nonpolar flavors for the aqueous phase and therefore decreases their partition coefficient (K_{GL}). By a similar argument, reverse micelles in an oil phase are capable of solubilizing polar flavor molecules and therefore decreasing their partition coefficient (K_{GL}).

The partitioning of flavor compounds between a surfactant solution and the gas above it can be described by the following equation:

$$\frac{1}{K_{GL}} = \frac{\phi_M}{K_{GM}} + \frac{\phi_C}{K_{GC}} \tag{9.10}$$

where ϕ_M is the volume fraction of the micelles, ϕ_C is the volume fraction of the continuous phase surrounding the micelles ($\phi_M + \phi_C = 1$), K_{GM} ($= c_G/c_M$) is the gas–micelle partition coefficient, and K_{GC} ($= c_G/c_C$) is the gas–continuous phase partition coefficient. It is difficult to directly measure the partitioning of a flavor between gas and micelles, and so it is more convenient to rewrite the above equation in the following form:

$$K_{GL} = \frac{K_{GC}}{[1 + \phi_M (K_{MC} - 1)]} \tag{9.11}$$

where K_{MC} ($= c_M/c_C$) is the partition coefficient of the flavor molecules between the micelles and the continuous phase. Equation 9.11 indicates that the higher the affinity of the flavor molecules for the micelles ($K_{MC} \gg 1$), the smaller the concentration of flavor above the solution.

9.2.1.5. Partitioning in Emulsions in the Absence of an Interfacial Membrane

In an emulsion, which consists of two immiscible liquids, we must consider the partitioning of the additive between the emulsion droplets and the surrounding liquid, as well as into the vapor phase (Figure 9.3). We therefore have to define three different partition coefficients:

$$K_{DC} = \frac{c_D}{c_C} \qquad K_{GC} = \frac{c_G}{c_C} \qquad K_{GD} = \frac{c_G}{c_D} \tag{9.12}$$

where K is the partition coefficient; c is the concentration; and the subscripts D, C, and G refer to the dispersed, continuous, and gas phases, respectively. The partitioning of the additive between the oil and aqueous phases depends on the relative strength of its molecular interactions with its surroundings in the two different phases. Thus nonpolar molecules will tend to exist preferentially in the oil phase, while polar molecules will tend to exist preferentially in the aqueous phase. Thus, for oil-in-water emulsions, K_{DC} is large for nonpolar molecules and small for polar molecules, while the opposite is true for water-in-oil emulsions.

The partition coefficient of a flavor molecule between an emulsion and vapor phase is given by (Overbosch et al. 1991):

$$\frac{1}{K_{GE}} = \frac{\phi_D}{K_{GD}} + \frac{\phi_C}{K_{GC}} \tag{9.13}$$

where $\phi_D + \phi_C = 1$. Thus the partition coefficient between an emulsion and its vapor can be predicted from a knowledge of K_{GD} and K_{GC}. Experiments with flavor compounds dispersed

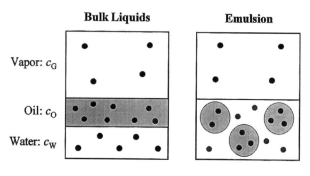

$$K_{OW} = c_O/c_W \qquad K_{GW} = c_G/c_W \qquad K_{GO} = c_G/c_O$$

FIGURE 9.3 In a two-liquid system, the flavor partitions between the oil, water, and vapor phases according to the partition coefficients.

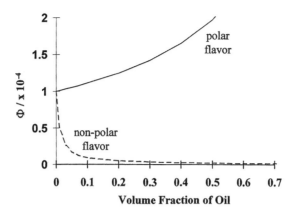

FIGURE 9.4   Influence of dispersed-phase volume fraction on the concentration of a polar (K_{DC} = 0.01) and a nonpolar (K_{DC} = 100) flavor in the vapor phase of an oil-in-water emulsion (V_G = 10 cm³, V_E = 100 cm³).

in oil-in-water emulsions have shown that this equation gives a good description of the behavior of emulsions, provided the flavor does not interact with the interface or any free emulsifier in the aqueous phase (Guyot et al. 1996).

Predictions of the mass fraction of flavor in the vapor phase of oil-in-water emulsions with the same overall flavor concentration but different dispersed-phase volume fractions are shown in Figure 9.4. There is a decrease in the fraction of a nonpolar flavor in the vapor phase as the oil content increases, whereas the amount of a polar flavor is relatively unaffected. Thus the nonpolar flavors in an emulsion become more odorous as the fat content is decreased, whereas the polar flavors remain relatively unchanged. This has important consequences when deciding the type and concentration of flavors to use in low-fat analogs of existing emulsion-based food products.

One possible limitation of Equation 9.13 is that it does not take into account the droplet size. The assumption that the partitioning of additives is independent of particle size is likely to be valid for emulsions which contain fairly large droplets (i.e., d > 1 μm), because the influence of the curvature of a droplet on the solubility of the material within it is not significant (Hunter 1986). However, when the droplet diameter falls below a critical size, there is a large increase in the solubility of the material within it because of the increased Laplace pressure. Thus one would expect the flavor concentration in the continuous phase and in the vapor phase to increase as the size of the droplets in an emulsion decreased. These predictions are supported by experimental evidence which indicates that the partition coefficient of a nonpolar additive in an emulsion is less than in a macroscopic two-phase system with the same overall composition (Matsubara and Texter 1986, Texter et al. 1987, Wedzicha 1988).

9.2.1.6.   Partitioning in Emulsions in the Presence of an Interfacial Membrane

Even though the interfacial region constitutes only a small fraction of the total volume of an emulsion, it can have a pronounced influence on the partitioning of surface-active molecules, especially when they are present at low concentrations, which is usually the case for food flavors. The influence of the interfacial membrane can be highlighted by a simple calculation of the amount of a surface-active additive which can associate with it (Wedzicha 1988). Assume that the additive occupies an interfacial area of 1 m²/mg, which is typical of many surface-active components (Dickinson 1992). The interfacial area per unit volume of an emulsion is given by the following relationship: $A_S = 6\phi/d_{VS}$, where d_{VS} is the volume–surface mean diameter (McClements and Dungan 1993). If we assume that the additive is present in 100 cm^3 of an emulsion with a dispersed-phase volume fraction of 0.1 and a mean droplet diameter of 1 μm, then the total interfacial area of the droplets is 60 m^2. It would therefore take about 60 mg of additive to completely saturate the interface, which corresponds to a concentration of approximately 0.1 wt%. Many flavors are used at concentrations which are considerably less than this value, and therefore their ability to accumulate at an interface has a large influence on their partitioning within an emulsion.

The accumulation of a flavor at an interface reduces its concentration in the oil, water, and gaseous phases by an amount which depends on the interfacial area, the flavor concentration, and the affinity of the flavor for the interface.

Reversible Binding. When the binding between the flavor and the interface is reversible, we can define a number of additional partition coefficients:

$$K_{ID} = \frac{c_I}{c_D} \qquad K_{IC} = \frac{c_I}{c_C} \qquad K_{IG} = \frac{c_I}{c_G} \tag{9.14}$$

In this case, the partition coefficient between the gas and the emulsion is given by:

$$\frac{1}{K_{GE}} = \frac{\phi_D}{K_{GD}} + \frac{\phi_C}{K_{GC}} + \frac{\phi_L}{K_{GI}} \tag{9.15}$$

In practice, it is difficult to directly measure the partition coefficient between the gas and interfacial region (K_{GI}), and so it is better to express the equation in terms of properties which are simpler to measure (i.e., $K_{GI} = K_{GC}/K_{IC}$). In addition, the properties of the interface are usually better expressed in terms of the interfacial area rather than the volume fraction. Equation 9.15 can therefore be expressed in the following manner:

$$\frac{1}{K_{GE}} = \frac{\phi_D}{K_{GD}} + \frac{\phi_C}{K_{GC}} + \frac{A_S K_{IC}^*}{K_{GC}} \tag{9.16}$$

where $K_{IC}^* = \Gamma_I/c_C$ is the partition coefficient between the interface and the continuous phase and Γ_I is the mass of the additive per unit interfacial area. Thus the partition coefficient of an emulsion (K_{GE}) can be predicted from experimental measurements which are all relatively simple to carry out (i.e., K_{GC}, K_{GD}, and K_{IC}). This equation assumes that the concentration of flavor at the interface is well below the saturation level. Once the interfacial region becomes saturated with flavor, the remainder will be distributed between the bulk phases.

The influence of the interface on the volatility of a surface-active flavor molecule is shown in Figure 9.5. As the size of the droplets in the emulsion is decreased, the interfacial area increases, and therefore a greater amount of flavor associates with the interface, thereby reducing its volatility. Nevertheless, it should be stated that this type of behavior is only likely to occur when there is no free emulsifier in the aqueous phase, either as individual molecules or micelles. In practice, there is often free emulsifier in the aqueous phase and the flavor molecules may bind to it as well as to the interfacial region. In these systems, the flavor volatility is more likely to depend on the total concentration of emulsifier in the system rather than on the droplet size. Another factor which must be considered is that the concentration

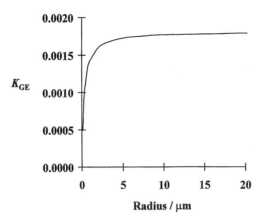

FIGURE 9.5 Influence of droplet size on the volatility of surface-active flavor molecules which associate with the interfacial membrane.

of a flavor in the continuous and gas phases may actually increase with decreasing droplet size because of the influence of the droplet curvature on solubility (Section 9.2.1.5).

Irreversible Binding. When the binding of the flavor to the interface is irreversible, then its concentration in the vapor phase is only determined by the amount of free flavor in the emulsion ($K_{GE} = c_G/c_{E,F}$). Under these circumstances, it is usually more convenient to use an *effective* partition coefficient, which is equal to the concentration of flavor in the vapor phase relative to the *total* amount of flavor in the emulsion ($K_{GE}^e = c_G/c_E$):

$$\frac{1}{K_{GE}^e} = \left(\frac{c_E}{c_E - A_S\Gamma_I} \right) \left(\frac{\phi_D}{K_{GD}} + \frac{\phi_C}{K_{GC}} \right) \tag{9.17}$$

where, Γ_I is now the amount of flavor that is irreversibly bound to the interface per unit surface area. If the flavor does not interact with the interface, and the interfacial region has a negligible volume, then this equation reduces to Equation 9.13, but if the flavor binds to the interface, its concentration in the vapor phase is reduced.

Recent studies using oil-in-water emulsions that contain different types of flavor compounds have indicated that amphiphilic flavors, such as butyric acid, bind strongly to the interface of a droplet and thus reduce the partition coefficient (K_{GL}) (Guyot et al. 1996). Nevertheless, a great deal of systematic research is still needed to determine the factors which influence the volatility of different flavor compounds in food emulsions. Special emphasis should be placed on establishing the molecular basis of this process so that predictions about the flavor profile of a food can be made from a knowledge of its composition and the type of flavor components present. This type of information could then be used by flavor chemists to formulate foods with specific flavor profiles.

9.2.2. Flavor Release

Flavor release is the process whereby flavor molecules move out of a food and into the surrounding saliva or vapor phase during mastication (McNulty 1987, Overbosch et al. 1991). The release of the flavors from a food material occurs under extremely complex and dynamic conditions (Land 1996). A food usually spends a relatively short period (typically 1 to 30 s) in the mouth before being swallowed. During this period, it is diluted with saliva, experiences temperature changes, and is subjected to a variety of mechanical forces. Mastication may therefore cause dramatic changes in the structural characteristics of a food.

During mastication, nonvolatile flavor molecules must move from within the food, through the saliva, to the taste receptors on the tongue and the inside of the mouth, whereas volatile flavor molecules must move from the food, through the saliva, and into the gas phase, where they are carried to the aroma receptors in the nasal cavity (Thomson 1986). The two major factors which determine the rate at which these processes occur are the equilibrium partition coefficient (because this determines the magnitude of the flavor concentration gradients at the various boundaries) and the mass transfer coefficient (because this determines the speed at which the molecules move from one location to another). In this section, we examine some of the simple models which have been developed to describe the complex processes which occur during the release of both nonvolatile and volatile flavor components from foods.

9.2.2.1. Release of Nonvolatile Compounds (Taste)

Ideally, we would like to know the maximum amount of flavor which can be released from a food and the time taken for this release to occur.

Maximum Amount Released. A relatively simple model, based on the equilibrium partition coefficient of the flavor between oil and water, has been used to describe the maximum amount of flavor which can be released by an oil-in-water emulsion when it is placed in the mouth (McNulty 1987). The model assumes that the food is initially at equilibrium, so that the distribution of the flavor between the droplets and continuous phase is given by the equilibrium partition coefficient (K_{ow}). When the food is placed in the mouth, it is diluted by saliva (Figure 9.6). Immediately after dilution, the concentration of flavor in the aqueous phase is reduced, and so there is a thermodynamic driving force which favors the release of flavor from the droplets until the equilibrium flavor distribution is restored.

The potential extent of the flavor release can be characterized by the ratio of the flavor in the aqueous phase once equilibrium has been reestablished to that immediately after dilution:

$$E_F = \frac{c_{We}}{c_{Wd}} = \frac{[\phi(K_{ow} - 1) + 1](DF - \phi)}{[\phi(K_{ow} - 1) + DF](1 - \phi)} \tag{9.18}$$

where DF is the dilution factor of the emulsion ($= V_f/V_i$), V_i and V_f are the emulsion volume before and after dilution, ϕ is the dispersed-phase volume fraction of the initial emulsion, c_{Wd} is the concentration of flavor in the aqueous phase immediately after dilution, and c_{We} is the concentration in the aqueous phase once equilibrium has been reestablished. The higher the

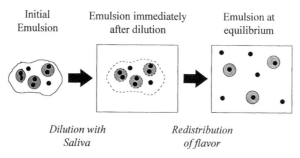

Initial   Emulsion immediately   Emulsion at
Emulsion   after dilution   equilibrium

Dilution with   Redistribution
Saliva   of flavor

FIGURE 9.6   The flavor in a food is initially distributed according to the partition coefficients. When it is diluted with saliva, the equilibrium is upset, and flavor is released from the droplets.

value of E_F, the greater the potential for flavor release. Despite its simplicity, this model can be used to make some valuable predictions about the factors which determine the flavor release from foods (McNulty 1987) (e.g., the extent of flavor release on dilution increases as either ϕ or K_{ow} increases). The major limitation of this model is that it provides no information about the rate at which the flavor is released from the droplets.

Kinetics of Flavor Release. The taste of an emulsion depends on the rate at which the flavor molecules move from the food to the receptors on the tongue and inside of the mouth. Flavor molecules may be located in either the oil or water phase, although it is widely believed that taste perception is principally a result of those molecules which are present in the water phase (McNulty 1987), because the flavor must cross an aqueous membrane before reaching the taste receptors (Thomson 1986). An indication of the kinetics of flavor release can therefore be obtained from a knowledge of the time dependence of the flavor concentration in the aqueous phase.

When an oil-in-water emulsion is diluted with saliva, some of the flavor molecules in the droplets move into the aqueous phase. A mathematical model has been developed to describe the rate at which a solute is released from a spherical droplet surrounded by a finite volume of a well-stirred liquid (Crank 1975):

$$\frac{M_t}{M_\infty} = 1 - \sum_{n=1}^{\infty} \frac{6\alpha(\alpha + 1) \exp(1 - Dq_n^2 t / r^2)}{9 + 9\alpha + q_n^2\alpha^2} \tag{9.19}$$

where M_t is the total amount of solute which has left the sphere by time t, M_∞ is the total amount of solute which has left the sphere once equilibrium has been established, D is the translational diffusion coefficient of the flavor within the droplets, t is the time, r is the droplet radius, $\alpha = 3V/(4\pi r^3)K_{DC}$, V is the volume of the continuous phase, and q_n are the nonzero roots of the relation, $\tan q_n = 3q_n/(1+\alpha q_n^2)$. This equation assumes that the concentration of solute (flavor) in the aqueous phase is initially zero, and therefore this equation is only strictly applicable to emulsions which are diluted with high concentrations of saliva. Nevertheless, it does provide some useful insights into the rate of flavor release from oil droplets.

The influence of droplet radius on the flavor release rate from droplets in a typical oil-in-water emulsion is shown in Figure 9.7. The release rate increases as the size of the droplets decreases. For the system shown in Figure 9.7, the time required for half of the flavor to leave the emulsion droplets is given by $t_{1/2} = (0.162r)^2/D$. The variation of $t_{1/2}$ with droplet radius

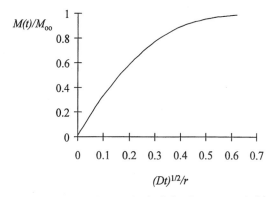

FIGURE 9.7   Kinetics of flavor release from spherical droplets suspended in a liquid.

TABLE 9.2
**Influence of Droplet Size on the
Time It Takes for the Flavor to
Diffuse Out of the Droplets**

Radius (μm)	$t_{1/2}$ (s)
0.1	6.6×10^{-7}
0.5	1.6×10^{-5}
1	6.6×10^{-5}
5	1.6×10^{-3}
10	6.6×10^{-3}
50	1.6×10^{-1}

is shown in Table 9.2. Flavor release is extremely rapid in emulsions that contain droplets less than about 10 μm, but becomes appreciably slower at large sizes. It therefore seems that the movement of flavor molecules from the droplets to the aqueous phase will not be the rate-limiting step in flavor release provided the emulsion is well-agitated and the droplets are relatively small. Instead, it is more likely that the breakdown of the structure and the diffusion of the molecules through the aqueous phase to the receptors are rate limiting. If the droplets are coated by interfacial membranes, the release rate may be slowed considerably and would be dependent on the type of emulsifier present, although little work has been carried out in this area (Harvey et al. 1995).

9.2.2.2. Release of Volatile Compounds (Aroma)

The release of volatile compounds from a food involves the mass transfer of the compounds through the emulsion and into the vapor phase (Overbosch et al. 1991). We are therefore interested in the change in the concentration of the flavor in the vapor phase with time.

Flavor Release from Homogeneous Liquids. The release rate of volatile flavors from a homogeneous liquid (e.g., oil or water) has been modeled by assuming that static diffusion conditions apply (Overbosch et al. 1991). A finite thickness of the liquid containing the flavor is assumed to be in contact with an infinite gas phase. The flavor moves from the liquid into the gas phase in order to try and establish the equilibrium partition coefficient (which can never be achieved because of the infinite extent of the gas phase). The flux of the flavor across the boundary separating the liquid and the gas and the change in the flavor concentration in the liquid with time are given by:

$$J = F c_L^0 \sqrt{D_L / \pi t} \tag{9.20}$$

$$M(t) = 2 F c_L^0 \sqrt{D_L t / \pi} \tag{9.21}$$

where

$$F = \frac{K_{GL} \sqrt{D_G / D_L}}{1 + K_{GL} \sqrt{D_G / D_L}} \tag{9.22}$$

Here, D_L and D_G are the translational diffusion coefficients of the flavor in the liquid and the gas phase, respectively; t is the time; K_{GL} ($= c_G/c_L$) is the equilibrium partition coefficient between the gas and liquid; and c_L^0 is the initial concentration of flavor in the liquid. The factor F is the driving force for the diffusion process: its value varies from 0 (low flavor release rate) to 1 (high flavor release rate). The value of F is determined principally by the equilibrium partition coefficient of the flavor between the gas and the liquid (K_{GL}), because there are much larger variations in this parameter than in diffusion coefficients for different flavors.

These equations indicate that the rate of flavor release from a good solvent (K_{GL} low) is much slower than the rate from a poor solvent (K_{GL} high). Thus a nonpolar flavor will be released more slowly from a nonpolar solvent than a polar solvent and vice versa. This accounts for the fact that the ranking of the release rates of flavors from water is opposite to that from oil (Overbosch et al. 1991). It also indicates that anything which decreases the diffusion coefficient of the flavor molecules in the liquid will decrease the release rate (e.g., increasing viscosity or molecular size).

Flavor release does not occur under static conditions, and therefore the above model has limited applicability in practice. In reality, there will be a flow of gas over the food in the mouth due to respiration and swallowing (Thomson 1986). When the flavor is constantly swept away from the surface of the food, the concentration gradient is increased, and so the rate of flavor loss is more rapid than under static conditions. Overbosch et al. (1991) developed a model to take this convective diffusion process into account. A numerical solution of this model indicated that the release rate due to convective diffusion is much greater than that due to static diffusion when K_{GL} is small, because the rate-limiting step is the movement of the flavor compound from the liquid surface into the gas. At large K_{GL} values, the difference between the two theories is much smaller because the rate-limiting step is the diffusion of the molecules through the liquid, rather than from the surface into the gas. Under convective conditions, the rate of flavor release therefore depends on the equilibrium partition coefficient of the flavor, as well as the flow rate of the gas through the mouth and into the nose.

Influence of Ingredient Interactions. A number of ingredients commonly found in food emulsions are capable of decreasing the rate of flavor release because of their ability to either bind flavors or retard their mass transfer (e.g., proteins, carbohydrates, and surfactant micelles).

The effect of flavor binding on the rate of flavor release can be accounted for using the same approach as for static diffusion in homogeneous liquids (i.e., Equations 9.20 to 9.22) (Overbosch et al. 1991). For reversible binding, D_L is replaced by an effective diffusion coefficient $D_L^e = D_L/(K^* +1)$, and K_{GL} is replaced by an effective partition coefficient $K_{GL}^e = K_{GL}/(K^* +1)$, where K^* is the binding coefficient ($K^* = c_{L,B}/c_{L,F}$). The release rate is reduced because D_L^e and K_L^e are smaller than D_L and K_{GL}, although the total amount of flavor released when the process is allowed to go to completion is unchanged. For irreversible binding, c_L^0 is replaced by $c_{L,F}$ in Equations 9.20 to 9.22 because the rest of the flavor is "lost." In addition, the total amount of flavor released when the process is allowed to go to completion is reduced by a factor $c_{L,F}/c_L^0$.

The release rate may also be reduced because of the ability of certain food ingredients to retard the movement of flavor molecules to the surface of the liquid, which may be due to an enhanced viscosity or due to structural hindrance (Kokini 1987). The diffusion coefficient of a molecule is inversely proportional to the viscosity of the surrounding liquid, and so increasing the viscosity of the liquid will decrease the rate of flavor release because the movement of the flavor molecules is reduced. The presence of a network of aggregated biopolymer molecules may provide a physical barrier through which the molecules cannot directly pass. Instead, they may have to take a tortuous path through the network, which increases the time taken for them to reach the surface. If the flavor molecules are associated with surfactant micelles, their release rate will depend on the diffusion coefficient of the micelles, as well as the kinetics of micelle breakdown (Section 4.5).

Highly volatile flavors (high K_{GL}) are most affected by viscosity or structural hindrance effects because the rate-limiting step in their release from a food is the movement through the liquid rather than the movement away from the liquid surface. On other hand, low-volatility flavors (low K_{GL}) are affected less, because the rate-limiting step in their release is the movement away from the liquid surface rather than through the liquid (Roberts et al. 1996).

A great deal of research has been carried out to establish the relative importance of binding and retarded mass transfer mechanisms. Many experimental studies have shown that increasing the biopolymer concentration decreases the rate of flavor release, but have been unable to establish the relative importance of the two mechanisms (Hau et al. 1996, Guichard 1996, Roberts et al. 1996). The importance of rheology has been demonstrated by studies which have shown that the intensity of flavors decreases as the viscosity of a biopolymer solution or the strength of a biopolymer gel increases (Baines and Morris 1989, Carr et al. 1996). On the other hand, solutions with the same viscosity often have different flavor release rates, which may be because they have different microstructures or because the flavors bind to them differently (Guichard 1996). It is clear that more systematic research is needed to establish the role of biopolymers and other ingredients in retarding flavor release. The influence of ingredient interactions on release rates has important consequences for the formulation of many food products. For example, it may be necessary to incorporate more flavor into a food to achieve the same flavor intensity when the biopolymer concentration is increased.

Flavor Release from Emulsions. Overbosch et al. (1991) used the same mathematical approach as for homogeneous liquids to describe the rate of flavor release from emulsions. For static diffusion conditions, the flux and time dependence of the mass of flavor in an emulsion are given by:

$$J = F_E c_E^o \sqrt{D_E / \pi t} \tag{9.23}$$

$$M(t) = 2 F c_E^o \sqrt{D_E t / \pi} \tag{9.24}$$

where

$$F_E = \frac{K_{GE} \sqrt{D_G / D_E}}{1 + K_{GE} \sqrt{D_G / D_E}} \qquad\qquad c_E = (1 - \phi)c_C + \phi c_D$$

$$D_E = \frac{(1 - \phi)D_C + \phi D_D}{1 + (K_{DC} - 1)\phi} \qquad\qquad K_{GE} = \frac{c_G}{c_E} = \frac{K_{GD} K_{DC}}{1 + (K_{DC} - 1)\phi}$$

These equations provide some useful insights into the major factors which determine flavor release in emulsions. They indicate that the release rate depends on the diffusion and partition coefficients of the flavor in the oil, water, and gas phases. The larger the value of K_{GE} or D_E, the more rapid the flavor release. Similar conclusions were recently drawn by Harrison et al. (1997) using another mathematical model based on penetration theory.

The above equations predict that the release rate from an oil-in-water emulsion is the same as that from a water-in-oil emulsion with the same composition, because they are symmetrical with respect to the physical properties of the two phases (Overbosch et al. 1991). This will only be true if the flavor in the droplets and continuous phase is in equilibrium. Some studies have indicated that the taste perception of oil-in-water and water-in-oil emulsions of the same composition is approximately the same (Barylko-Pikielna et al. 1994, Brossard et al. 1996). Nevertheless, other studies have shown that the release rate is faster from oil-in-water emulsions than from water-in-oil emulsions, which suggests that the emulsion cannot be simply treated as a homogeneous liquid with "averaged" properties (Overbosch et al. 1991, Bakker and Mela 1996).

The above equations are most likely to be suitable for describing systems in which the rate-limiting step is the movement of the flavor molecules from the emulsion surface to the gas phase (i.e., when K_{GE} is small), rather than those where the movement of the flavor molecules through the emulsion is rate limiting (i.e., when K_{GE} is large), especially for emulsions with relatively large droplet sizes.

It is possible to develop more sophisticated theories to describe the kinetics of flavor release from emulsions which take into account the partitioning and diffusion of the flavor molecules in the oil, water, gas, and interfacial membrane. Nevertheless, these theories are much more complex and can usually only be solved numerically.

9.2.3. Measurements of Partition Coefficients and Flavor Release

A variety of experimental techniques can be used to measure equilibrium partition coefficients and the kinetics of flavor release in emulsions.

9.2.3.1. Head Space Analysis

The concentration of flavor in the vapor phase above a liquid can be determined by head space analysis (Franzen and Kinsella 1975, Overbosch et al. 1991, O'Neill 1996, Landy et al. 1996, Guyot et al. 1996). The liquid is placed in a sealed container which is stored under conditions of constant temperature and pressure. Samples of the gas phase are removed from the head space using a syringe which is inserted through the lid of the sealed container, and the concentration of flavor is measured, usually by gas chromatography or high-performance liquid chromatography. The concentration of volatile flavors in many foods is too low to be detected directly by conventional chromatography techniques, and therefore it is necessary to concentrate the samples prior to analysis. Head space analysis can be carried out over time to monitor the kinetics of flavor release or after the sample has been left long enough for equilibrium to be attained to determine equilibrium partition coefficients.

9.2.3.2. Concentration Analysis in Static Binary Liquids

The equilibrium partition coefficient of a flavor between a bulk oil and bulk water phase can be determined by measuring the concentration of flavor in the two phases after the system has been left long enough to attain equilibrium (McNulty 1987, Guyot et al. 1996, Huang et al. 1997). The flavor is usually added to one of the liquids first, and then the water phase is poured into the container and the oil phase is poured on top. The container is sealed and stored in a temperature-controlled environment until equilibrium is achieved, which can be a considerable period (a few days or weeks), although this time can be shortened by mild agitation of the sample. The method used to determine the concentration depends on the nature of the flavor molecule. The most commonly used techniques are spectrophotometry, chromatography, and radiolabeling. In some systems, it is possible to analyze the solutions directly, whereas in others it is necessary to extract the flavors first using appropriate solvents.

The partition coefficient of flavors in emulsions can be determined using a similar procedure. The emulsion to be analyzed is placed into a container and a known amount of the flavor is added to the continuous phase. The container is filled to the top and sealed to prevent any of the flavor from partitioning into the vapor phase. It is then stored in a temperature-controlled environment until it reaches equilibrium. Equilibrium is attained much more rapidly than in a nonemulsified system because the molecules only have to diffuse a short distance through the droplets. The emulsion is centrifuged to separate the droplets from the continuous phase, and then a sample of the continuous phase is removed for analysis of the flavor concentration. The partition coefficient can then be determined from a knowledge of the overall flavor concentration: $K_{DC} = c_D/c_C = (c_{total} - c_C)/c_C$. One important limitation of this technique is that it cannot distinguish between the flavor which is contained within the droplets and that which is associated with the interfacial membrane.

9.2.3.3. Concentration Analysis in Stirred Diffusion Cells

Flavor release occurs under highly dynamic conditions within the mouth (Land 1996), and so a number of workers have developed experimental techniques which attempt to mimic these conditions. McNulty and Karel (1973) developed a stirred diffusion cell for monitoring the mass transport of flavor compounds between an oil and aqueous phase under shear conditions. The flavor compound is initially dissolved in either the oil or aqueous phase. A known volume of the aqueous phase is then poured into the vessel, and a known volume of oil is poured on top. The oil and aqueous phases are sheared separately using a pair of stirrers, and samples are extracted periodically using syringes which protrude into each of the liquids (Figure 9.8). These samples are then analyzed to determine the concentration of the flavor within them. The liquids are stirred at a rate which ensures a uniform flavor distribution, without significantly disturbing the air–water or oil–water interfaces. By carrying out the measurements over a function of time, it is possible to determine the kinetics of flavor transport between the oil and water.

A similar system can be used to study the movement of flavor from a solution or emulsion to the vapor phase above it. The liquid to be analyzed is placed in a sealed vessel and stirred at a constant rate. A syringe is used to withdraw samples from the head space above the liquid as a function of time. Alternatively, a continuous flow of gas can be passed across the stirred liquid and the concentration of flavor within it determined by chromatography (Roberts and Acree 1996). Thus it is possible to simulate the agitation of the food within the mouth, as well as the flow of the gas across the food during manufacture.

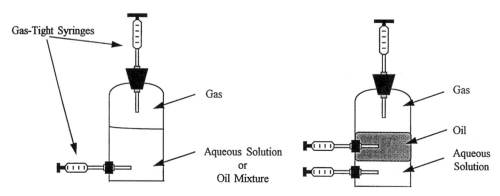

FIGURE 9.8   Diagram of stirred diffusion cells used to measure the kinetics of flavor release in liquids and emulsions.

9.2.3.4. *Sensory Analysis*

The ultimate test of the flavor profile of a food is its acceptance by consumers. Analytical tests carried out in a laboratory help to identify the most important factors which determine flavor release, but they cannot model the extreme complexity of the human sensory system. For this reason, many researchers use sensory analysis by human subjects to assess the overall flavor profile of a food sample (Buttery et al. 1973, Williams 1986, Barylko-Pikielna et al. 1994, Guyot et al. 1996).

9.3. EMULSION APPEARANCE

The first impression that a consumer usually has of a food emulsion is a result of its appearance (Francis and Clydesdale 1975, Hutchings 1994). Appearance therefore plays an important role in determining whether or not a consumer will purchase a particular product, as well as his or her perception of the quality once the product is consumed. A number of different characteristics contribute to the overall appearance of a food emulsion, including its opacity, color, and homogeneity. These characteristics are the result of interactions between light waves and the emulsion. The light which is incident upon an emulsion may be reflected, transmitted, scattered, absorbed, and refracted before being detected by the human eye (Francis and Clydesdale 1975, Farinato and Rowell 1983, Hutchings 1994, Francis 1995). A better understanding of the relationship between the appearance of emulsions and their composition and microstructure will aid in the design of foods with improved quality. This section highlights some of the most important factors which contribute to the overall appearance of emulsions.

9.3.1. Interaction of Light Waves with Emulsions

9.3.1.1. *Transmission, Reflection, and Refraction*

When an electromagnetic wave is incident upon a boundary between two homogeneous nonabsorbing materials, it is partly reflected and partly transmitted (or refracted) (Figure 9.9). The relative importance of these processes is determined by the refractive indices of the two materials, the surface topography, and the angle at which the light meets the surface (Hutchings 1994).

The reflection of an electromagnetic wave from a surface may be either *specular* or *diffuse*. Specular reflectance occurs when the angle of reflection is equal to the angle of incidence ($\phi_{reflection} = \phi_{incidence}$) and is the predominant form of reflection from optically smooth surfaces.

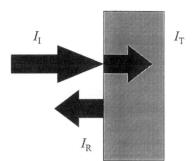

FIGURE 9.9 When a light wave encounters a planar interface between two materials, it is partly reflected and partly transmitted.

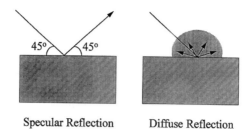

Specular Reflection   Diffuse Reflection

FIGURE 9.10   Comparison of specular and diffuse reflectance.

Diffuse reflectance occurs when the light is reflected over many different angles and is the most important form of reflection from optically rough surfaces (Figure 9.10).

When the angle of incidence is perpendicular to the surface of a material, the fraction of light which is specularly reflected is given by the reflection coefficient (R)

$$R = \left(\frac{n - 1}{n + 1} \right)^2 \tag{9.25}$$

where n is the relative refractive index ($= n_2/n_1$), and n_1 and n_2 are the refractive indices of the two materials (Hutchings 1994). The greater the difference in refractive index between the two materials, the larger the fraction of light which is reflected. About 0.1% of light is reflected from an interface between oil ($n_O = 1.43$) and water ($n_W = 1.33$), while about 2% is reflected from an interface between water and air ($n_A = 1$). These reflection coefficients may seem small, but the total amount of energy reflected from a concentrated emulsion becomes significant because a light wave encounters a huge number of different droplets and is reflected from each one of them.

When a light wave encounters a smooth material at an angle, part of the wave is reflected at an angle equal to that of the incident wave, while the rest is refracted (transmitted) at an angle which is determined by the relative refractive indices of the two materials and the angle of incidence: $\sin(\phi_{\text{refraction}}) = \sin(\phi_{\text{incidence}})/n$.

9.3.1.2. Absorption

Absorption is the process whereby a photon of electromagnetic energy is transferred to an atom or molecule (Atkins 1994, Penner 1994a, Pomeranz and Meloan 1994). The primary cause of absorption of electromagnetic radiation in the visible region is the transition of outer-shell electrons from lower to higher electronic energy levels. A photon is only absorbed when it has an energy which exactly corresponds to the difference between the energy levels involved in the transition, that is $\Delta E = h v = hc/\lambda$, where h is Planck's constant, v is the frequency of the electromagnetic wave, c is the velocity of the wave, and λ is the wavelength.

The visible region consists of electromagnetic radiation with wavelengths between about 380 and 750 nm, which corresponds to energies of between about 120 and 230 kJ mol^{-1} in water (Penner 1994a). Substances which can absorb electromagnetic energy in this region are usually referred to as *chromophores* (Patterson 1967). The most common type of chromophoric groups which are present in food emulsions are attached to organic molecules which contain conjugated unsaturated bonds or aromatic ring structures (Hutchings 1994). Single unsaturated bonds tend to absorb in the ultraviolet rather than the visible region (Pomeranz and Meloan 1994).

Absorption causes a reduction in the intensity of a light wave as is passes through a material, which can be described by the following equation (Penner 1994b, Pomeranz and Meloan 1994):

$$T = \frac{I_S}{I_0} \tag{9.26}$$

where T is the *transmittance*, I_S is the intensity of the light which travels directly through the sample, and I_0 is the intensity of the incident wave. In practice, I_S is also reduced because of reflections from the surfaces of the cell and due to absorption by the solvent and cell (Penner 1994b). These losses can be taken into account by comparing I_S with the intensity of a wave which has traveled through a reference cell (which usually contains pure solvent), rather than with I_0:

$$T = \frac{I_S}{I_R} \tag{9.27}$$

where I_R is the intensity of the light which has traveled directly through the reference cell. The transmittance of a substance decreases exponentially with increasing chromophore concentration or sample length (Pomeranz and Meloan 1994). For this reason, it is often more convenient to express the absorption of light in terms of an *absorbance* (A) because this is proportional to the chromophore concentration and the sample length:

$$A = \log \frac{I_S}{I_R} = \alpha c l \tag{9.28}$$

where α is a constant of proportionality known as the *absorptivity, c* is the chromophore concentration, and l is the sample path length. The linear relationship between absorbance and concentration holds over the chromophore concentrations used in most food emulsions. The color intensity of a particular chromophore therefore depends on its concentration and absorptivity.

The appearance of an emulsion to the human eye is determined by the interactions between it and electromagnetic radiation in the visible region (Francis and Clydesdale 1975, Hutchings 1994). It is therefore important to measure the absorbance over the whole range of visible wavelengths (390 to 750 nm). A plot of absorbance versus wavelength is referred to as an *absorption spectrum* (Figure 9.11). An absorption peak occurs at a wavelength which depends on the difference between the energy levels of the electronic transitions in the chromophores ($\lambda = ch/\Delta E$). Absorption peaks are fairly broad in the visible region because transitions occur between the different vibrational and rotational energy levels within the electronic energy levels and because of interactions between neighboring molecules (Penner 1994b).

9.3.1.3. Scattering

Scattering is the process whereby a wave which is incident upon a particle is directed into directions which are different from that of the incident wave (Farinato and Rowell 1983, Hiemenz 1986). The extent of light scattering by an emulsion is determined mainly by the relationship between the droplet size and wavelength and by the difference in the refractive index between the droplets and the surrounding liquid.

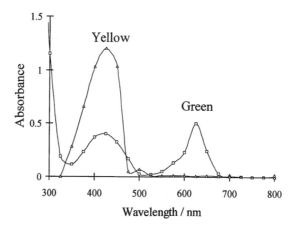

FIGURE 9.11   Absorption spectrum of colored liquids.

Light scattering causes a reduction in the intensity of an electromagnetic wave as it passes through an emulsion, which is characterized by the *transmittance* (Farinato and Rowell 1983):

$$T = \frac{I_S}{I_R} = \exp(-\tau l) \tag{9.29}$$

where τ is the turbidity of the emulsion. In dilute emulsions (<0.05%), the turbidity is proportional to the droplet concentration, but in concentrated emulsions, the turbidity falls below the expected value because of multiple scattering effects (Ma et al. 1990).

Turbidity is a desirable attribute of many food emulsions because it imparts a natural-looking character and appeal (Hernandez and Baker 1991, Hernandez et al. 1991). For example, low concentrations of oil droplets consisting of vegetable or flavor oils are used to provide a turbid appearance to many types of fruit beverage (Tan 1990, Dickinson 1994). The turbidity of these emulsions depends principally on the size, concentration, and relative refractive index of the droplets.

The turbidity of an emulsion is related to the characteristics of the droplets which it contains by the following relationship (Hernandez and Baker 1991):

$$\tau = \frac{3\pi c}{4\rho_1}\left(\frac{Q}{r}\right) \tag{9.30}$$

where c is the droplet concentration, ρ_1 is the density of the continuous phase, r is the droplet radius, and Q is the scattering cross-section of the individual particles:

$$Q = 2 - \frac{4}{\beta}\sin\beta + \frac{4}{\beta^2}(1 - \cos\beta) \tag{9.31}$$

where

$$\alpha = \frac{2\pi rn}{\lambda} \qquad \beta = 2\alpha|n - 1|$$

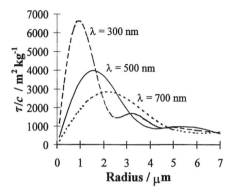

FIGURE 9.12   The turbidity of an emulsion depends on the droplet size and wavelength.

These equations have to be modified for polydisperse emulsions to take into account the fact that each droplet has its own scattering profile (Hernandez and Baker 1991). It should be noted that Equation 9.31 is only applicable to emulsions in which the relative refractive index is close to unity and the size of the droplets is greater than the wavelength of light.

The variation of emulsion turbidity with droplet size and wavelength is illustrated in Figure 9.12. The dependence of scattering on droplet size can be conveniently divided into three regions: long-wavelength regime ($r \ll \lambda$), intermediate-wavelength regime ($r \approx \lambda$), and short-wavelength regime ($r \gg \lambda$). The turbidity is greatest in the intermediate-wavelength regime, where the size of the droplets is similar to that of the wavelength. The wavelength of light varies from about 0.3 to 0.7 μm, and therefore droplets of this size would be expected to scatter most strongly.

Experiments with citrus oil-in-water emulsions have shown that there is a maximum in their turbidity (at 650 nm) when the droplets have a diameter of just under 1 μm (Hernandez and Baker 1991). This has important implications for the formulation of food beverages because it means that the characteristic turbidity of a product can be obtained using a lower concentration of oil when the droplet diameter is optimized to give the maximum amount of scattering. Nevertheless, other factors which depend on particle size also have to be considered, such as the stability of the product to creaming or sedimentation.

The influence of the relative refractive index on the scattering cross-section of an emulsion is illustrated in Figure 9.13. The scattering is minimum when the droplets have a refractive index equal to the surrounding liquid, but increases as the refractive index moves to higher or lower values. The refractive index of pure water is 1.33 (Walstra 1968), while that of most food oils ranges between 1.4 and 1.5 (Formo 1979).* The refractive index of aqueous solutions depends on the type and concentration of solutes present (e.g., proteins, sugars, alcohols, and salts) (Figure 9.14). At high concentrations of these components, it is possible for the refractive index of the droplets to be matched to that of the surrounding liquid, which greatly reduces the degree of scattering by the droplets and therefore causes the emulsion to appear transparent (Taisne et al. 1996). This effect can clearly be observed by diluting an oil-in-water emulsion in a series of aqueous solutions that contain different concentrations of sucrose. An emulsion which appears optically opaque when the droplets are dispersed in water becomes transparent when they are dispersed in a concentrated sugar solution (≈60%), even though the droplet size and concentration are unchanged.

---

* It should be noted that the refractive indices of substances vary with wavelength (Walstra 1968).

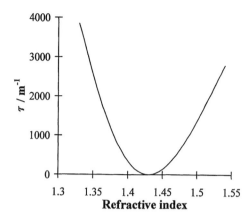

FIGURE 9.13   Influence of refractive index on the degree of scattering from emulsions. The scattering is smallest in the region where the droplet and continuous phase have similar refractive indices, but increases at lower or higher values.

9.3.1.4. Absorption and Scattering in Concentrated Emulsions

The optical characteristics of concentrated emulsions can be described by the Kubelka–Munk theory (Francis and Clydesdale 1975, Hutchings 1994). This theory enables one to determine the relative amounts of scattering and absorption in a concentrated emulsion from measurements of the light reflected from it:

$$F_{KM} = \frac{K}{S} = \frac{(1 - R_\infty)^2}{2R_\infty} \tag{9.32}$$

where F_{KM} is the Kubelka–Munk parameter, K and S are the absorption and scattering coefficients, and R_∞ is the reflectance from an infinitely thick sample at a given wavelength. When an emulsion is highly colored, the absorption of light dominates (high F_{KM}) and the reflectance is low. As scattering effects become more important (low F_{KM}), the reflectance increases and the emulsion appears "lighter" in color. The scattering coefficient is proportional to the droplet concentration, while the absorption coefficient is proportional to the chromophore concentration. The appearance of an emulsion therefore depends on the concentration of the various components within it.

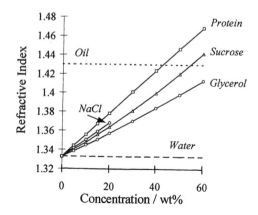

FIGURE 9.14   Dependence of the refractive index of various aqueous solutions on solute concentration.

The Kubelka–Munk theory can be used to predict the influence of different types of scatterers or chromophores on the overall appearance of an emulsion. It can also be related to the scattering cross-sections of the individual droplets in an emulsion using diffuse scattering theory. It therefore provides a valuable mathematical link between the physicochemical characteristics of the droplets and the appearance of an emulsion.*

9.3.2.   Overall Appearance

The overall appearance of an emulsion is a result of the various types of interactions between light waves and foods mentioned in the previous section.

9.3.2.1.   Opacity

An object which allows all of the light to pass through it is referred to as being *transparent*, whereas an object which scatters or absorbs all of the light is referred to as being *opaque* (Clydesdale 1975). Many dilute emulsions fall somewhere between these two extremes and are therefore referred to as being *translucent*. The opacity of most food emulsions is determined mainly by the scattering of light from the droplets: the greater the scattering, the greater the opacity (Hernandez and Baker 1991, Dickinson 1994). The extent of scattering is determined by the concentration, size, and relative refractive index of the droplets (Section 9.3.1.3). An emulsion becomes more opaque as the droplet concentration or scattering cross-section increases (Equation 9.29).

The scattering of light accounts for the opacity of emulsions that are made from two liquids which are themselves optically transparent (e.g., water and a mineral oil). When a light wave impinges on an emulsion, all of the different wavelengths are scattered by the droplets, and so the light cannot penetrate very far into the emulsion. As a consequence, the emulsion appears to be optically opaque (Farinato and Rowell 1983).

9.3.2.2.   Color

It is extremely difficult for human beings to objectively describe the colors of materials using everyday language (Hutchings 1994). For this reason, a number of standardized methods have been developed to measure and specify color in a consistent way (Francis and Clydesdale 1975, Hutchings 1994). The underlying principle of these methods is that all colors can be simulated by combining three selected colored lights (red, green, and blue) in the appropriate ratio and intensities. This trichromatic principle means that it is possible to describe any color in terms of just three mathematical variables (e.g., hue, value, and chroma) (Francis and Clydesdale 1975). Experimental techniques for measuring these variables are discussed briefly in Section 9.3.3.

The color of an emulsion is determined by the absorption and scattering of light waves from both the droplets and continuous phase (Dickinson 1994). The absorption of light depends on the type and concentration of chromophores present, while the scattering of light depends on the size, concentration, and relative refractive index of any particulate matter. Whether an emulsion appears "red," "orange," "yellow," "blue," etc. depends principally on its absorption spectra. Under normal viewing conditions, an emulsion is exposed to white light from all directions.** When this light is reflected, transmitted, or scattered by the

---

* Recently, it has been shown how the color of an emulsion can be predicted using the Kubelka–Munk theory from a knowledge of the droplet and dye characteristics (McClements et al. 1998a).

** If an emulsion is placed in a dark room and a white light beam is directed through it, it appears blue when observed from the side and red when observed from the back, because blue light has a lower wavelength than red light and is therefore scattered to a wider angle (Farinato and Rowell 1983).

emulsion, some of the wavelengths are absorbed by the chromophores present. The color of the light which reaches the eye is a result of the nonabsorbed wavelengths (e.g., an emulsion appears red if it absorbs all of the other colors except the red) (Francis and Clydesdale 1975). The color of an emulsion is modified by the presence of the droplets or any other particulate matter. As the concentration or scattering cross-section of the particles increases, an emulsion becomes lighter in appearance because the scattered light does not travel very far through the emulsion and is therefore absorbed less by the chromophores. It is therefore possible to modify the color of an emulsion by altering the characteristics of the emulsion droplets or other particulate matter.

9.3.2.3. Homogeneity

The quality of a food emulsion is often determined by the uniformity of its appearance over the whole of its surface. An emulsion may have a heterogeneous appearance for a number of reasons: (1) it contains particles which are large enough to be resolved by the human eye (>0.1 mm) or (2) it contains particles which have moved to either the top or the bottom of the container because of gravitational separation. Methods of retarding creaming and sedimentation were considered in Chapter 7.

9.3.3. Experimental Techniques

9.3.3.1. Spectrophotometry

A variety of different types of spectrophotometers have been developed to measure the transmission and reflection of light from objects as a function of wavelength in the visible region (Clydesdale 1975, Francis and Clydesdale 1975, Pomeranz and Meloan 1994, Hutchings 1994). These instruments usually consist of a light source, a wavelength selector, a sample holder, and a light detector (Figure 9.15).

Transmission Spectrophotometer. A beam of white light, which contains electromagnetic radiation across the whole of the visible spectrum, is passed through a wavelength selector, which isolates radiation of a specific wavelength (Penner 1994b). This monochromatic wave is then passed through a cell containing the sample, and the intensity of the transmitted wave is measured using a light detector. By comparing the intensity of the light transmitted by the sample with that transmitted by a reference material, it is possible to determine the transmittance of the sample (Equation 9.27). A transmittance spectrum is obtained by carrying out this procedure across the whole range of wavelengths in the visible region. Transmission measurements can only be carried out on emulsions which allow light to pass through, and therefore they cannot be used to analyze concentrated emulsions.

Reflection Spectrophotometer. In these instruments, the intensity of light reflected from the surface of a sample is measured (Francis and Clydesdale 1975). The reflectance (R) of a material is defined as the ratio of the intensity of the light reflected from the sample (R_S) to the intensity of the light reflected from a reference material of known reflectance (R_R): $R = R_S/R_R$. The precise nature of the experimental device depends on whether the reflection is specular or diffuse. For specular reflection, the intensity of the reflected light is usually measured at an angle of 90° to the incident wave, whereas for diffuse reflection, the sum of the intensity of the reflected light over all angles is measured using a device called an *integrating sphere* (Figure 9.15). A reflectance spectrum is obtained by carrying out this procedure across the whole range of wavelengths in the visible region.

The transmittance and reflectance spectra obtained from a sample can be used to calculate the relative magnitudes of the absorption and scattering of light by an emulsion as a function of wavelength. Alternatively, the color of a product can be specified in terms of trichromatic

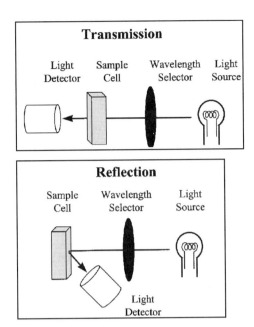

FIGURE 9.15 Examples of transmission and reflection spectrophotometers.

coordinates by analyzing the spectra using appropriate mathematical techniques (McClements et al. 1998). The details of these techniques have been described elsewhere and are beyond the scope of this book (Francis and Clydesdale 1975, Hutchings 1994).

9.3.3.2. Light Scattering

Light-scattering techniques are used principally to determine the size distribution of the droplets in an emulsion (Chapter 10). A knowledge of the droplet size distribution enables one to predict the influence of the droplets on light scattering and therefore on the turbidity of an emulsion (Hernandez and Baker 1991, Dickinson 1994). Alternatively, the experimentally determined scattering pattern — the intensity of scattered light versus scattering angle — can be used directly to describe the scattering characteristics of the emulsion droplets.

The scattering of light from dilute emulsions is sometimes characterized by a device known as a nephelometer (Hernandez et al. 1991). This device measures the intensity of light which is scattered at an angle of 90° to the incident beam. The intensity of light scattered by a sample is compared with that scattered by a standard material of known scattering characteristics (e.g., formazin) (Hernandez et al. 1991). Because small droplets scatter light more strongly at wide angles than large droplets, the nephelometer is more sensitive to the presence of small droplets than are turbidity measurements.

9.3.3.3. Colorimetry

A large number of instruments have been developed to characterize the color of materials which are based on the trichromatic principle mentioned previously (Francis and Clydesdale 1975, Hutchings 1994, Francis 1995). A simple colorimeter consists of a light source, the sample to be analyzed, a set of three filters (red, green, and blue), and a photocell to determine the light intensity (Figure 9.16). Colorimetry measurements can be carried out in either a transmission or a reflection mode. In the transmission mode, the intensity of a light beam is measured after it has been passed through the sample and one of the color filters. This

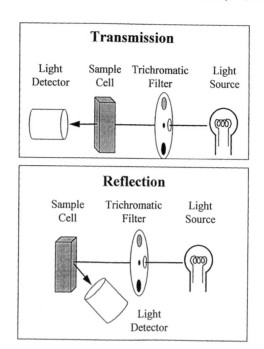

FIGURE 9.16   Schematic diagram of a simple colorimeter.

procedure is then repeated for the other two filters, and the sample is characterized by the intensity of the light wave passing through the red, green, and blue filters. In the reflection mode, the light beam is passed through a filter and is then reflected from the surface of a sample. For specular reflection, a photocell is usually positioned at 90° to the incident wave, while for diffuse reflection, an integrating sphere is used to determine the intensity of the reflected light. This procedure is carried out for each of the three color filters, so that the color of the sample is characterized in terms of the intensities of the red, green, and blue lights. Reflectance measurements are most suitable for determining the color of concentrated emulsions, while transmission measurements are more suitable for characterizing dilute emulsions.

9.3.3.4.   Sensory Analysis

Ultimately, the appearance of an emulsion must be acceptable to the consumer, and therefore it is important to carry out sensory tests (Hutchings 1994). These tests can be carried out using either trained specialists or untrained individuals. Each individual is given one or more samples to analyze and asked either to compare the appearance of the samples with each other or to rank certain attributes of the appearance of each sample. Sensory tests must be carried out in a room where the light source is carefully controlled to obtain reproducible measurements which correspond to the conditions a person might experience when consuming the product (Francis and Clydesdale 1975).

10 Characterization of Emulsion Properties

10.1. INTRODUCTION

Previous chapters stressed the relationship between the bulk physicochemical characteristics of food emulsions and their colloidal properties. The more efficient production of high-quality food products depends on an improved understanding of this relationship, as well as on the successful implementation of this knowledge in practice. Advances in these areas rely on the availability of analytical techniques to characterize emulsion properties. These techniques are needed in the laboratory for research and development and in the food factory to monitor the properties of foods before, during, and after production.

10.1.1. Research and Development

The properties of foods can be considered at a number of different levels, ranging from the molecular to the bulk physicochemical and organoleptic characteristics of the final product (Baianu 1992, Eads 1994). The quality attributes of a food emulsion are therefore the result of a complex set of interactions between and within the different levels of structural organization (e.g., molecular, colloidal, microscopic, macroscopic, and organoleptic). This relationship is made even more complex because of the dynamic nature of food emulsions. The chemical structure or spatial organization of the components within an emulsion may change with time or in response to changes in environmental conditions. A major concern of food scientists is identifying the most important factors which determine the properties at each level and determining the relationship between the different levels. To elucidate these relationships, it is necessary to use a combination of experimental techniques and theoretical understanding. An array of different experimental techniques are available for studying emulsion properties, ranging from the molecular to the organoleptic. By using a combination of these techniques, it is possible to obtain a greater understanding of the relationship between the different levels of structural organization. This type of information can then be used to systematically design and manufacture foods with improved properties.

10.1.2. Quality Control

The information obtained from research and development is used to select the most appropriate ingredients and processing operations required to manufacture a product with the desired quality attributes (Kokini et al. 1993). Nevertheless, there are always inherent variations in the raw materials (e.g., composition, quality) and processing conditions (e.g., temperatures, pressures, times, flow rates) used to manufacture food products, which inevitably leads to variability in the properties of the final product. To minimize these variations, it is necessary to understand the contribution that each ingredient makes to the overall properties and the influence of processing conditions on these properties. This type of information comes from the fundamental studies discussed in Section 10.1.1.

Analytical techniques are also needed to monitor the properties of foods at each stage of manufacturing (Nielsen 1994). The data obtained from these techniques are used to optimize processing conditions, thus improving product quality, reducing waste, improving energy efficiency, and minimizing variations in the properties of the final product.

As mentioned earlier, the overall quality of food emulsions is determined by a wide range of different factors, ranging from the molecular to the organoleptic. In this chapter, the focus is on those experimental techniques which measure properties that are unique to emulsions (i.e., emulsifier efficiency, dispersed-phase volume fraction, droplet size distribution, droplet crystallinity, and droplet charge). In previous chapters, experimental techniques were reviewed for characterizing interfacial properties (Chapter 5), emulsion stability (Chapter 7), and emulsion rheology (Chapter 8). Despite their importance in understanding the properties of food emulsions, we have not considered techniques for measuring molecular, chemical, enzymatic, microbiological, or organoleptic properties, because these techniques are common to all foods and are therefore beyond the scope of this book.

10.2. TESTING EMULSIFIER EFFICIENCY

One of the most important decisions a food manufacturer must make when developing an emulsion-based food product is the selection of the most appropriate emulsifier (Fisher and Parker 1985, Charalambous and Doxastakis 1989, Dickinson 1992, Hasenhuettl 1997). A huge number of emulsifiers are available as food ingredients, and each has its own unique characteristics and optimum range of applications (Hasenhuettl and Hartel 1997). The efficiency of an emulsifier is governed by a number of characteristics, including the minimum amount required to produce a stable emulsion, its ability to prevent droplets from aggregating over time, the speed at which it adsorbs to the droplet surface during homogenization, the interfacial tension, and the thickness and viscoelasticity of the interfacial membrane. These characteristics depend on the food in which the emulsifier is present and the prevailing environmental conditions (e.g., pH, ionic strength, ion type, oil type, ingredient interactions, temperature, and mechanical agitation) (Sherman 1995). For this reason, it is difficult to accurately predict the behavior of an emulsifier from a knowledge of its chemical structure (although some general prediction about its functional properties is usually possible). Instead, it is often better to test the efficiency of an emulsifier under conditions which are similar to those found in the actual food product in which it is going to be used (Sherman 1995). A number of procedures commonly used to test emulsifier efficiency are discussed in this section.

10.2.1. Emulsifying Capacity

It is often important for a food manufacturer to know the minimum amount of an emulsifier that can be used to create a stable emulsion. The *emulsifying capacity* of a water-soluble emulsifier is defined as the maximum amount of oil that can be dispersed in an aqueous solution that contains a specific amount of the emulsifier without the emulsion breaking down or inverting into a water-in-oil emulsion (Sherman 1995). Experimentally, it is determined by placing an aqueous emulsifier solution into a vessel and continuously agitating using a high-speed blender as small volumes of oil are titrated into the vessel (Swift et al. 1961, Das and Kinsella 1990).* The end point of the titration occurs when the emulsion breaks down or inverts, which can be determined by optical, rheological, or electrical conductivity measurements. The greater the volume of oil which can be incorporated into the emulsion before it

---

* The emulsifying capacity of an oil-soluble emulsifier can be determined in the same way, except that the water is titrated into the oil phase.

breaks down, the higher the emulsifying capacity of the emulsifier. Although this test is widely used to characterize emulsifiers, it has a number of drawbacks which limit its application as a standard procedure (Sherman 1995, Dalgleish 1996a). The main problem with the technique is that the amount of emulsifier required to stabilize the emulsion is governed by the oil–water interfacial area rather than by the oil concentration, and so the emulsifying capacity depends on the size of the droplets produced during agitation. As a consequence, the results are particularly sensitive to the type of blender and blending conditions used in the test. In addition, the results of the test have also been found to depend on the rate at which the oil is titrated into the vessel, the method used to determine the end point, the initial emulsifier concentration, and the measurement temperature (Sherman 1995). The emulsifying capacity should therefore be regarded as a qualitative index which depends on the specific conditions used to carry out the test. Nevertheless, it is useful for comparing the efficiency of different emulsifiers under the same experimental conditions.

A more reliable means of characterizing the minimum amount of emulsifier required to form an emulsion is to measure the *surface load* (Γ_S), which corresponds to the mass of emulsifier required to cover a unit area of droplet surface (Dickinson 1992). A stable emulsion is prepared by homogenizing known amounts of oil, water, and emulsifier. The mass of emulsifier adsorbed to the surface of the droplets per unit volume of emulsion (C_a/kg m^{-3}) is equal to the initial emulsifier concentration minus that remaining in the aqueous phase after homogenization (which is determined by centrifuging the emulsion to remove the droplets and then analyzing the emulsifier concentration in the serum). The total droplet surface area covered by the adsorbed emulsifier is given by $S = 6\phi V_e/d_{32}$, where V_e is the emulsion volume and d_{32} is the volume–surface mean droplet diameter. Thus the surface load can be calculated: $\Gamma_S = C_a V_e/S = C_a d_{32}/6\phi$, which is typically a few milligrams per meter squared. A knowledge of the surface load enables one to calculate the minimum amount of emulsifier required to prepare an emulsion that contain droplets of a given size and concentration. In practice, an excess of emulsifier is usually needed because it does not all adsorb to the surface of the droplets during homogenization due to the finite time it takes for an emulsifier to reach the oil–water interface and because there is an equilibrium between the emulsifier at the droplet surface and that in the continuous phase (Hunt and Dalgleish 1994, Dalgleish 1996a). In addition, the surface load is often dependent on environmental conditions, such as pH, ionic strength, temperature, and protein concentration (Dickinson 1992, Hunt and Dalgleish 1994, Dalgleish 1996a).

10.2.2. Emulsion Stability Index

An efficient emulsifier produces an emulsion in which there is no visible separation of the oil and water phases over time. Phase separation may not become visible to the human eye for a long time, even though some emulsion breakdown has occurred. Consequently, it is important to have analytical tests which can be used to detect the initial stages of emulsion breakdown, so that their long-term stability can be predicted.

One widely used test is to centrifuge an emulsion at a given speed and time and observe the amount of creaming and/or oil separation which occurs (Smith and Mitchell 1976, Tornberg and Hermannson 1977, Aoki et al. 1984, Das and Kinsella 1990). This test can be used to predict the stability of an emulsion to creaming using relatively low centrifuge speeds or to coalescence by using speeds which are high enough to rupture the interfacial membranes. The greater the degree of creaming or oil separation that occurs, the greater the instability of an emulsion and the less efficient the emulsifier.

An alternative approach which can be used to accelerate emulsion instability is to measure the degree of droplet coalescence when an emulsion is subjected to mechanical agitation (Britten and Giroux 1991, Dickinson et al. 1993b, Dickinson and Williams 1994). The droplet size distribution of the emulsions can be measured either as a function of time as the emulsions are agitated at a constant stirring speed or as a function of stirring speed after the emulsions have been agitated for a fixed time. The faster the increase in droplet size with time, the greater the instability of the emulsion and the lower the efficiency of the emulsifier.

Although these tests are widely used and can be carried out fairly rapidly, they do have a number of important limitations: the rate of creaming or coalescence in a centrifugal field or during agitation may not be a good indication of emulsion instability under normal storage conditions, and it does not take into account chemical or biochemical reactions that might alter emulsion stability over extended periods.

A more quantitative method of determining emulsifier efficiency is to measure the change in the particle size distribution of an emulsion with time using one of the analytical techniques discussed in Section 10.3. An efficient emulsifier produces emulsions in which the particle size distribution does not change over time, whereas a poor emulsifier produces emulsions in which the particle size increases due to coalescence and/or flocculation. The kinetics of emulsion stability can be established by measuring the rate at which the particle size increases with time. These tests should be carried out under conditions similar to those found in the final product (e.g., pH, ionic strength, composition, temperature, etc.).

The analytical instruments used for measuring particle size distributions are often fairly expensive and are not available in many small institutions. In this case, a measure of the flocculation or coalescence in an emulsion can be obtained using a simple UV-visible spectrophotometer (Walstra 1968, Pearce and Kinsella 1978, Reddy and Fogler 1981, Pandolfe and Masucci 1984). The turbidity of light is measured at a single wavelength or over a range of wavelengths, and the mean particle size is estimated using light-scattering theory. This technique should be used with caution, because the relationship between particle size and turbidity is fairly complex in the region where the droplet radius is the same order of magnitude as the wavelength of light used (see Figure 9.12).

10.2.3. Interfacial Tension

One of the most valuable means of obtaining information about the characteristics of an emulsifier is to measure the reduction in the surface (or interfacial) tension when it adsorbs to a surface (or an interface). Surface tension measurements can be used to determine the kinetics of emulsifier adsorption, the packing of emulsifier molecules at an interface, critical micelle concentrations, surface pressures, and competitive adsorption (Chapter 5). A variety of analytical instruments are available for measuring the surface or interfacial tension of liquids, as discussed in Section 5.10.

10.2.4. Interfacial Rheology

The stability of food emulsions to creaming and droplet coalescence depends on the rheological characteristics of the interfacial membranes which surround the droplets (Chapter 7), and so it is often important for food scientists to be able to quantify the rheological characteristics of interfaces. In addition, interfacial rheology measurements can also be used to provide valuable information about other characteristics of emulsifiers, such as adsorption kinetics, competitive adsorption, and interfacial interactions. Interfacial rheology is the two-dimensional equivalent of bulk rheology (Chapter 8), and consequently many of the principles and concepts are directly analogous.* An interface can be viscous, elastic, or viscoelastic depending on the type, concentration, and interactions of the molecules present (Murray and Dickinson

---

* Although it is convenient to consider interfacial rheology in two dimensions, it must be remembered that the interfacial membrane has a finite thickness (usually a few nanometers) in reality, and this may also influence its rheological characteristics.

1996). Two types of deformation are particularly important at an interface: shear and dilatation. The shear behavior of an interface is characterized by its resistance to the "sliding" of neighboring regions past one another, without any change in the overall interfacial area. The dilatational behavior of an interface is characterized by its resistance to the expansion or contraction of its surface area. Instruments for measuring the shear and dilatational rheology of interfaces were reviewed in Section 5.11.

10.3. MICROSTRUCTURE AND DROPLET SIZE DISTRIBUTION

10.3.1. Microscopy

The unaided human eye can resolve objects which are greater than about 0.1 mm (100 µm) apart (Aguilera and Stanley 1990). Many of the structural components in food emulsions are smaller than this lower limit and therefore cannot be observed directly by the eye (e.g., emulsion droplets, surfactant micelles, fat crystals, gas bubbles, and protein aggregates) (Dickinson 1992). Our normal senses must therefore be augmented by microscopic techniques which enable us to observe tiny objects (Aguilera and Stanley 1990, Kalab et al. 1995, Smart et al. 1995). A number of these techniques are available to provide information about the structure, dimensions, and organization of the components in food emulsions (e.g., optical microscopy, scanning and transmission electron microscopy, and atomic force microscopy) (Kirby et al. 1995, Kalab et al. 1995, Smart et al. 1995). These techniques have the ability to provide information about structurally complex systems in the form of "images" which are relatively easy to comprehend by human beings (Kirby et al. 1995). Each microscopic technique works on different physicochemical principles and can be used to examine different levels and types of structural organization. Nevertheless, any type of microscope must have three qualities if it is going to be used to examine the structure of small objects: resolution, magnification, and contrast (Aguilera and Stanley 1990). *Resolution* is the ability to distinguish between two objects which are close together. *Magnification* is the number of times that the image is greater than the object being examined. *Contrast* determines how well an object can be distinguished from its background.

10.3.1.1. Conventional Optical Microscopy

Although the optical microscope was developed over a century ago, it is still one of the most valuable tools for observing the microstructure of emulsions (Mikula 1992; Hunter 1986, 1993). An optical microscope contains a series of lenses which direct the light through the specimen and magnify the resulting image (Figure 10.1). The resolution of an optical microscope is determined by the wavelength of light used and the mechanical design of the instrument (Franklin 1977, Hunter 1986). The theoretical limit of resolution of an optical microscope is about 0.2 µm, but in practice it is difficult to obtain reliable measurements below about 1 µm (Hunter 1993). This is because of technical difficulties associated with the design of the instrument and because the Brownian motion of small particles causes images to appear blurred. The optical microscope therefore has limited application to many food emulsions because they contain structures with sizes below the lower limit of resolution. Nevertheless, it can provide valuable information about the size distribution of droplets in emulsions which contain larger droplets and can be used to distinguish between flocculation and coalescence (Mikula 1992), which is often difficult using instrumental techniques based on light scattering, electrical pulse counting, or ultrasonics.

The natural contrast between the major components in food emulsions is often fairly poor (because they have similar refractive indices or color), which makes it difficult to reliably distinguish them from each other using conventional bright-field optical microscopy. For this

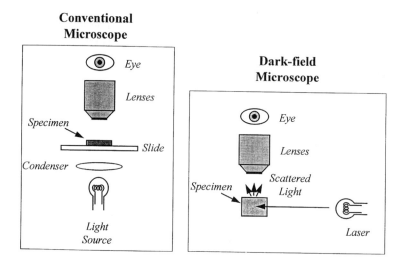

FIGURE 10.1   Comparison of conventional bright-field and dark-field microscopes.

reason, the technique has been modified in a number of ways to enhance the contrast, improve the image quality, and provide more detailed information about the composition and micro-structure of food emulsions. Various types of chemical stains are available which bind to particular components within an emulsion (e.g., the proteins, polysaccharides, or lipids) and therefore enable specific structural features to be highlighted (Gurr 1961, Smart et al. 1995). These stains must be used with caution because they can alter the structure being examined. In addition, they are often difficult to incorporate into concentrated or semisolid emulsions.

The contrast between different components can be improved without using chemical stains by modifying the design of the optical microscope (e.g., by using *phase contrast* or *differential interference contrast* microscopy) (Aguilera and Stanley 1990). These techniques improve contrast by using special lenses which convert small differences in refractive index into differences in light intensity.

The structure of optically anisotropic food components, such as fat crystals, starch granules, and muscle fibers, can be studied using *birefringent* microscopy (Aguilera and Stanley 1990). Anisotropic materials cause plane polarized light to be rotated, whereas isotropic materials do not, and so it is possible to distinguish anisotropic and isotropic materials using "crossed polarizers." This technique is particularly useful for monitoring phase transitions of fat and for determining the location and morphology of fat crystals in emulsion droplets (Walstra 1967, Boode 1992).

The characteristics of particles with sizes less than a micrometer can be observed by dark-field illumination using an instrument known as an ultramicroscope (Shaw 1980, Farinato and Rowell 1983, Hunter 1986). A beam of light is passed through the specimen at a right angle to the eyepiece (Figure 10.1). In the absence of any particles, the specimen appears completely black, but when there are particles present, they scatter light and the image appears as a series of bright spots against a black background. This technique can be used to detect particles as small as 10 nm; however, the particles appear as blurred spots rather than well-defined images whose size can be measured directly. The particle size is inferred from the brightness of the spots or from measurements of their Brownian motion. The ultramicroscope can also be used to determine the number of particles in a given volume or to monitor the motion of particles in an electric field (Section 10.6).

Certain food components either fluoresce naturally or can be made to fluoresce by adding fluorescent dyes which bind to them (Aguilera and Stanley 1990, Kalab et al. 1995).

Fluorescent materials adsorb electromagnetic radiation at one wavelength and emit it at a higher wavelength (Skoog et al. 1994). A conventional bright-field optical microscope can be modified to act as a fluorescence microscope by adding two filters (or other suitable wavelength selectors). One filter is placed before the light enters the sample and produces a monochromatic *excitation* beam. The other filter is placed after the light beam has passed through the sample and produces a monochromatic *emission* beam. A variety of fluorescent dyes (fluorophores) are available which bind to specific components within a food (e.g., proteins, fats, or carbohydrates) (Larison 1992). Fluorescence microscopes usually use an ultraviolet light source to illuminate the specimen (which is therefore invisible to the human eye), whereas the light emitted by the fluorescent components within a specimen is in the visible part of the electromagnetic spectrum, and so they appear as bright objects against a black background. Fluorescence microscopy is a very sensitive technique that is particularly useful for studying structures that are present at such small concentrations that they cannot be observed using conventional optical microscopy. In addition, it can be used to highlight specific structures within an emulsion by selecting fluorescent dyes which bind to them.

One of the major drawbacks of optical microscopy is the possibility that sample preparation alters the structure of the specimen being analyzed (Aguilera and Stanley 1990, Kalab et al. 1995). Sample preparation may be a simple procedure, such as spreading an emulsion across a slide, or a more complex procedure, such as fixing, embedding, slicing, and staining a sample (Smart et al. 1995). Even a procedure as simple as spreading a specimen across a slide may alter its structural properties and should therefore be carried out carefully and reproducibly. Other disadvantages of optical microscopy are that measurements are often time consuming and subjective, it is often necessary to analyze a large number of different regions within a sample to obtain statistically reliable data, and it is limited to studying structures greater than about 1 μm (Mikula 1992). Many modern optical microscopes now have the capability of being linked to personal computers which can rapidly store and analyze images and thus enhance their ease of operation (Klemaszeski et al. 1989, Mikula 1992).

10.3.1.2. *Laser Scanning Confocal Microscopy*

This is a fairly recent development in optical microscopy which can provide extremely valuable information about the microstructure of food emulsions (Blonk and van Aalst 1993, Brooker 1995, Smart et al. 1995, Vodovotz et al. 1996). Laser scanning confocal microscopy (LSCM) provides higher clarity images than conventional optical microscopy and allows the generation of three-dimensional images of structures without the need to physically section the specimen. The LSCM focuses an extremely narrow laser beam at a particular point in the specimen being analyzed, and a detector measures the intensity of the resulting signal (Figure 10.2). A two-dimensional image is obtained by carrying out measurements at different points in the *x*–*y* plane, either by moving the specimen (and keeping the laser beam stationary) or by moving the laser beam (and keeping the specimen stationary). An image is generated by combining the measurements from each individual point. Three-dimensional images are obtained by focusing the laser beam at different depths into the sample and then scanning in the horizontal plane. Observation of the microstructure of multicomponent systems is often facilitated by using the natural fluorescence of certain components or by using fluorescent dyes that bind selectively to specific components (e.g., proteins, lipids, or polysaccharides) (Larison 1992). The LSCM technique suffers from many of the same problems as conventional optical microscopy. Nevertheless, it has a slightly better resolution and sensitivity, and the sample preparation is often less severe.

LSCM has been used to study the size, concentration, and organization of droplets in emulsions (Jokela et al. 1990); to examine the microstructure of butter, margarines, and low-

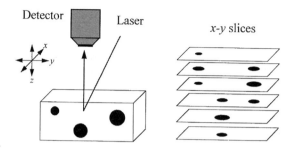

FIGURE 10.2   Laser scanning confocal microscopy. A laser beam is scanned across a particular *x–y* plane within a sample. By combining successive *x–y* planes, it is possible to create a three-dimensional image.

fat spreads (Blonk and van Aalst 1993); to monitor creaming of droplets (Brakenoff et al. 1988); and to follow the desorption of proteins from the surface of emulsion droplets.

10.3.1.3.   *Electron Microscopy*

Electron microscopy is widely used to examine the microstructure of food emulsions, particularly those that contain structural components which are smaller than the lower limit of resolution of optical microscopes (ca. <1 μm) (e.g., protein aggregates, small emulsion droplets, fat or ice crystals, micelles, and interfacial membranes) (Chang et al. 1972, Tung and Jones 1981, Hunter 1986, Aguilera and Stanley 1990, Heertje and Paques 1995). It can be used to provide information about the concentration, dimensions, and spatial distribution of structural entities within a specimen, provided the microstructure of the specimen is not significantly altered by the sample preparation (Heertje and Paques 1995). With suitable sample preparation, electron microscopy can be used to analyze both oil-in-water and water-in-oil emulsions that are either liquid or solid. Electron microscopes are fairly large pieces of equipment which are relatively expensive to purchase and maintain (Smart et al. 1995). For this reason, they tend to be available only at fairly large research laboratories or food companies.

Electron microscopes use electron beams, rather than light beams, to provide information about the structure of materials (Aguilera and Stanley 1990, Heertje and Paques 1995). These beams are directed through the microscope using a series of magnetic fields rather than optical lenses. Electron beams have much smaller wavelengths than light and so can be used to examine much smaller objects. In principle, the smallest size that can be resolved is about 0.2 nm, but in practice, it is usually about 1 nm due to limitations in the stability and performance of the magnetic lenses. Two types of electron microscope are commonly used to examine the structure of food systems: transmission electron microscopy (TEM) and scanning electron microscopy (SEM). In both of these techniques, it is necessary to keep the microscope under high vacuum because electrons are easily scattered by atoms or molecules in a gas, and this would cause a deterioration in the image quality. This also means that specimens must be prepared so that they are free of all volatile components that could evaporate (e.g., water and organic molecules).

Transmission Electron Microscopy.   A cloud of electrons, produced by a tungsten cathode, is accelerated through a small aperture in a positively charged plate to form an electron beam (Figure 10.3). This beam is focused and directed through the specimen by a series of magnetic lenses. Part of the electron beam is either adsorbed or scattered by the specimen, while the rest is transmitted. The beam of transmitted electrons is magnified by a magnetic

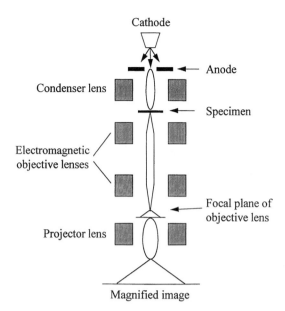

FIGURE 10.3   Schematic diagram of TEM. An electron beam is directed through the sample, and the intensity of the transmitted beam is measured.

lens and then projected onto a fluorescent screen to create an image of the specimen. The fraction of electrons transmitted by a substance depends on its electron density: the lower the electron density, the greater the fraction of electrons transmitted and the more intense the image. Components with different electron densities therefore appear as regions of different intensity on the image. Images are typically 100 to 500,000 times larger than the portion of specimen being examined, which means that structures as small as 0.4 nm can be observed (Hunter 1993).

Electrons are highly attenuated by most materials, and therefore specimens must be extremely thin in order to allow enough of the electron beam through to be detected (Aguilera and Stanley 1990, Heertje and Paques 1995). Specimens used in TEM are therefore much thinner ($\approx$0.05 to 0.1 μm) than those used in light microscopy ($\approx$1 μm to a few millimeters). The difference in electron densities of food components is quite small, and therefore it is difficult to distinguish between them. For this reason, the density contrast between components is usually enhanced by selectively staining the sample with various heavy metal salts (which have high electron densities), such as lead, tungsten, or uranium. These salts bind selectively to specific components, which enables them to be distinguished from the rest of the sample. The metal salts may bind to the component itself (positive staining) or to the surrounding material (negative staining). In positive staining, a dark specimen is seen against a light background, whereas in negative staining, an illuminated specimen is seen against a dark background. The need to have very thin and dehydrated specimens, that often require staining, means that sample preparation is considerably more time consuming, destructive, and cumbersome than for other forms of microscopy.

The information produced by TEM is usually in the form of a two-dimensional image that represents a thin slice of the specimen (Figure 10.4). Nevertheless, it is possible to obtain some insight about the three-dimensional structure of a sample using a technique called *metal shadowing* (Figure 10.5). A vapor of a heavy metal, such as platinum, is sprayed onto the surface of a sample at an angle (Hunter 1993). The sample is then dissolved away using a strong acid, which leaves a metal replica of the sample. When a beam of electrons is

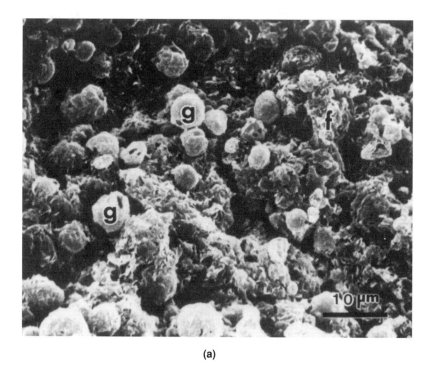

(a)

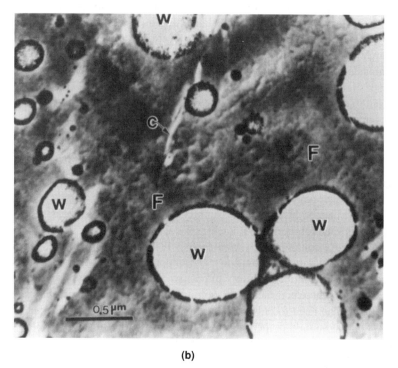

(b)

FIGURE 10.4 Comparison of electron micrographs of emulsions produced by (a) SEM (of butter) and (b) TEM (of margarine). TEM normally produces a two-dimensional image of a sample, whereas SEM produces a more three-dimensional image. The SEM of butter: g = fat droplet, f = fat crystals. The TEM of margarine: W = water droplets, F = fat continuous matrix, c = fat crystals. (Photographs kindly provided by I. Heertje.)

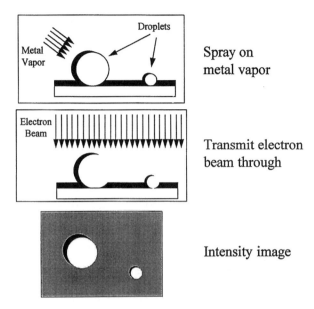

FIGURE 10.5   TEM can be used to obtain images of the surface topography of a sample using a technique known as *metal shadowing*.

transmitted through the metal replica, the "shadows" formed by the specimen are observed as illuminated regions which have characteristic patterns from which the topography of the specimen can be deduced.

Scanning Electron Microscopy.   SEM is used to provide images of the surface topography of specimens (Heertje and Paques 1995). It relies on the measurement of secondary electrons generated by a specimen when it is bombarded by an electron beam, rather than the electrons which have traveled through the specimen. A focused electron beam is directed at a particular point on the surface of a specimen. Some of the energy associated with the electron beam is absorbed by the material and causes it to generate secondary electrons, which leave the surface of the sample and are recorded by a detector. An image of the specimen is obtained by scanning the electron beam in an *x–y* direction over its surface and recording the number of electrons generated at each location. Because the intensity at each position depends on the angle between the electron beam and the surface, the electron micrograph has a three-dimensional appearance (Figure 10.4).

Sample preparation for SEM is considerably easier and tends to produce fewer artifacts than TEM. Because an image is produced by secondary electrons generated at the surface of a specimen, rather than by an electron beam that travels through a specimen, it is not necessary to use ultrathin samples. Even so, specimens often have to be cut, fractured, fixed, and dehydrated, which may alter their structures. The resolving power of SEM is about 3 to 4 nm, which is an order of magnitude worse than TEM but about three orders of magnitude better than optical microscopy. Another major advantage of SEM over optical microscopy is the large *depth of field,* which means that images of relatively large structures are all in focus (Hunter 1993).

Electron microscopy is widely used by food scientists to determine the size of emulsion droplets; the dimensions and structure of flocs; the surface morphology of droplets and air bubbles; the size, shape, and location of fat crystals; and the microstructure of three-dimensional networks of aggregated biopolymers (Chang et al. 1972, Kalab 1981, Tung and Jones

1981, Hermansson 1988, Aguilera and Stanley 1990, Bucheim and Dejmek 1990, Lee and Morr 1992, Heertje and Paques 1995). TEM and SEM electron micrographs of two water-in-oil emulsions are shown in Figure 10.4. TEM gives a two-dimensional cross-section of the sample, whereas SEM gives a more three-dimensional image. As mentioned earlier, the major limitation of the technique is the difficulty in preparing samples without altering their structure. In addition, the use of high-energy electron beams can change the structure of delicate specimens. Many of these problems are being overcome as the result of recent advances in the design of electron microscopes and sample preparation techniques (Smart et al. 1995).

10.3.1.4. *Atomic Force Microscopy*

Atomic force microscopy (AFM) has the ability to provide information about structures at the atomic and molecular levels and is therefore complementary to the other forms of microscopy mentioned above (Miles and McMaster 1995). The technique has only recently been developed as a commercial instrument, although these instruments are still fairly expensive to purchase. For this reason, the application of AFM to foods is still largely in its infancy (Kirby et al. 1995) and is only used by a small number of research laboratories. Nevertheless, the technique has the potential to provide a wealth of valuable information about the structure and organization of molecules in food emulsions and will increase our fundamental understanding of these complex systems.

The AFM creates an image by scanning a tiny probe (similar to the stylus of a record player, but only a few micrometers in size) across the surface of the specimen being analyzed (Figure 10.6). When the probe is held extremely close to the surface of a material, it experiences a repulsive force, which causes the cantilever to which it is attached to be bent away from the surface. The extent of the bending is measured using an extremely sensitive optical system. By measuring the deflection of the stylus as it is moved over the surface of the material, it is possible to obtain an image of its structure. In practice, it is more common to measure the force required to keep the deflection of the stylus constant, as this reduces the possible damage caused by a stylus as it moves across the surface of a sample. The resolution of AFM depends principally on the size and shape of the probe and the accuracy to which it can be positioned relative to the sample. Samples to be analyzed are usually dissolved in a suitable solvent and then dried onto the surface of an extremely flat plate such as mica.

Atomic force measurements have been used to observe the structure of individual and aggregated polysaccharide and protein molecules (e.g., xanthan, pectin, acetan, starch, col-

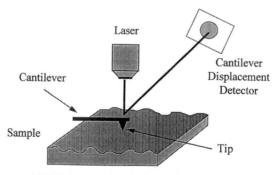

FIGURE 10.6 Principles of AFM. A probe is scanned over the surface of a material, and its displacement is measured using an accurate laser detection system. The force required to keep the probe in the same location, or the deflection of the probe, is measured.

lagen, myosin, and bovine serum albumin) (Miles and McMaster 1995, Kirby et al. 1995). This type of study is useful for examining the relationship between the structure and inter-actions of biopolymer molecules and the type of gels they form. AFM has also been used to study the molecular organization of emulsifiers at planar oil–water and air–water interfaces (Kirby et al. 1995).

10.3.2.   Static Light Scattering

10.3.2.1.   Principles

Static light scattering is used to determine particle sizes between about 0.1 and 1000 μm and is therefore suitable for characterizing the droplets in most food emulsions. When a beam of light is directed through an emulsion, it is scattered by the droplets (Dickinson and Stainsby 1982, Farinato and Rowell 1983, Hiemenz 1986, Hunter 1986, Everett 1988). A measurement of the degree of scattering can be used to provide information about the droplet size distribution and concentration. Analytical instruments based on this principle have been commercially available for many years (Mikula 1992) and are widely used in the food industry for research, development, and quality control purposes. Most of these instruments are fully automated, simple to use, and provide an analysis of an emulsion within a few minutes. Even so, they tend to be fairly expensive to purchase, which has limited their application somewhat.

The interaction of an electromagnetic wave with an emulsion is characterized by a *scattering pattern,* which is the angular dependence of the intensity of the light emerging from the emulsion, $I(\Phi)$ (Figure 10.7). The size and concentration of droplets in an emulsion are ascertained from the scattering pattern using a suitable theory (Farinato and Rowell 1983). Theories that relate light-scattering data to droplet size distributions are based on a mathematical analysis of the propagation of an electromagnetic wave through an ensemble of particles (van de Hulst 1957, Bohren and Huffman 1983). A number of theories are available, which vary according to their mathematical complexity and the type of systems to which they can be applied. The interaction between light waves and emulsion droplets can be conveniently divided into three regimes, according to the relationship between the droplet radius (r) and the wavelength (λ): (1) long-wavelength regime ($r < \lambda/20$), (2) intermediate-wavelength regime ($\lambda/20 < r < 20\lambda$), and (3) short-wavelength regime ($r > 20\lambda$). A characteristic scattering pattern is associated with each of these regimes (Figure 10.8).

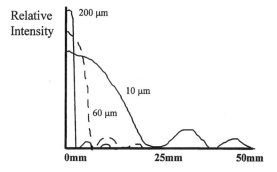

FIGURE 10.7   The scattering pattern from an emulsion is characterized by measuring the intensity of the scattered light, $I(\Phi)$, as a function of angle (Φ) between the incident and scattered beams. The scattering pattern is particularly sensitive to particle size relative to the wavelength of light.

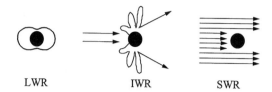

LWR   IWR   SWR

FIGURE 10.8 Typical scattering patterns for droplets in the long-, intermediate-, and short-wavelength regimes.

Long-Wavelength Regime. The simplest equation for relating the characteristics of the particles in a suspension to the scattering pattern produced when a monochromatic light beam passes through it was derived by Lord Rayleigh over a century ago and is applicable in the long-wavelength regime (Sherman 1968b):

$$\frac{I(\Phi)}{I_i} = \frac{6\pi^3 \phi r^3}{\lambda^4 R^2} \frac{\left(n_1^2 - n_0^2\right)^2}{\left(n_1^2 + 2n_0^2\right)^2} (1 + \cos^2 \Phi) \tag{10.1}$$

where I_i is the initial intensity of the light beam in the surrounding medium, ϕ is the dispersed-phase volume fraction, R is the distance between the detector and the scattering droplet, and n_0 and n_1 are the refractive indices of the continuous phase and droplets, respectively. This equation cannot be used to interpret the scattering patterns of most food emulsions because the size of the droplets (typically between 0.1 and 100 μm) is of the same order or larger than the wavelength of light used (typically between 0.2 and 1 μm). Nevertheless, it can be applied to suspensions which contain smaller particles, such as surfactant micelles or protein molecules (Hiemenz 1986). Equation 10.1 also provides some useful insights into the factors which influence the scattering profile of emulsions. It indicates that the degree of scattering from an emulsion is linearly related to the droplet concentration and increases as the refractive indices of the materials become more dissimilar.

Intermediate-Wavelength Regime. As mentioned earlier, most food emulsions contain droplets which are in the intermediate-wavelength regime. The scattering profile in this regime is extremely complex because light waves scattered from different parts of the same droplet are out of phase and therefore constructively and destructively interfere with one another (Shaw 1980, Hiemenz 1986). For the same reason, the mathematical relationship between the scattering pattern and the particle size is much more complex. A mathematician called Mie developed a theory which can be used to interpret the scattering patterns of dilute emulsions that contain spherical droplets of any size (van de Hulst 1957, Kerker 1969). The Mie theory is fairly complicated, but it can be solved rapidly using modern computers. This theory gives excellent agreement with experimental measurements and is used by most commercial particle-sizing instruments. It should be pointed out that the Mie theory assumes that the light waves are only scattered by a single particle, and so it is only strictly applicable to dilute emulsions. In more concentrated emulsions, a light beam scattered by one droplet may subsequently interact with another droplet, and this alters the scattering pattern (Ma et al. 1990). For this reason, emulsions must be diluted prior to analysis to a concentration where multiple scattering effects are negligible (i.e., $\phi < 0.05\%$).

Short-Wavelength Regime. When the wavelength of light is much smaller than the particle diameter, the droplet size distribution can be determined directly by optical microscopy (Section 10.3.1).

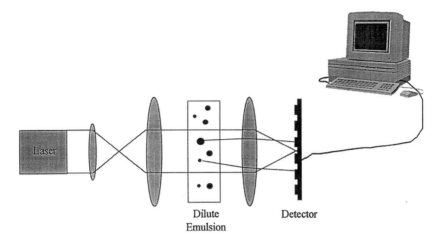

FIGURE 10.9 Design for a particle-sizing instrument that utilizes angular scattering.

10.3.2.2. Measurement Techniques

Angular Scattering Methods. Most modern particle-sizing instruments, based on static light scattering, measure the angular dependence of the scattered light (Farinato and Rowell 1983, Mikula 1992). The sample to be analyzed is diluted to an appropriate concentration and then placed in a glass measurement cell (Figure 10.9). A laser beam is generated by a helium–neon laser (λ = 632.8 nm) and directed through the measurement cell, where it is scattered by the emulsion droplets. The intensity of the scattered light is measured as a function of scattering angle using an array of detectors located around the sample. The scattering pattern recorded by the detectors is sent to a computer, where it is stored and analyzed. The droplet size distribution that gives the best fit between the experimental measurements of $I(\Phi)$ versus Φ and those predicted by the Mie theory is calculated. The data are then presented as a table or graph of droplet concentration versus droplet size (Section 1.3.2). Once the emulsion sample has been placed in the instrument, the measurement procedure is fully automated and only takes a few minutes to complete. Before analyzing a sample, the instrument is usually *blanked* by measuring the scattering profile from the continuous phase in the absence of emulsion droplets. This scattering pattern is then subtracted from that of the emulsion to eliminate extraneous scattering from background sources other than the droplets (e.g., dust or optical imperfections). The range of droplet radii that can be detected using this type of experimental arrangement is about 0.1 to 1000 µm. A number of instrument manufacturers have developed methods of determining droplet sizes down to 0.01 µm, but there is still much skepticism about the validity of data at these low sizes. This is highlighted by the fact that measurements on exactly the same emulsion using a number of different commercial instruments often give very different particle size distributions (Coupland and McClements 1998).

Spectroturbidimetric Methods. These techniques measure the turbidity of a dilute emulsion as a function of wavelength (Walstra 1968, Pearce and Kinsella 1978, Reddy and Fogler 1981, Pandolfe and Masucci 1984). The turbidity (τ) is determined by comparing the intensity of light which has traveled directly through an emulsion (I) with that which has traveled directly through the pure continuous phase (I_i): $\tau = -\ln(I/I_i)/d$, where d is the sample path length. The greater the scattering of light by an emulsion, the lower the intensity of the transmitted wave, and therefore the larger the turbidity. The turbidity of a dilute emulsion is linearly related to the dispersed-phase volume fraction, and so turbidity measurements can be

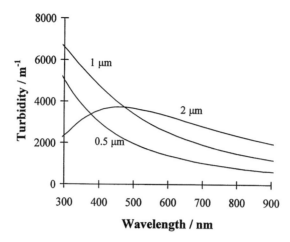

FIGURE 10.10   Turbidity versus wavelength spectra for emulsions with different droplet sizes.

used to determine ϕ if the droplet size remains constant. Turbidity versus wavelength spectra for emulsions with different droplet sizes are shown in Figure 10.10.

The emulsion to be analyzed is placed in a cuvette, and its turbidity is measured over a range of wavelengths (typically between 200 to 1000 nm). The droplet size distribution is then determined by finding the best fit between the experimental measurements of turbidity versus wavelength and those predicted by the Mie theory. The spectroturbidimetric technique can be carried out using the UV-visible spectrophotometers found in most research laboratories, which may circumnavigate the need to purchase one of the expensive commercial light-scattering instruments mentioned above. Nevertheless, some samples adsorb light strongly in the UV-visible region, which interferes with the interpretation of the turbidity spectra. In addition, the refractive indices of the dispersed and continuous phases must be known, and these vary with wavelength (Walstra 1968).

Reflectance Methods.   An alternative method of determining the droplet size of emulsions is to measure the light reflected back from the emulsion droplets ($\Phi = 180°$). The intensity of backscattered light is related to the size of the droplets in an emulsion (Lloyd 1959, Sherman 1968b). This technique has been used much less frequently than the spectroturbidimetric or angular scattering techniques but may prove useful for the study of more concentrated emulsions which are opaque to light.

10.3.2.3.   Applications

The principal application of static light-scattering techniques in the food industry is to determine droplet size distributions (Dickinson and Stainsby 1982). A knowledge of the particle size of an emulsion is useful for predicting its long-term stability to creaming, flocculation, coalescence, and Ostwald ripening (Chapter 7). Measurements of the time dependence of the particle size distribution can be used to monitor the kinetics of these processes (Chapter 7). Light-scattering instruments are widely used in research and development laboratories to investigate the influence of droplet size on physicochemical properties, such as stability, appearance, and rheology. They are also used in quality control laboratories to ensure that a product meets the relevant specifications for droplet size.

It should be noted that the data from any commercial light-scattering technique should be treated with some caution. To determine the droplet size distribution of an emulsion, commercial instruments have to make some *a priori* assumption about the shape of the distribution in order to solve the scattering theory in a reasonable time. In addition, the solution of the scattering theory is often particularly sensitive to the refractive indices and adsorptivities of the continuous and dispersed phases, and these values are often not known accurately (Zhang and Xu 1992). Finally, the mechanical design of each commercial instrument is different. All of these factors mean that the same emulsion can be analyzed using instruments from different manufacturers (or even from the same manufacturer) and quite large variations in the measured droplet size distributions can be observed, even though they should be identical (Coupland and McClements 1998). For this reason, commercial light-scattering instruments are often more useful for following qualitative changes rather than giving absolute values.

One must be especially careful when using light scattering to determine the particle size distribution of flocculated emulsions. The theory used to calculate the size distribution assumes that the particles are isolated homogeneous spheres. In flocculated emulsions, the droplets aggregate into heterogeneous "particles" which have an ill-defined refractive index and shape. Consequently, the particle size distribution determined by light scattering gives only an approximate indication of the true size of the flocs. In addition, emulsions often have to be diluted in continuous phase (to eliminate multiple scattering effects) and stirred (to ensure they are homogeneous) prior to measurement. Dilution and stirring are likely to disrupt any weakly flocculated droplets but leave strongly flocculated droplets intact. For these reasons, commercial light-scattering techniques can only give a qualitative indication of the extent of droplet flocculation. Another important limitation of light-scattering techniques is that they cannot be used to analyze optically opaque or semisolid emulsions *in situ* (e.g., butter, margarine, and ice cream).

10.3.3. Dynamic Light Scattering

10.3.3.1. Principles

Dynamic light scattering is used to determine the size of particles which are below the lower limit of detection of static light-scattering techniques (e.g., small emulsion droplets, protein aggregates, and surfactant micelles) (Hallet 1994, Dalgleish and Hallet 1995, Horne 1995). Instruments based on this principle are commercially available and are capable of analyzing particles with diameters between about 3 nm and 3 μm. Dynamic light-scattering techniques utilize the fact that the droplets in an emulsion continually move around because of their Brownian motion (Hunter 1986). The translational diffusion coefficient (D) of the droplets is determined by analyzing the interaction between a laser beam and the emulsion, and the size of the droplets is then calculated using the Stokes–Einstein equation (Horne 1995):

$$r = \frac{kT}{6\pi\eta_1 D} \tag{10.2}$$

where η_1 is the viscosity of the continuous phase.

10.3.3.2. Measurement Techniques

A number of experimental techniques have been developed to measure the translational diffusion coefficient of colloidal particles (Hunter 1993). The two most commonly used methods in commercial instruments are photon correlation spectroscopy (PCS) and Doppler shift spectroscopy (DSS).

Photon Correlation Spectroscopy. When a laser beam is directed through an ensemble of particles, a scattering pattern is produced which is a result of the interaction between the electromagnetic waves and the particles (Horne 1995). The precise nature of this scattering pattern depends on the relative position of the particles in the measurement cell. If the scattering pattern is observed over very short time intervals (approximately microseconds), one notices that there are slight variations in its intensity with time, which are caused by the change in the relative position of the particles due to their Brownian motion. The frequency of these fluctuations depends on the speed at which the particles move, and hence on their size. The change in the scattering pattern is monitored using a detector that measures the intensity of the photons, $I(t)$, which arrive at a particular scattering angle (or range of scattering angles) with time. If there is little change in the position of the particles within a specified time interval (τ), the scattering pattern remains fairly constant and $I(t) \approx I(t + \tau)$. On the other hand, if the particles move an appreciable distance within the time interval, the scattering pattern is altered significantly and $I(t) \neq I(t + \tau)$. The correlation between the scattering patterns can be expressed mathematically by an *autocorrelation function* (Horne 1995):

$$C(\tau) = \frac{1}{N} \sum_{i=1}^{N} I(t_i)I(t_i + \tau)$$

(10.3)

where N is the number of times this procedure is carried out, which is typically of the order of 10^5 to 10^6. As the time interval (τ) between which the two scattering patterns are compared is increased, the autocorrelation function decreases from a high value, where the scattering patterns are highly correlated, to a constant low value, when all correlation between the scattering patterns is lost (Figure 10.11). For a monodisperse emulsion, the autocorrelation function decays exponentially with a relaxation time (t_c) that is related to the translational diffusion coefficient of the particles: $t_c = (2Q^2D)^{-1}$, where Q is the scattering vector, which describes the strength of the interaction between the light wave and the particles (Horne 1995):

$$Q = \frac{4\pi n}{\lambda} \sin \frac{\theta}{2}$$

(10.4)

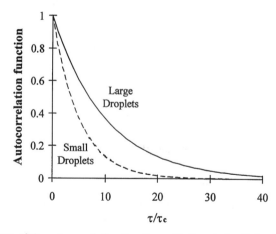

FIGURE 10.11   Decay of the autocorrelation function with time interval due to Brownian motion of the particles. The decay occurs more rapidly for small particles because they move faster.

Here, n is the refractive index of the medium, λ is the wavelength of light, and θ is the scattering angle. By measuring the decay of the autocorrelation function with time, it is possible to determine the diffusion coefficient and hence the particle size from the Stokes–Einstein equation (Equation 10.2).

PCS is suitable for accurate determination of particle size in suspensions that are monodisperse or that have a narrow size distribution. It is less reliable for suspensions with broad size distributions because of difficulties associated with interpreting the more complex autocorrelation decay curves (Horne 1995). PCS can be used to monitor the flocculation of particles in suspensions because aggregation causes them to move more slowly (Dalgleish and Hallet 1995). It can also be used to determine the thickness of adsorbed layers of emulsifier on spherical particles (Dalgleish and Hallet 1995). The radius of the spherical particles is measured in the absence of emulsifier (provided they are stable to aggregation) and then in the presence of emulsifier. The difference in radius is equal to the thickness of the adsorbed layer, although this value may also include the presence of any solvent molecules associated with the emulsifier. PCS is restricted to the analysis of dilute suspensions of particles ($\phi < 0.1\%$).

Doppler Shift Spectroscopy. When a laser beam is scattered from a moving particle, it experiences a shift in frequency, known as a Doppler shift (Trainer et al. 1992, Horne 1995). The frequency of the scattered wave increases slightly when the particle moves toward the laser beam and decreases slightly when it moves away. Consequently, there is a symmetrical distribution of Doppler shifts around the original frequency of the laser beam. The magnitude of the Doppler shift increases with particle velocity, and hence with decreasing particle size. Consequently, there is a broad distribution of Doppler shifts in a polydisperse suspension that contains particles moving at different velocities (Figure 10.12). The particle size distribution is determined by analyzing this Doppler shift spectrum. Two types of measurement techniques are commonly used in DSS: homodyne and heterodyne. Homodyne techniques determine the Doppler shifts by making use of the interference of light scattered from one particle with that scattered from all the other particles, whereas heterodyne techniques determine frequency shifts by comparing the frequency of the light scattered from the particles with a reference beam of fixed frequency (Trainer et al. 1992). The heterodyne technique is capable of analyzing suspensions with much higher particle concentrations than is possible using the homodyne technique, and so only it will be considered here.

A typical experimental arrangement for making Doppler shift measurements is shown in Figure 10.13. A laser beam is propagated along an optical waveguide which is immersed in the sample being analyzed. Part of the laser beam is reflected from the end of the waveguide and returns to the detector, where it is used as a reference beam because its frequency is not

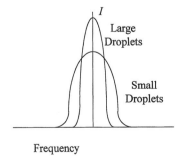

FIGURE 10.12 The distribution of frequency shifts can be related to the particle size distribution using suitable theories: small particles move more rapidly and therefore cause a greater frequency shift.

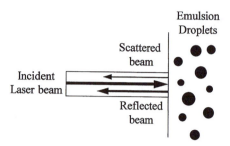

FIGURE 10.13 Experimental technique for measuring the Doppler shift of particles based on the heterodyne measurement technique.

Doppler shifted. The remainder of the laser beam propagates into the sample, and part of it is scattered back up the waveguide by the particles in its immediate vicinity (≈100 μm path length). This backscattered light is frequency shifted by an amount that depends on the velocity of the particles. The difference in frequency between the reflected and scattered waves is equivalent to the Doppler shift.

The variation of the power, $P(\omega)$, of the scattered waves with angular frequency (ω) has the form of a Lorentzian function:

$$P(\omega) = I_0 \langle I_S \rangle \frac{\omega_0}{\omega^2 + \omega_0^2} \tag{10.5}$$

where I_0 and $<I_S>$ are the reference and scattered light intensities, and ω_0 is a characteristic frequency that is related to the scattering efficiency and diffusion coefficient of the particles: $\omega_0 = DQ^2$. Particle size is determined by measuring the variation of $P(\omega)$ with ω, and then finding the value of the diffusion coefficient which gives the best fit between the measured spectra and that predicted by Equation 10.5. For polydisperse systems, it is necessary to take into account that there is a distribution of particle sizes. The calculation of particle size requires that the analyst input the refractive indices of the continuous and dispersed phases and the viscosity of the continuous phase. Because the path length of the light beam in the sample is so small (about 100 μm), it is possible to analyze much more concentrated emulsions (up to 40%) than with static light scattering or PCS techniques (Trainer et al. 1992), although some form of correction factor usually has to be applied to take into account the influence of the droplet interactions on the diffusion coefficient (Horne 1995). Commercial instruments are available that are easy to use and which can analyze an emulsion in a few minutes. The major limitations of dynamic light-scattering techniques are that they can only be used to analyze droplets with sizes smaller than about 3 μm and that the viscosity of the aqueous phase must be Newtonian, which is not the case for many food emulsions, especially those that contain thickening agents.

10.3.4.   Electrical Pulse Counting

Electrical pulse counting techniques are capable of determining the size distribution of particles with diameters between about 0.6 and 400 μm (Hunter 1986, Mikula 1992) and are therefore suitable for analyzing most food emulsions. These instruments have been commercially available for many years and are widely used in the food industry. The emulsion to be analyzed is placed in a beaker that has two electrodes dipping into it (Figure 10.14). One of

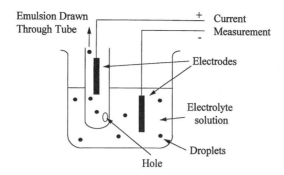

FIGURE 10.14 An electrical pulse counter. The decrease in current is measured when an oil droplet passes through a small hole in a glass tube.

the electrodes is contained in a glass tube which has a small hole in it, through which the emulsion is sucked. When an oil droplet passes through the hole, it causes a decrease in the current between the electrodes because oil has a much lower electrical conductivity than water. Each time a droplet passes through the hole, the instrument records a decrease in current, which it converts into an electrical pulse. The instrument controls the volume of liquid that passes through the hole, and so the droplet concentration can be determined by counting the number of electrical pulses in a known volume. When the droplets are small compared to the diameter of the hole, the droplet size is simply related to the height of the pulses: $d^3 = kP$, where d is the droplet diameter, P is the pulse height, and k is an instrument constant which is determined by recording the pulse height of a suspension of monodisperse particles of known diameter.

To cover the whole range of droplet sizes from 0.6 to 400 μm, it is necessary to use glass tubes with different sized holes. Typically, droplets between about 4 and 40% of the diameter of the hole can be reliably analyzed. Thus a tube with a hole of 16 μm can be used to analyze droplets with diameters between about 0.6 and 6 μm.

There are a number of practical problems associated with this technique which may limit its application to certain systems. First, it is necessary to dilute an emulsion considerably before analysis so that only one particle passes through the hole at a time; otherwise, a pair of particles would be counted as a single larger particle. As mentioned earlier, dilution of an emulsion may alter the structure of any flocs present, especially when the attraction between the droplets is small. Second, the emulsion droplets must be suspended in an electrolyte solution (typically 5 wt% salt) to ensure that the electrical conductivity of the aqueous phase is sufficiently large to obtain accurate measurements. The presence of an electrolyte alters the nature of the colloidal interactions between charged droplets, which may change the extent of flocculation from that which was present in the original sample. Third, any weakly flocculated droplets may be disrupted by the shear forces generated by the flocs as they pass through the small hole. Fourth, if an emulsion contains a wide droplet size distribution, it is necessary to use a number of glass tubes with different hole sizes to cover the full size distribution. Nevertheless, in those emulsions where it can be applied, electrical pulse counting is usually considered to be more reliable and accurate than light scattering.

10.3.5. Sedimentation Techniques

Sedimentation techniques can be used to determine the size distribution of particles between about 1 nm and 1 mm, although a number of different types of instruments have to be used to cover the whole of this range (Hunter 1986, 1993). Particle size is determined by measuring the velocity at which droplets sediment (or cream) in a gravitational or centrifugal field.

10.3.5.1. Gravitational Sedimentation

The velocity that an isolated rigid spherical particle suspended in a Newtonian liquid moves due to gravity is given by Stokes' equation:

$$v = \frac{2(\rho_2 - \rho_1)gr^2}{9\eta_1} \tag{10.6}$$

where ρ_2 is the density of the droplets, ρ_1 and η_1 are the density and viscosity of the continuous phase, g is the acceleration due to gravity, and r is the droplet radius. The droplet size can therefore be determined by measuring the velocity at which the particle moves through the liquid once the densities of both phases and the viscosity of the continuous phase are known. The movement of the droplets could be monitored using a variety of experimental methods, including visual observation, optical microscopy, light scattering, nuclear magnetic resonance, ultrasound, and electrical measurements (Mikula 1992, Pal 1994, Dickinson and McClements 1995). The Stokes equation cannot be used to estimate droplet sizes less than about 1 μm because of their Brownian motion (Hunter 1986). The gravitational forces acting on a droplet cause it to move in a certain direction, either up or down, depending on the particle density relative to the continuous phase. On the other hand, Brownian motion tends to randomize the spatial distribution of the particles. This effect is important when the distance a particle moves due to Brownian motion is comparable to the distance it moves due to sedimentation or creaming in the same time. In a quiescent system, the average distance (root mean square displacement, x_{rms}) that a sphere moves due to Brownian motion is given by (Hunter 1993):

$$x_{rms} = \sqrt{\frac{kTt}{3\pi\eta r}} \tag{10.7}$$

Thus, a 1-μm oil droplet ($\rho_2 = 920$ kg m^{-3}) moves about 0.7 μm s^{-1} due to Brownian motion when suspended in pure water, whereas it will move about 2 μm in the same time due to creaming. Thus, even for a droplet of this size, the thermal motion can have a pronounced influence on its creaming rate. In addition, temperature gradients within a sample can lead to convective currents that interfere with the droplet movement. For these reasons, gravitational sedimentation techniques have limited application for determining droplet sizes in emulsions. Instead, the droplets are made to move more rapidly by applying a centrifugal force, so that the problems associated with Brownian motion and convection are overcome.

10.3.5.2. Centrifugation

When an emulsion is placed in a centrifuge and rotated rapidly, it is subjected to a centrifugal force that causes the droplets to move inward when they have a lower density than the surrounding liquid (e.g., oil-in-water emulsions) or outward when they have a higher density (e.g., water-in-oil emulsions). The velocity, $v(x)$, at which the droplets move through the surrounding liquid depends on the angular velocity (ω) at which the tube is centrifuged and their distance from the center of the rotor (x) The droplet motion can be conveniently characterized by a *sedimentation coefficient,* which is independent of the angular velocity and location of the droplets (Hunter 1993):

$$S = \frac{v(x)}{\omega^2 x} \tag{10.8}$$

The radius of an isolated spherical particle in a fluid is related to the sedimentation coefficient by the following equation:

$$r = \sqrt{\frac{9\eta_1 S}{2(\rho_2 - \rho_1)}} \tag{10.9}$$

By measuring the droplet velocity at different positions within a centrifuge tube, it is possible to determine S and therefore the droplet radius.

The sedimentation coefficient is related to the distance from the rotor center: $\ln(x_2/x_1) = S\omega^2(t_2 - t_1)$, where x is the position of the particle at time t (Hunter 1993). Thus a plot of $\ln(x)$ versus time at a fixed rotation speed can be used to determine S. For oil-in-water emulsions with sizes between 0.1 and 100 μm, S varies between about 2 ns and 2 ms. The position of the droplets is usually determined by optical microscopy or light scattering. Modern instruments are capable of measuring the full droplet size distribution of emulsions by analyzing the velocity at which a number of different droplets move. Instruments that utilize this principle are commercially available and are used in the food industry.

10.3.6. Ultrasonic Spectrometry

10.3.6.1. Principles

Ultrasonic spectrometry utilizes interactions between ultrasonic waves and particles to obtain information about the droplet size distribution of an emulsion (McClements 1991, 1996; Dukhin and Goetz 1996; Povey 1995, 1997). It can be used to determine droplet sizes between about 10 nm and 1000 μm. Particle-sizing instruments based on ultrasonic spectrometry have recently become available commercially and are likely to gain wide acceptance in the food industry in the near future because they have a number of important advantages over alternative technologies (e.g., they can be used to analyze emulsions which are concentrated and optically opaque, without the need for any sample preparation).

As an ultrasonic wave propagates through an emulsion, its velocity and attenuation are altered due to its interaction with the droplets. These interactions may take a number of different forms: (1) some of the wave is *scattered* into directions which are different from that of the incident wave; (2) some of the ultrasonic energy is converted into heat due to various *absorption* mechanisms (e.g., thermal conduction and viscous drag); and (3) there is interference between waves which travel through the droplets, waves which travel through the surrounding medium, and waves which are scattered. The relative importance of these different mechanisms depends on the thermophysical properties of the component phases, the frequency of the ultrasonic wave, and the concentration and size of the droplets.

As with light scattering, the scattering of ultrasound by a droplet can be divided into three regimes according to the relationship between the particle size and the wavelength of the radiation: (1) the long-wavelength regime ($r < \lambda/20$), (2) the intermediate-wavelength regime ($\lambda/20 < r < 20\lambda$), and (3) the short-wavelength regime ($r > 20\lambda$). The wavelength of ultrasound (10 μm to 10 mm) is much greater than that of light (0.2 to 1 μm), and so ultrasonic measurements are usually made in the long-wavelength regime, whereas light-scattering measurements are made in the intermediate-wavelength regime. As a consequence, light-scattering results are often more difficult to interpret reliably because of the greater complexity of the scattering theory in the intermediate-wavelength regime.

In the long-wavelength regime, the ultrasonic properties of fairly dilute emulsions ($\phi <$ 15%) can be related to their physicochemical characteristics using the following equation:

$$\left(\frac{K}{k_1}\right)^2 = \left(1 - \frac{3\phi iA_0}{k_1^3 r^3}\right)\left(1 - \frac{9\phi iA_1}{k_1^3 r^3}\right)$$

(10.10)

where K is the complex propagation constant of the emulsion ($= \omega/c_e + i\alpha_e$), k_1 is the complex propagation constant of the continuous phase ($= \omega/c_1 + i\alpha_1$), ω is the angular frequency, c is the ultrasonic velocity, α is the attenuation coefficient, $i = \sqrt{-1}$, and r is the droplet radius. The A_0 and A_1 terms are the monopole and dipole scattering coefficients of the individual droplets, which depend on the adiabatic compressibility, density, specific heat capacity, thermal conductivity, cubical expansivity, and viscosity of the component phases, as well as the frequency and droplet size (Epstein and Carhart 1953). Equation 10.10 can be used to calculate the ultrasonic velocity and attenuation coefficient of an emulsion as a function of droplet size and concentration, once the thermophysical properties of the component phases are known, since $c_e = \omega/\text{Re}(K_e)$ and $\alpha_e = \text{Im}(K_e)$.

The dependence of the ultrasonic velocity and attenuation coefficient of an emulsion on droplet size and concentration is shown in Figure 10.15. The ultrasonic velocity increases with droplet size, while the attenuation per wavelength ($\alpha\lambda$) has a maximum value at an intermediate droplet size. It is this dependence of the ultrasonic properties of an emulsion on droplet size that enables ultrasound to be used as a particle-sizing technology. An emulsion is analyzed by measuring its ultrasonic velocity and/or attenuation spectra and finding the droplet size and concentration which give the best fit between the experimental data and Equation 10.10. For polydisperse emulsions, the equation has to be modified to take into account the droplet size distribution (McClements 1991). One of the major limitations of the ultrasonic technique is the fact that a great deal of information about the thermophysical properties of the component phases is needed to interpret the measurements, and this information often is not readily available in the literature (Coupland and McClements 1998). In addition, for droplet concentrations greater than about 15%, it is necessary to extend the above equation to take into account interactions between the droplets, which has been done recently (Hemar et al. 1997, McClements et al. 1998b).

10.3.6.2. Measurement Techniques

The ultrasonic properties of materials can be measured using a number of different experimental techniques (e.g., pulse echo, through transmission, and interferometric) (McClements

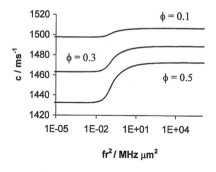
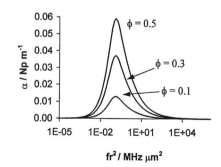

FIGURE 10.15 Dependence of the ultrasonic properties of an emulsion on its droplet size and dispersed-phase volume fraction.

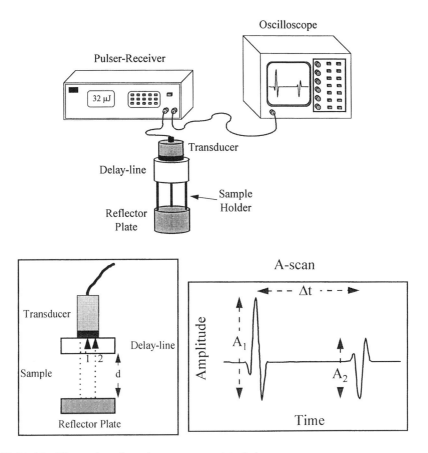

FIGURE 10.16 Ultrasonic pulse-echo measurement technique.

1996). The major difference between them is the form in which the ultrasonic energy is applied to the sample and the experimental configuration used to carry out the measurements. In this section, we shall only consider the pulse-echo technique because it is one of the simplest and most widely used techniques to analyze emulsions. The sample to be analyzed is contained within a thermostated measurement cell which has an ultrasonic transducer fixed to one side (Figure 10.16). An electrical pulse is applied to the transducer, which stimulates it to generate a pulse of ultrasound that is directed into the sample. This pulse travels across the sample, is reflected from the back wall of the measurement cell, travels back through the sample, and is then detected by the same transducer (Figure 10.16). The ultrasonic pulse received by the transducer is converted back into an electrical pulse which is digitized and saved for analysis.

The ultrasonic velocity and attenuation coefficient of a sample are determined by measuring the time of flight (t) and amplitude (A) of the ultrasonic pulse which has traveled through it. The ultrasonic velocity is equal to the distance traveled by the pulse ($2d$) divided by its time of flight: $c = 2d/t$. The attenuation coefficient is calculated by comparing the amplitude of a pulse which has traveled through the sample with that of a pulse which has traveled through a material whose attenuation coefficient is known: $\alpha_2 = \alpha_1 - \ln(A_2/A_1)/2d$, where the subscripts 1 and 2 refer to the properties of the reference material and the material being tested, respectively. The through-transmission technique is very similar to the pulse-echo technique, except that separate transducers are used to generate and receive the ultrasonic pulse (McClements 1996).

To determine the droplet size distribution of an emulsion, it is necessary to measure the frequency dependence of its ultrasonic properties. Two different approaches can be used to measure the dependence of the ultrasonic properties on frequency using the pulse-echo technique. In the first approach, a *broadband* ultrasonic pulse is used, which is a single pulse that contains a wide range of frequencies. After the pulse has traveled through the sample, it is analyzed by a Fourier transform algorithm to determine the values of t and A (and therefore c and α) as a function of frequency (McClements and Fairley 1991, 1992). To cover the whole frequency range (0.1 to 100 MHz), a number of transducers with different frequencies have to be used. In the second approach, a *tone-burst* ultrasonic pulse is used, which is a single pulse that contains a number of cycles of ultrasound at a particular frequency (McClements 1996). The transducer is tuned to a particular frequency, and a measurement of the ultrasonic velocity and attenuation coefficient is made. The transducer is then tuned to another frequency and the process is repeated. Because measurements are carried out separately at a number of different frequencies, this approach is more time consuming and laborious than the one which uses broadband pulses. Both of these methods are used to determine the frequency dependence of the ultrasonic properties of emulsions in commercial ultrasonic particle sizers.

10.3.6.3. Applications

Ultrasound has major advantages over many other particle-sizing technologies because it can be used to measure droplet size distributions in concentrated and optically opaque emulsions *in situ*. In addition, it can be used as an on-line sensor for monitoring the characteristics of food emulsions during processing, which gives food manufacturers much greater control over the quality of the final product. The possibility of using ultrasound to measure droplet sizes in real foods has been demonstrated for casein micelles (Griffin and Griffin 1990), sunflower-oil-in-water emulsions (McClements and Povey 1989), corn-oil-in-water emulsions (Coupland and McClements 1998), salad creams (McClements et al. 1990), and milk fat globules (Miles et al. 1990).

The two major disadvantages of the ultrasonic technique are the large amount of thermophysical data required to interpret the measurements and the fact that small air bubbles can interfere with the signal from the emulsion droplets (McClements 1991, 1996).

10.3.7. Nuclear Magnetic Resonance

Instrumental techniques based on nuclear magnetic resonance (NMR) utilize interactions between radio waves and the nuclei of hydrogen atoms to obtain information about the properties of materials.* An NMR technique has been developed to measure the droplet size distribution of emulsions (Callaghan et al. 1983, van den Enden et al. 1990, Li et al. 1992, Soderman et al. 1992), which is sensitive to particle sizes between about 0.2 and 100 μm (Dickinson and McClements 1995). This technique relies on measurements of the *restricted diffusion* of molecules within emulsion droplets.

The principles of the technique are fairly complex and have been described in detail elsewhere (Dickinson and McClements 1995, Soderman and Bailnov 1996). Basically, the sample to be analyzed is placed in a static magnetic field gradient and a series of radio-frequency pulses are applied to it. These pulses cause some of the hydrogen nuclei in the sample to be excited to higher energy levels, which leads to the generation of a detectable NMR signal. The amplitude of this signal depends on the movement of the nuclei in the sample: the farther the nuclei move during the experiment, the greater the reduction in the

---

* NMR techniques can also be used to study the nuclei of certain other isotopes, but these are not widely used for particle sizing.

amplitude. A measurement of the signal amplitude can therefore be used to study molecular motion.

In a bulk liquid, the distance that a molecule can move in a certain time is governed by its translational diffusion coefficient, $x_{rms} = (\sqrt{2Dt})$. When a liquid is contained within an emulsion droplet, its diffusion may be restricted because of the presence of the interfacial boundary. If the movement of a molecule in a droplet is observed over relatively short times ($t << d^2/2D$), the diffusion is unrestricted, but if it is observed over longer times, its diffusion is restricted because it cannot move farther than the diameter of the droplet. By measuring the attenuation of the NMR signal at different times, it is possible to identify when the diffusion becomes restricted and thus estimate the droplet size. Because this technique relies on the movement of molecules within droplets, it is independent of droplet flocculation.

This technique has been used to determine the droplet size distribution of a variety of oil-in-water and water-in-oil emulsions, including margarine, cream, and cheese (Callaghan et al. 1983, van den Enden et al. 1990, Li et al. 1992, Soderman et al. 1992). Like ultrasonic spectrometry, it is nondestructive and can be used to analyze emulsions which are concentrated and optically opaque (Dickinson and McClements 1995). It therefore seems likely that NMR instruments specifically designed to measure the particle size distribution of emulsions will be developed and become commercially available in the near future.

10.3.8.   Neutron Scattering

Neutron scattering techniques utilize interactions between a beam of neutrons and an emulsion to determine the droplet size distribution (Dickinson and Stainsby 1982, Eastoe 1995). They can also be used to provide information about the thickness of interfacial layers and the spatial distribution of droplets. These techniques have a couple of special features which make them particularly suitable for studying food emulsions (Eastoe 1995). First, the scattering of neutrons from emulsion droplets is very weak, and therefore multiple scattering effects are not appreciable, which means that concentrated emulsions can be analyzed without dilution. Second, the scattering of neutrons from heterogeneous materials depends on the "contrast" between the different components, which can be manipulated by the experimenter. Thus it is possible to selectively highlight specific structural features within an emulsion (see below). Despite its ability to generate information which is difficult to obtain using other techniques, the application of neutron scattering to food emulsions is limited because a nuclear reactor is needed to generate the neutron beam. There are only a small number of neutron-scattering facilities in the world which are generally accessible, and beam time is rather limited and must be scheduled many months in advance of the proposed experiment (Stothart 1995).

In many respects, the measurement principle of neutron scattering is similar to that of static light scattering, except that a beam of neutrons is used instead of light. The sample to be analyzed is placed into a cuvette which is inserted between a source of neutrons and a neutron detector (Stothart 1995). A beam of neutrons is passed through the emulsion, and the intensity of the scattered neutrons is measured as a function of scattering angle (and/or wavelength). Information about the properties of the emulsion is then obtained by interpreting the resulting spectra using an appropriate neutron-scattering theory (Eastoe 1995).

Each type of atomic nuclei scatters neutrons to a different extent, which is characterized by a "scattering cross-section" (Lovsey 1984). The scattering of neutrons from a heterogenous material, such as an emulsion, depends on the contrast between the scattering cross-sections of the different components: the greater the contrast, the more intense the scattering. One of the most important attributes of neutron scattering is the ability to alter the scattering cross-section of molecules that contain hydrogen atoms (e.g., water, proteins, fats, and carbohydrates) (Eastoe 1995, Stothart 1995). Normal hydrogen (^1H) and deuterium (^2H) have

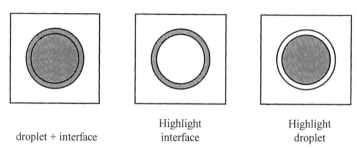

| droplet + interface | Highlight interface | Highlight droplet |

FIGURE 10.17 Concept of contrast matching of emulsions in neutron-scattering experiments.

significantly different scattering cross-sections, and so by varying the $^1H:^2H$ ratio of a particular type of molecule, it is possible to increase or decrease its contrast with respect to the other components in an emulsion. As a consequence, it is possible to emphasize specific structural components within an emulsion.

Consider an oil-in-water emulsion (Figure 10.17) which consists of oil droplets covered by an interfacial layer and suspended in an aqueous phase. By altering the ratio of water to deuterated water in the aqueous phase, it is possible to match the aqueous phase to either the interfacial layer or the oil within the droplets. Alternatively, the interfacial layer could be matched to the droplet by partial deuteration of either the oil or emulsifier molecules. Thus it is possible to obtain information about the thickness of the interfacial layer, the oil droplet, or the droplet + interfacial layer.

10.3.9. Dielectric Spectroscopy

This technique depends on the dielectric response of an emulsion during the application of an electromagnetic wave (Clausse 1983, Asami 1995, Sjoblom et al. 1996). The possibility of using dielectric spectroscopy to determine the droplet size distribution of concentrated emulsions was recently demonstrated (Garrouch et al. 1996). The dielectric permittivity of an emulsion is measured over a wide range of electromagnetic frequencies, and the resulting spectra are analyzed using a suitable theory to determine the droplet size distribution. The major limitation of this technique is that it can only be used to determine droplet size distributions in emulsions that contain charged particles. Even so, it can be used to simultaneously measure the zeta potential and droplet size distribution, and it can be used to analyze emulsions which are concentrated and optically opaque without the need for any sample dilution. It may therefore have some important applications in the food industry. Nevertheless, dielectric spectroscopy is still in its infancy, and a lot more research is still required before the technique becomes more widely accepted and utilized.

10.3.10. Electroacoustics

The droplet size distribution of emulsions that contain charged droplets can be determined using electroacoustic techniques (O'Brien et al. 1995, Carasso et al. 1995, Dukhin and Goetz 1996). These techniques use a combination of electric and acoustic phenomena to determine both the size and zeta potential of emulsion droplets (Section 10.6.3). Instruments based on this principle have recently become commercially available and are capable of analyzing droplets with sizes between about 0.1 and 10 µm (Hunter 1998). These instruments are capable of analyzing emulsions with high droplet concentrations (<40%) without any sample dilution. The major limitation of the electroacoustic technique is that measurements rely on the droplets having an electrical charge, the surrounding liquid must be Newtonian, and there must be a significant density contrast between the droplets and the surrounding liquid. These conditions are not always met for food emulsions, which means that electroacoustics may only have limited application within the food industry.

10.4. DISPERSED-PHASE VOLUME FRACTION

10.4.1. Proximate Analysis

The concentration of droplets in an emulsion can be determined using many of the standard analytical methods developed to determine the composition of foods (Nielsen 1994, Pomeranz and Meloan 1994). A variety of solvent-extraction techniques are available for measuring fat content (Min 1994, Pal 1994). The sample to be analyzed is mixed with a nonpolar organic solvent which extracts the oil. The solvent is then physically separated from the aqueous phase, and the oil content is determined by evaporating the solvent and weighing the residual oil. A possible difficulty associated with applying this technique to oil-in-water emulsions is that the interfacial membrane may be resistant to rupture, and therefore all of the oil is not released. This problem has been overcome in a number of nonsolvent techniques developed to measure the fat content of dairy emulsions. In the *Gerber* and *Babcock* methods, an emulsion is placed in a specially designed bottle and then mixed with sulfuric acid, which digests the interfacial membrane surrounding the droplets and thus causes coalescence (Min 1994). The bottle is centrifuged to facilitate the separation of the oil and aqueous phases, and the percentage of oil in the emulsion is determined from the calibrated neck of the bottle. A similar procedure is involved in the *detergent* method, except that rather than sulfuric acid, a surfactant is added to promote droplet coalescence.

The water content of an emulsion can be determined using a variety of proximate analysis techniques (Mikula 1992, Pal 1994, Pomeranz and Meloan 1994, Bradley 1994). The simplest of these involves weighing an emulsion before and after the water has been evaporated, which may be achieved by conventional oven, vacuum oven, microwave oven, or infrared light. The moisture content can also be determined by distillation. The emulsion is placed in a specially designed flask which has a calibrated side arm. An organic solvent is mixed with the emulsion, and the flask is heated to cause the water to evaporate and collect in the side arm. This procedure is continued until all of the water has evaporated, and then its volume is determined from the calibrations on the side arm.

Many of these techniques are labor intensive, time consuming, and destructive and therefore are unsuitable for rapid quality assurance tests. Instruments based on infrared absorption are becoming increasingly popular for rapid and nondestructive analysis of food composition (Wehling 1994, Wilson 1995). Once calibrated, these instruments are capable of simultaneously determining the concentration of fat, water, protein, and carbohydrate. These instruments are likely to find widespread use in the food industry, particularly for on-line measurements.

10.4.2. Density Measurements

10.4.2.1. *Principles*

One of the simplest methods of determining the dispersed-phase volume fraction of an emulsion is to measure its density (Pal 1994). The density of an emulsion (ρ_e) is related to the densities of the continuous (ρ_1) and dispersed phases (ρ_2): $\rho_e = \phi\rho_2 + (1 - \phi)\rho_1$. The dispersed-phase volume fraction of an emulsion can therefore be determined by measuring its density and knowing the density of the oil and aqueous phases:

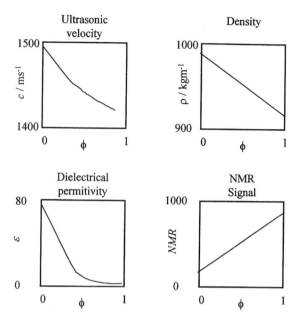

FIGURE 10.18 Dependence of the physicochemical properties of various oil-in-water emulsions on their dispersed-phase volume fraction.

$$\phi = \frac{\rho_e - \rho_1}{\rho_2 - \rho_1} \tag{10.11}$$

The densities of the dispersed and continuous phases in a food emulsion are appreciably different, about 900 kg m^{-3} for liquid oils and about 1000 kg m^{-3} for aqueous phases. It is possible to measure the density of an emulsion to about 0.2 kg m^{-3}, and so the dispersed-phase volume fraction can be determined to within 0.002 (0.2%) using this technique. The fact that the density of liquid oils is lower than that of water means that the density of an emulsion usually decreases with increasing oil content (Figure 10.18).

10.4.2.2. Measurement Techniques

Density Bottles. A liquid sample is poured into a glass bottle of known mass and volume (Pomeranz and Meloan 1994). The bottle and liquid are allowed to equilibrate to the measurement temperature and are then weighed using an accurate balance. The mass of emulsion ($m_{emulsion}$) required to completely fill the container at a given temperature is measured. The internal volume of the container is determined by measuring the mass of distilled water (a material whose density is known accurately) it takes to fill the bottle ($V_{bottle} = m_{water}/\rho_{water}$). Thus, the density of the emulsion can be determined: $\rho_{emulsion} = m_{emulsion}/V_{bottle}$. The density of an emulsion is particularly sensitive to temperature, and so it is important to carefully control the temperature when accurate measurements are required. It is important to ensure that the container is clean and dry prior to weighing and that it contains no gas bubbles. Gas bubbles reduce the volume of liquid in the bottle without contributing to the mass and therefore lead to an underestimate of the density.

Hydrometers. Several methods for measuring the density of liquids are based on the Archimedes principle, which states that the upward buoyant force exerted on a body immersed in a liquid is equal to the weight of the displaced liquid (Pomeranz and Meloan 1994). This principle is used in *hydrometers,* which are graduated hollow glass bodies that float on the liquid to be tested. The depth that the hydrometer sinks into a liquid depends on the density of the liquid. The hydrometer sinks to a point where the mass of the displaced liquid is equal to the mass of the hydrometer. Thus the depth to which a hydrometer sinks increases as the density of the liquid decreases. The density of a liquid is read directly from graduated calibrations on the neck of the hydrometer. Hydrometers are less accurate than density bottles, but much more rapid and convenient to use.

Oscillating U-tubes. The density of fluids can be measured rapidly and accurately using an instrument called an *oscillating U-tube densitometer* (Pal 1994). The sample to be analyzed is placed in a glass U-tube, which is forced to oscillate sinusoidally by the application of an alternating mechanical force. The resonant frequency of the U-tube is related to its mass and therefore depends on the density of the material contained within it. The density of a fluid is determined by measuring the resonant frequency of the U-tube and relating it to the density using an appropriate mathematical equation. The instrument must be calibrated with two fluids of accurately known density (usually distilled water and air). Density can be measured to within 0.1 kg m^{-3} in a few minutes using this technique. Recently, on-line versions of this technique have been developed for monitoring the density of fluids during processing. A small portion of a fluid flowing through a pipe is directed through an oscillating U-tube and its density is measured before being redirected into the main flow.

10.4.2.3. *Applications*

An accurate measurement of emulsion density can be carried out using inexpensive equipment that is available in many laboratories. The technique is nondestructive and can be used to analyze emulsions which are concentrated and optically opaque. One possible problem with the technique is that the physical state of the emulsion constituents may alter the accuracy of a measurement. For example, the density of solid fat is greater than that of liquid oil, and therefore the density of an emulsion depends on the solid fat content, as well as the total fat content. In these situations, it is necessary to heat the emulsion to a temperature where it is known that all of the fat crystals have melted and then measure its density.

10.4.3. Electrical Conductivity

10.4.3.1. *Principles*

The dispersed-phase volume fraction of an emulsion can be conveniently determined by measuring its electrical conductivity (ε) (Clausse 1983, Robin et al. 1994, Asami 1995). The electrical conductivity of water is much higher than that of oil, and so there is a decrease in ε as the oil content of an emulsion increases (Figure 10.18). In dilute emulsions, the dispersed-phase volume fraction is related to the electrical conductivity by the following equation (Clausse 1983):

$$\phi = \left(\frac{\varepsilon_e - \varepsilon_1}{\varepsilon_e + 2\varepsilon_1} \right)\left(\frac{\varepsilon_2 + 2\varepsilon_1}{\varepsilon_2 - \varepsilon_1} \right) \tag{10.12}$$

where the subscripts 1, 2, and *e* refer to the continuous phase, dispersed phase, and emulsion, respectively. More complex expressions have been derived to relate the dispersed-phase volume fraction of concentrated emulsions to their electrical properties.

10.4.3.2. Measurement Techniques

The electrical conductivity of an emulsion can simply be determined using a conductivity cell (Siano 1998). This cell consists of a couple of electrodes which are connected to electrical circuitry that is capable of measuring the electrical conductivity of the sample contained between the electrodes. The electrical conductivity of an aqueous phase depends on the concentration of electrolytes present, and so it is important to properly characterize the properties of the component phases.

10.4.3.3. Applications

The electrical conductivity technique can be used to determine the dispersed-phase volume fraction of concentrated and optically opaque emulsions without the need for any sample preparation. Measurements are independent of the size of the emulsion droplets, which is an advantage when the droplet size distribution is unknown (Robin et al. 1994). The electrical conductivity of an emulsion is dependent on the physical state of its constituents, and therefore it may be necessary to heat an emulsion to a temperature where all of the crystals have melted before making a measurement. The electrical conductivity is also sensitive to the ionic strength of the aqueous phase, and therefore it is necessary to take this into account when carrying out the analysis (Skodvin et al. 1994).

10.4.4. Alternative Techniques

The dispersed-phase volume fraction can be measured using many of the techniques used to determine droplet size distributions (Section 10.3). Light-scattering and electrical pulse counting techniques can be used to determine dispersed-phase volume fractions in dilute emulsions (ϕ < 0.1%), whereas Doppler shift spectroscopy, ultrasonic, electroacoustic, dielectric, neutron-scattering, and NMR techniques can be used to analyze much more concentrated emulsions. All of these techniques rely on there being a measurable change in some physicochemical property of an emulsion as its droplet concentration increases (e.g., the intensity of scattered or transmitted light, the attenuation or velocity of an ultrasonic wave, the amplitude or decay time of an NMR signal) (Figure 10.18). Some of these techniques can be used to simulta-neously determine the droplet size distribution and dispersed-phase volume fraction, whereas others can be used to determine ϕ independently of a knowledge of the droplet size.

10.5. DROPLET CRYSTALLINITY

10.5.1. Dilatometry

10.5.1.1. Principles

Dilatometry has been used for many years to monitor the crystallinity of both the dispersed and continuous phases in emulsions (Turnbull and Cormia 1961, Skoda and van den Tempel 1963, Phipps 1964). The technique is based on measurements of the density change which occurs when a material melts or crystallizes. The density of the solid state of a material is usually greater than that of the liquid state because the molecules are able to pack more efficiently.* Consequently, there is a decrease in the density of a material when it melts and an increase when it crystallizes. The fraction of crystalline droplets in an emulsion can be determined from the following equation:

---

* With the important exception of water near its freezing point.

$$\phi_c \cdot = \frac{\rho_e - \rho_{eL}}{\rho_{eS} - \rho_{eL}}$$ (10.13)

where, ρ_e is the density of an emulsion that contains partially crystalline droplets, and ρ_{eL} and ρ_{eS} are the densities of the same emulsion when the droplets are either completely liquid or completely solid, respectively. The values of ρ_{eL} and ρ_{eS} are usually determined by extrapolating density measurements from higher and lower temperatures into the region where the droplets are partially crystalline.

10.5.1.2. Measurement Techniques

In principle, dilatometry can be carried out using any experimental technique that is capable of measuring density changes. In practice, dilatometry is often performed using a specially designed piece of apparatus (Figure 10.19). A known mass of sample is placed into a glass bulb which is connected to a calibrated capillary tube. A liquid, such as mercury or colored water, is poured into the capillary tube above the fat. The change in volume of the sample when it crystallizes or melts is then determined by observing the change in height of the liquid in the capillary tube. These measurements can be carried out either as the temperature of the sample is varied in a controlled way or as the sample is held at a constant temperature over time (Turnbull and Cormia 1961). The data can be presented as a volume change or as a density change, depending on which is most convenient.

10.5.1.3. Applications

The temperature dependence of the density of an oil-in-water emulsion that contains droplets which undergo a phase transition is shown in Figure 10.20. When the emulsion is heated from a temperature where the droplets are completely solid, there is a sharp decrease in density when the droplets melt. Conversely, when an emulsion is cooled from a temperature where the droplets are completely liquid, there is a sudden increase in density when the droplets crystallize. The crystallization temperature is considerably lower than the melting temperature because of supercooling (Section 4.2). In food emulsions, oil droplets normally melt over a much wider temperature range than that shown for a pure oil in Figure 10.20 because they contain a mixture of different triacylglycerols, each with its own melting point.

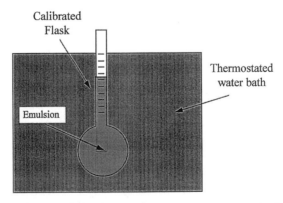

FIGURE 10.19 Schematic diagram of a simple dilatometer used for monitoring phase transitions of materials.

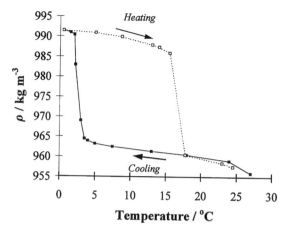

FIGURE 10.20   Temperature dependence of the density of a hexadecane oil-in-water emulsion. For food oils, melting and crystallization normally occur over a wider range of temperatures.

10.5.2.   Nuclear Magnetic Resonance

10.5.2.1.   Principles

NMR has been used to measure the solid content of food emulsions for many years (Walstra and van Beresteyn 1975, van Boekel 1981, Dickinson and McClements 1995). NMR instruments are capable of rapidly analyzing emulsions that are concentrated and optically opaque, without the need for any sample preparation. For this reason, they have largely replaced the more cumbersome and time-consuming dilatometry method in laboratories which can afford the relatively high initial cost of an NMR instrument. The NMR technique utilizes interactions between radio waves and the nuclei of hydrogen atoms to obtain information about the solid content of a material. The fundamental principles of this application have been described elsewhere (Dickinson and McClements 1995), and so only a simplified description of the technique will be given here. Basically, a radio-frequency pulse is applied to an emulsion, which causes some of the hydrogen nuclei to move into an excited state, which leads to the generation of a detectable NMR signal. The frequency, amplitude, and decay time of this signal depend on the ratio of solid to liquid material in the sample. Thus, by analyzing the characteristics of the NMR signal, it is possible to obtain information about the solid content of the material.

10.5.2.2.   Measurement Techniques

A variety of NMR instruments which can be used to measure the solid content of emulsions are commercially available. These instruments vary in their operating principles, the types of information they are capable of providing, and their cost. The more sophisticated instruments can measure a large number of different physicochemical characteristics of emulsions, but because they are often very expensive and require highly trained operators, their application is restricted to a small number of research laboratories. A number of less sophisticated instruments are available that have a more limited range of applications but which are considerably less expensive. These instruments are the most widely used to determine the solid content of foods, and therefore only they will be considered here.

A typical pulsed NMR instrument consists of a *static magnet*, a *probe coil*, *electronics* to generate and receive electromagnetic pulses, and a *computer* to regulate the measurement procedure and analyze and store the data (Figure 10.21). Nuclei in the sample are excited to

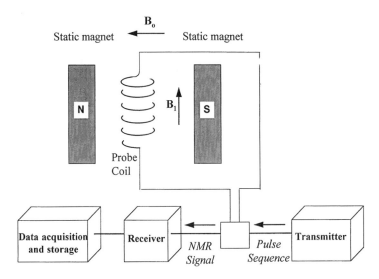

FIGURE 10.21 A typical arrangement for making NMR measurements of the solid content of an emulsion.

a higher energy level by applying a radio-frequency pulse via the probe coil, and the excited nuclei are detected by the same coil once the radio-frequency pulse is removed. The detected signal is digitized by a digital-to-analog convertor and stored in the computer for data analysis. The solid content is determined by analyzing the decay rate of the detected signal after the application of the radio-frequency pulse.

The decay of the NMR signal is much more rapid for a solid than a liquid (Figure 10.22). Consequently, the solid and liquid phases in a sample can be differentiated by measuring the decay of the NMR signal with time. Immediately after the radio-frequency pulse is switched off, the NMR signal (S_0) is proportional to the total number of nuclei in the liquid and solid phases. After a certain time, the contribution from the solid phase has completely decayed, and the signal (S_t) is then proportional to the number of nuclei in the liquid phase. Assuming that the NMR signal per gram of the solid and liquid phases is similar, the solid content can simply be determined: $\phi_c = (S_0 - S_t)/S_0$. In practice, it is not possible to measure the signal at zero time because the NMR receiver takes a short time to recover after the radio-frequency

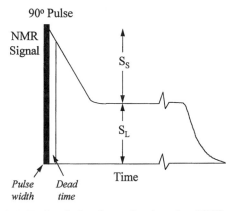

FIGURE 10.22 Decay of NMR signal after the application of an NMR pulse is much more rapid for solid than liquid fat.

pulse is applied. Consequently, it is necessary to use a correction factor that accounts for the slight decay of the signal before the first measurement is made. The solid content can also be determined by only making measurements of the signal from the liquid phase, since S_t is proportional to the mass of liquid phase present (M_L). The mass of liquid in a sample is deduced by measuring S_t and using a previously prepared calibration curve of M_L versus S_t. If the total mass of the sample analyzed (M_T) is known, then $\phi_c = (M_T - M_L)/M_T$. The solid content can also be determined by measuring S_t for a partially crystalline sample and then heating it to a temperature where all of the solid phase melts and measuring it again: $\phi_c = (S_t' - S_t)/S_t'$, where the prime refers to the measurement at the higher temperature. This value has to be corrected to take into account the temperature dependence of the NMR signal of the liquid phase.

10.5.2.3. Applications

NMR has been used to determine the extent of droplet crystallization in emulsions in a number of fundamental studies and commercial products (Walstra and van Beresteyn 1975, Waddington 1980, van Boekel 1981, Dickinson and McClements 1995). Benchtop instruments are available which are extremely simple and rapid to use. A tube containing the sample is placed in the NMR instrument and the solid content is given out in a few minutes or less. The technique is therefore particularly useful for quality control purposes when many samples have to be rapidly analyzed. Attempts have been made to develop on-line versions of these NMR instruments, but their application is limited because the sample cannot be analyzed within a metal pipe because of its distorting influence on the magnetic field. An additional limitation is that the technique cannot be used to accurately determine the solid contents when the degree of crystallization is small (<5%).

Before leaving this section, it should be noted that there are a variety of other NMR techniques which can also be used to determine droplet solid contents (Dickinson and McClements 1995). The most powerful of these is *NMR imaging,* which can be used to determine the solid content at any location within a material, so that it is possible to obtain a three-dimensional image of the droplet crystallinity (Simoneau et al. 1991, 1993). These imaging techniques provide food scientists with an extremely powerful method of monitoring and predicting the long-term stability of food emulsions, which will undoubtedly help to identify the most important factors that determine emulsion properties. Nevertheless, these techniques are extremely expensive and require highly skilled operators and therefore are currently available to a small number of research laboratories.

10.5.3. Thermal Analysis

10.5.3.1. Principles

Differential thermal analysis (DTA) and differential scanning calorimetry (DSC) are two thermal analysis techniques that can be used to monitor melting and crystallization of emulsion droplets (Walstra and van Beresteyn 1975, McClements et al. 1993a, Davis 1994a). These techniques are based on measurements of the heat released or adsorbed by a sample when it is subjected to a controlled temperature program. A material tends to release heat when it crystallizes and absorb heat when it melts. The major difference between the two techniques is the method used to measure the heat adsorbed or released by the sample.

10.5.3.2. Measurement Techniques

Differential Thermal Analysis. DTA records the difference in *temperature* between a substance and a reference material when they are heated or cooled at a controlled rate. A DTA

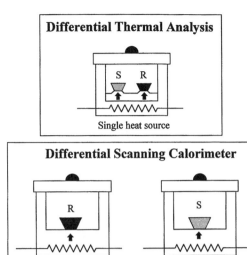

FIGURE 10.23 DTA and DSC instruments.

instrument consists of two sample holders which are connected to the same furnace, which is a device for varying the temperature in a controlled fashion (Figure 10.23). A few milligrams of sample is accurately weighed into a small aluminum pan and placed into one of the sample holders. A reference, usually an empty aluminum pan or a pan containing a material that does not undergo a phase transition over the temperature range studied, is placed into the other sample holder. The sample and reference pans are then heated or cooled together at a controlled rate. The difference in temperature (ΔT) between the two pans is recorded by thermocouples located below each of the pans. The output from the instrument is therefore a plot of ΔT versus T. If the sample *absorbs* heat (an endothermic process), then it will have a slightly *lower* temperature than the reference, but if it *releases* heat (an exothermic process), it will have a slightly *higher* temperature. Thus measurements of the difference in temperature between the sample and reference pans can be used to provide information about physical and chemical changes which occur in the sample. An exothermic process leads to the generation of a negative peak, and an endothermic process leads to the generation of a positive peak (Figure 10.24).

Differential Scanning Calorimetry. DSC records the *energy* necessary to establish a zero temperature difference between a sample and a reference material which are either heated or cooled at a controlled rate. Thermocouples constantly measure the temperature of each pan, and two heaters below the pans supply heat to one of the pans so that they both have exactly the same temperature. If a sample were to undergo a phase transition, it would either absorb or release heat. To keep the temperature of the two pans the same, an equivalent amount of energy must be supplied to either the test or reference cells. Special electrical circuitry is used to determine the amount of energy needed to keep the two sample pans at the same temperature. DSC data are therefore reported as the rate of energy absorption (Q) by the sample relative to the reference material as a function of temperature.

10.5.3.3. Applications

Thermal analysis can be used to determine the temperature range of a phase transition, as well as the amount of material involved in a phase transition (Figure 10.24). When an emulsion

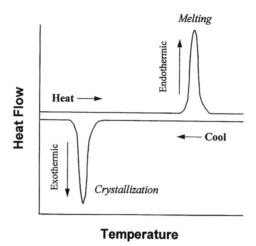

FIGURE 10.24   Thermal analysis of the melting and crystallization of oil droplets in an oil-in-water emulsion.

that contains solid droplets is heated above the melting point of the dispersed phase, the droplets melt and an exothermic peak is observed. The melting temperature can therefore be ascertained by measuring the position of the peak. The area under the peak is proportional to the amount of material which undergoes the phase transition: $A = k\Delta H_f m$, where ΔH_f is the heat of fusion per gram, m is the mass of material undergoing the phase transition in grams, and k is a constant which depends on the instrument settings used to make the measurement. The value of k is determined by measuring the peak areas of a series of samples of known mass and heat of fusion. Thus the mass of material which melts can be determined by measuring the peak area. When an emulsion that contains liquid droplets is cooled, an endothermic peak is observed when the droplets crystallize. The crystallization temperature and the amount of material which has crystallized can be determined in the same way as for the melting curve. As mentioned earlier, the droplets tend to crystallize at a much lower temperature than they melt because of supercooling. In addition, the melting range of edible fats is much wider than that shown in Figure 10.24 because they contain a mixture of triacylglycerols.

Thermal analysis has been used to monitor the influence of oil type, emulsifier type, cooling rates, catalytic impurities, polymorphic changes, and droplet size on the nucleation and crystallization of droplets in oil-in-water emulsions (Dickinson and McClements 1995). It has also been used to study the factors which influence the melting and crystallization of water droplets in water-in-oil emulsions (Clausse 1985) and to study phase transitions in the continuous phase of emulsions.

10.5.4.   Ultrasonics

10.5.4.1.   Principles

The ultrasonic properties of a material change significantly when it melts or crystallizes, and so ultrasound can be used to monitor phase transitions in emulsions (Dickinson et al. 1990, 1991c; McClements 1991). The temperature dependence of the ultrasonic velocity of an oil-in-water emulsion in which the droplets crystallize is shown in Figure 10.25. As the emulsion is cooled to a temperature where the droplets crystallize, the ultrasonic velocity increases steeply, because the velocity of ultrasound is larger in solid fat than in liquid oil. As the

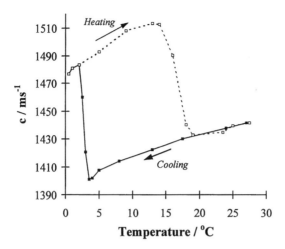

FIGURE 10.25 Crystallization and melting of oil droplets in an emulsion can be monitored by ultrasonic velocity measurements.

emulsion is heated to a temperature where the droplets melt, the velocity decreases rapidly. The droplets crystallize at a temperature which is much lower than the melting point of the bulk oil because of supercooling effects (Section 4.2).

To a first approximation, the fraction of crystalline material (ϕ_{SFC}) in an emulsion can be determined using the following equation (Dickinson et al. 1991):

$$\phi_{SFC} = \left(\frac{1}{c_e^2} - \frac{1}{c_{eL}^2} \right) \left(\frac{1}{c_{eS}^2} - \frac{1}{c_{eL}^2} \right)^{-1} \tag{10.14}$$

where c_{eL} and c_{eS} are the ultrasonic velocities in the emulsion if all the droplets were either completely liquid or completely solid, respectively. These values are determined by extrapolating measurements from higher and lower temperatures into the region where the fat is partially crystalline or by using ultrasonic scattering theory to calculate their values.

10.5.4.2. Measurement Techniques

The ultrasonic properties of an emulsion can be measured using the same techniques as used for determining the droplet size distribution (Section 10.3). The measurements are usually carried out in one of two ways: isothermal or temperature scanning. In an isothermal experiment, the temperature of the emulsion is kept constant and the change in the ultrasonic velocity is measured as a function of time. In a temperature scanning experiment, the ultrasonic velocity is measured as the temperature is increased or decreased at a controlled rate.

10.5.4.3. Applications

The solid contents determined using ultrasound are in good agreement with those determined using traditional techniques such as dilatometry (Hussin and Povey 1984) and NMR (McClements and Povey 1988). Ultrasound has been used to monitor phase transitions in nonfood oil-in-water emulsions (Dickinson et al. 1990, 1991), triacylglycerol oil-in-water and water-in-oil emulsions (McClements 1989, Coupland et al. 1993), margarine and butter

(McClements 1989), shortening and meat (Miles et al. 1985), and various triacylglycerol/oil mixtures (McClements and Povey 1987, 1988).

It should be noted that in some systems, large increases in the attenuation coefficient and appreciable velocity dispersion have been observed during melting and crystallization because of a relaxation mechanism associated with the solid–liquid phase equilibrium (McClements et al. 1998). The ultrasonic wave causes periodic fluctuations in the temperature and pressure of the material, which perturb the phase equilibrium. When a significant proportion of the material is at equilibrium, a large amount of ultrasonic energy is absorbed due to this process. This effect depends on the ultrasonic frequency, the relaxation time for the phase equilibrium, and the amount of material undergoing phase equilibrium and can cause large deviations in both the velocity and attenuation coefficient. In systems where this phenomenon is important, it is not possible to use Equation 10.14 to interpret the data. Nevertheless, it may be possible to use ultrasound to obtain valuable information about the dynamics of the phase equilibrium.

10.6. DROPLET CHARGE

10.6.1. Electrophoresis

The emulsion to be analyzed is placed in a measurement cell, and a static electrical field (E) is applied across it via a pair of electrodes (Figure 10.26). This causes any charged emulsion droplets to move toward the oppositely charged electrode (Hunter 1986, 1993). The sign of the charge on the emulsion droplets can therefore be deduced from the direction they move. When an electrical field is applied across an emulsion, the droplets accelerate until they reach a constant velocity (v) where the electrical pulling force is exactly balanced by the viscous drag force exerted by the surrounding liquid. This velocity depends on the size and charge of the emulsion droplets and can therefore be used to provide information about these parameters. Experimentally, particle velocity is determined by measuring the distance they move in a known time or the time it takes to move a known distance. Droplet motion can be monitored using a number of different experimental methods. The movement of relatively large particles (>1 μm) can be monitored by optical microscopy or static light scattering, whereas the movement of smaller particles can be monitored by an ultramicroscope or dynamic light scattering.

Mathematical expressions have been derived to relate the movement of a droplet in an electric field to its zeta potential (Hunter 1986). These are based on a theoretical consideration of the forces that act on a particle when it has reached constant velocity (i.e., the electrical pulling force is balanced by the viscous drag force). The mathematical theory that describes this process depends on the droplet size and charge, the thickness of the Debye

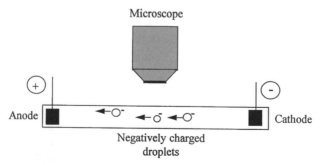

FIGURE 10.26 Electrophoretic mobility cell for measuring the zeta potential of emulsion droplets.

layer, the viscosity of the surrounding liquid, and the strength of the applied electric field. The general solution of this theory leads to a complicated expression that relates all of these parameters. Nevertheless, under certain experimental circumstances, it is possible to derive simpler expressions:

$$\xi = \frac{3\eta u}{2\varepsilon_0 \varepsilon_R} \qquad \kappa r \ll 1 \qquad (10.15)$$

$$\xi = \frac{\eta u}{\varepsilon_0 \varepsilon_R} \qquad \kappa r \gg 1 \qquad (10.16)$$

where η is the viscosity of the surrounding liquid, u is the electrophoretic mobility (= particle velocity divided by electric field strength), ε_0 is the dielectric constant of a vacuum, and ε_R is the relative dielectric constant of the material. In practice, the latter equation is the most applicable to emulsions, because the droplet size is much greater than the Debye length (κ^{-1}). Even so, there are many practical examples where the particle size is comparable to the Debye length, and so there are considerable deviations between Equation 10.16 and experimental measurements. In these cases, it is necessary to solve the full theory.

10.6.2. Zetasizer©

A more sophisticated instrument for measuring both the zeta potential and size of droplets in emulsions is the Zetasizer© developed by Malvern Instruments (Hunter 1986). Two coherent beams of light are made to intersect with each other at a particular position within a measurement cell so that they form an interference pattern which consists of regions of low and high light intensity. The charged emulsion droplets are made to move through the interference pattern by applying an electrical field across the cell. As the droplets move across the interference pattern, they scatter light in the bright regions, but not in the dark regions. The faster a droplet moves through the interference pattern, the greater the frequency of the intensity fluctuations. By measuring and analyzing the frequency of these fluctuations, it is possible to determine the particle velocity, which can then be mathematically related to the zeta potential (e.g., using Equation 10.16 for larger particles). The sign of the charge on the particles is ascertained from the direction they move in the electric field. The same instrument can also be used to determine droplet concentration and size distribution (from 10 nm to 3 μm) of an emulsion by a dynamic light-scattering technique. Consequently, it is possible to determine the droplet size, concentration, and charge using a single instrument, which is extremely valuable for predicting the stability and bulk physicochemical properties of emulsions.

10.6.3. Electroacoustics

Recently, analytical instruments based on electroacoustics have become commercially available for measuring the size, concentration, and zeta potential of droplets in emulsions (Hunter 1993, O'Brien et al. 1995, Carasso et al. 1995, Dukhin and Goetz 1996). The sample to be analyzed is placed in a measurement cell and an alternating electrical field is applied across it via a pair of electrodes. This causes any charged droplets to rapidly move backward and forward in response to the electrical field (Figure 10.27). An oscillating droplet generates a pressure wave with the same frequency as the alternating electric field which emanates from it and can be detected by an ultrasonic transducer. The amplitude of the signal received by the transducer is known as the electrokinetic sonic amplitude and is proportional to the

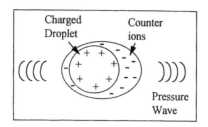

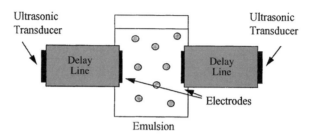

FIGURE 10.27 Principles of the electroacoustic technique.

dynamic mobility (μ_d) of the particles in the suspension. The dynamic mobility of the particles is related to their zeta potential and size:

$$\mu_d = \left(\frac{\varepsilon_0 \varepsilon_R \xi}{\eta} \right) G \left(\frac{\omega r^2 \rho}{\eta} \right) \tag{10.17}$$

Here, ε_R is the relative permittivity of the continuous phase, ξ is the zeta potential, ω is the angular frequency, r is the droplet radius, and η and ρ are the viscosity and density of the continuous phase, respectively. G is a function that varies from 1 at low frequencies to 0 at high frequencies. The frequency dependence of G is associated with the phase lag between the alternating electrical field and the oscillating pressure wave generated by the particles which arises because of their inertia. At low frequencies, the particles move in-phase with the electric field, and so the dynamic mobility is equal to the static mobility: $\mu_d = \varepsilon_0 \varepsilon_R \xi/\eta$ (i.e., $G = 1$). At extremely high frequencies, the electric field alternates so quickly that the particles have no time to respond and therefore remain stationary. Under these circumstances, no ultrasonic pressure wave is generated by the particles (i.e., $G = 0$). The transition from the low- to high-frequency regions occurs at a characteristic relaxation frequency (ω_R) which is related to the size of the particles, decreasing as the size (inertia) of the particles increases. By measuring the frequency dependence of the dynamic mobility, it is therefore possible to determine the droplet size. By measuring the dynamic mobility at low frequencies (where it is independent of droplet size), it is possible to determine the zeta potential.

Electroacoustic experiments can also be performed by applying an ultrasonic wave to an emulsion, which causes the droplets to oscillate backward and forward due to the density difference between themselves and the surrounding medium. An oscillating charged particle generates an alternating electric field that can be detected using suitable electrodes. The amplitude of the signal received by the electrodes is called the *colloid vibration potential*, and it can also be used to determine the size and zeta potential of the emulsion droplets. The major advantage of electroacoustic techniques is that they can be applied to concentrated emulsions ($\phi \leq 40\%$) without the need for any sample dilution. Nevertheless, they can only be used to

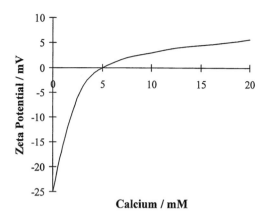

FIGURE 10.28 Dependence of the zeta potential of a 20% oil-in-water emulsion on the calcium concentration measured using an electroacoustic technique. (Data from Colloidal Dynamics Ltd., Application Note 203, Matec Applied Sciences, Hopkinton, MA.)

analyze emulsion droplets which are charged and have a significant density difference compared to the surrounding liquid, and therefore they have limited application to many food systems.

The use of electroacoustics for measuring the zeta potential of droplets in concentrated emulsions is illustrated in Figure 10.28. When calcium ions are added to a 20% oil-in-water emulsion stabilized by lecithin, they "bind" to the surface of the negatively charged droplets and reduce the zeta potential. Above a certain concentration, they actually cause the sign of the droplet charge to become reversed. One would therefore expect low concentrations of calcium to reduce the electrostatic repulsion between droplets, which may lead to flocculation (Chapters 3 and 7).

References

Agboola, S.O. and Dalgleish, D.G., Calcium-induced destabilization of oil-in-water emulsions stabilized by caseinate or by β-lactoglobulin, *Journal of Food Science,* **60**, 399, 1995.

Agboola, S.O. and Dalgleish, D.G., Kinetics of calcium-induced instability of oil-in-water emulsions: Studies under quiescent and shearing conditions, *Lebensmittal Wissenschaft unt Technologie,* **29**, 425 1996a.

Agboola, S.O. and Dalgleish, D.G., Enzymatic hydrolysis of milk proteins used for emulsion formation. 1. Kinetics of protein breakdown and storage stability of the emulsions, *Journal of Agriculture and Food Science,* **44**, 3631, 1996b.

Agboola, S.O. and Dalgleish, D.G., Enzymatic hydrolysis of milk proteins used for emulsion formation. 2. Effects of calcium, pH and ethanol on the stability of the emulsions, *Journal of Agriculture and Food Science,* **44**, 3637, 1996c.

Agboola, S.O. and Dalgleish, D.G., Effects of pH and ethanol on the kinetics of destabilization of oil-in-water emulsions containing milk proteins, *Journal of the Science of Food and Agriculture,* **72**, 448, 1996d.

Aguilera, J.M. and Kessler, H.G., Properties of mixed and filled-type dairy gels, *Journal of Food Science,* **54**, 1213, 1989.

Aguilera, J.M. and Stanley, D.W., *Microstructural Principles of Food Processing and Engineering,* Elsevier, Amsterdam, 1990.

Alaimo, M.H. and Kumosinski, T.F., Investigation of hydrophobic interactions in colloidal and biological systems by molecular dynamics simulations and NMR spectroscopy, *Langmuir,* **13**, 2007, 1997.

Alber, T., Stabilization energies of protein conformation, in *Prediction of Protein Structure and the Principles of Protein Conformation,* Fasman, G.D., Ed., Plenum Press, New York, 1989, 161.

Anandarajah, A. and Chen, J., Single correction function for computing retarded van der Waals attraction, *Journal of Colloid and Interface Science,* **176**, 293, 1995.

Aoki, H., Shirase, Y., Kato, J., and Watanabe, Y., Emulsion stabilizing properties of soy protein isolates mixed with sodium caseinates, *Journal of Food Science,* **49**, 212, 1984.

Arlauskas, R.A. and Weers, J.G., Sedimentation field-flow fractionation studies of composition ripening in emulsion mixtures, *Langmuir,* **12**, 1923, 1996.

Aronson, M.P., Surfactant induced flocculation of emulsions, in *Emulsions — A Fundamental and Practical Approach,* Sjoblom, J., Ed., Kluwer Academic Publishers, Netherlands, 1992, 75.

Asami, K., Evaluation of Colloids by Dielectric Spectroscopy, HP Application Note 380-3, Hewlett Packard, Palo Alto, CA, 1995.

Atkins, P.W., *Physical Chemistry,* 5th ed., Oxford University Press, Oxford, 1994.

Atkinson, P.J., Dickinson, E., Horne, D.S., and Richardson, R.M., Neutron reflectivity of adsorbed β-casein and β-lactoglobulin at the air/water interface, *Journal of the Chemical Society, Faraday Transactions,* **91**, 2847, 1995.

Aveyard, R., Binks, B.P., Clark, S., and Fletcher, P.D.I., Cloud points, solubilisation and interfacial tensions in systems containing nonionic surfactants, *Chemical Technology and Biotechnology,* **48**, 161, 1990.

Azzam, R.M. and Bachara, N.M., *Ellipsometry and Polarized Light,* North-Holland, Amsterdam, 1989.

Baianu, I.C., *Physical Chemistry of Food Processes,* Vol. 1, Van Nostrand Reinhold, New York, 1992.

Baines, Z.V. and Morris, E.R., Suppression of perceived flavor and taste by hydrocolloids, in *Food Colloids,* Bee, R.D., Richmond, P., and Mingins, J., Eds., Royal Society of Chemistry, Cambridge, 1989, 184.

Baker, R., *Controlled Release of Biologically Active Agents,* John Wiley & Sons, New York, 1987.

Baker, E.N. and Hubbard, R.E., Hydrogen bonding in globular proteins, *Progress in Biophysics and Molecular Biology,* **44**, 97, 1984.

Bakker, J., Flavor interactions with the food matrix and their effects on perception, in *Ingredient Interactions: Effects on Food Quality,* Gaonkar, A.G., Ed., Marcel Dekker, New York, 1995.

Bakker, J. and Mela, D.J., Effect of emulsion structure on flavor release and taste perception, in *Flavor–Food Interactions,* McGorrin, R.J. and Leland, J.V., Eds., American Chemical Society, Washington, D.C., 1996, chap. 4.

Baldwin, R.L., Temperature dependence of the hydrophobic interaction in protein folding, *Proceedings of the National Academy of Sciences,* **83**, 8069, 1986.

Baldwin, R.W., Cloninger, M.R., and Lindsay, R.C., Flavor thresholds for fatty acids in buffered solutions, *Journal of Food Science,* **38**, 528, 1973.

Banks, W. and Muir, D.D., Stability of alcohol containing emulsions, in *Advances in Food Emulsions and Foams,* Dickinson, E. and Stainsby, G., Eds., Elsevier Applied Science, London, 1988, chap. 8.

Barford, N.M., Methods for characterization of structure in whippable dairy based emulsions, in *Characterization of Foods: Emerging Techniques,* Gaonkar, A.G., Ed., Elsevier, Amsterdam, 1995, chap. 3.

Barford, N.M. and Krog, M., Destabilization and fat crystallization of whippable emulsions (toppings) studied by pulsed NMR, *Journal of the American Oil Chemists Society,* **64**, 112, 1987.

Barford, N.M., Krog, N., and Bucheim, W., Protein fat surfactant interactions in whippable emulsions, in *Food Emulsions and Foams*, Dickinson E., Ed., Royal Society of Chemistry, Cambridge, 1987, 141.

Barford, N.M., Krog, N., Larsen, G., and Bucheim, W., Effects of emulsifiers on protein–fat interaction in ice cream mix during ageing. I. Quantitative analysis, *Fat Science Technology,* **93**, 24, 1991.

Barnes, H.A., Rheology of emulsions — A review, *Colloids and Surfaces,* **91**, 89, 1994.

Barylko-Pikielna, N., Martin, A., and Mela, D.J., Perception of taste and viscosity of oil-in-water and water-in-oil emulsions, *Journal of Food Science,* **59**, 1318, 1994.

Basaran, T., Demetriades, K., and McClements, D.J., Ultrasonic imaging of gravitational separation in emulsions, *Colloids and Surfaces,* submitted, 1998.

Becher, P., *Emulsions: Theory and Practice*, Reinhold Publishing, New York, 1957.

Becher, P., *Encyclopedia of Emulsion Technology,* Vol. 1, Marcel Dekker, New York, 1983.

Becher, P., Hydrophile–lipophile balance: An updated bibliography, in *Encyclopedia of Emulsion Technology,* Vol. 2, Becher, P., Ed., Marcel Dekker, New York, 1985.

Becher, P., HLB: Update III, in *Encyclopedia of Emulsion Technology,* Vol. 4, Becher, P., Ed., Marcel Dekker, New York, 1996.

Bell, G.A., Molecular mechanisms of olfactory perception: Their potential for future technologies, *Trends in Food Science and Technology,* **7**, 425, 1996.

BeMiller, J.N. and Whistler, R.L., Carbohydrates, in *Food Chemistry,* 3rd ed., Fennema, O.R., Ed., Marcel Dekker, New York, 1996, 157.

Ben-Naim, A., *Hydrophobic Interactions,* Plenum Press, New York, 1980.

Bentley, B.J. and Leal, L.G., An experimental investigation of drop deformation and breakup in steady, two-dimensional linear flows, *Journal of Fluid Mechanics,* **167**, 241, 1986.

Bergenstahl, G., Physiochemical aspects of an emulsifier functionality, in *Food Emulsifiers and Their Applications,* Hasenhuettl, G.L. and Hartel, R.W., Eds., Chapman and Hall, New York, 1997, chap. 6.

Berger, K.G., Ice cream, in *Food Emulsions,* Friberg, S., Ed., Marcel Dekker, New York, 1976, chap. 4.

Bergethon, P.R. and Simons, E.R., *Biophysical Chemistry: From Molecules to Membranes,* Springer-Verlag, New York, 1990.

Besseling, N.A.M., Theory of hydration forces between surfaces, *Langmuir,* **13**, 2113, 1997.

Bibette, J., Depletion interactions and fractionated crystallization for polydisperse emulsion purification, *Journal of Colloid and Interface Science,* **147**, 474, 1991.

Bijsterbosch, B.H., Bos, M.T.A., Dickinson, E., van Opheusden, J.H.J., and Walstra, P., Brownian dynamics simulation of particle gel formation: From argon to yoghurt, *Faraday Discussions,* **101**, 51, 1995.

Binks, B.P., Emulsion type below and above the CMC in AOT microemulsion systems, *Colloids and Surfaces,* **71**, 167, 1993.

Birch, G.G. and Lindley, M.G., *Interactions of Food Components,* Elsevier Applied Science, London, 1986.

Birker, P.J.M.W.L. and Padley F.B., Physical properties of fats and oils, in *Recent Advances in Chemistry and Technology of Fats and Oils,* Hamilton, R.J. and Bhati, A., Eds., Elsevier Applied Science, London, 1987, chap. 1.

Blonk, J.C.G. and van Aalst, H., Confocal scanning light microscopy in food research, *Food Research International,* **26**, 297, 1993.

Boelens, M.H. and van Gemert, L.J., Physicochemical parameters related to organoleptic properties of flavour components, in *Developments in Food Flavours*, Birch, G.G. and Lindley, M.G., Eds., Elsevier Applied Science, London, 1986, 23.

Bohren, C.F. and Huffman, D.R., *Adsorption and Scattering of Light by Small Particles*, John Wiley & Sons, New York, 1983.

Boistelle, R., Fundamentals of nucleation and crystal growth, in *Crystallization and Polymorphism of Fats and Fatty Acids,* Garti, N. and Sato, K., Eds., Marcel Dekker, New York, 1988, chap. 5.

Bomben, J.L., Bruin, S., Thijssen, H.A.C., and Merson, R.L., Aroma recovery and retention in concentration and drying of foods, *Advances in Food Research,* **20**, 1, 1973.

Boode, K., Partial Coalescence in Oil-in-Water Emulsions, Ph.D. thesis, Wageningen Agricultural University, Wageningen, Netherlands, 1992.

Boode, K. and Walstra, P., Partial coalescence in oil-in-water emulsions. 1. Nature of the aggregation, *Colloids and Surfaces A,* **81**, 121, 1993a.

Boode, K. and Walstra, P., The kinetics of partial coalescence in oil-in-water emulsions, in *Food Colloids and Polymers: Stability and Mechanical Properties,* Dickinson, E. and Walstra, P., Eds., Royal Society of Chemistry, Cambridge, 1993b, 23.

Boode, K., Bispernick, C., and Walstra, P., Destabilization of O/W emulsions containing fat crystals by temperature cycling, *Colloids and Surfaces,* **61**, 55, 1991.

Boode, K., Walstra, P., and Degrootmostert, A.E.A., Partial coalescence in oil-in-water emulsions. 1. Influence of the properties of the fat, *Colloids and Surfaces A,* **81**, 139, 1993.

Bos, M., Nylander, T., Arnebrant, T., and Clark, D.C., Protein/emulsifier interactions, in *Food Emulsifiers and Their Applications,* Hasenhuettl, G.L. and Hartel, R.W., Eds., Chapman and Hall, New York, 1997, chap. 5.

Boudaud, N. and Dumont, J.P., Interaction between flavor components and β-lactoglobulin, in *Flavor–Food Interactions,* McGorrin, R.J. and Leland, J.V., Eds., American Chemical Society, Washington, D.C., 1996, chap. 4.

Bourne, M.C., *Food Texture and Viscosity: Concept and Measurement,* Academic Press, New York, 1982.

Bower, C., Washington, C., and Purewal, T.S., The use of image analysis to characterize aggregates in a shear field, *Colloids and Surfaces A,* **127**, 105, 1997.

Bradley, R.L., Moisture and total solids analysis, in *Introduction to the Chemical Analysis of Foods,* Nielsen, S.S., Ed., Bartlett and Jones, Boston, 1994, chap. 7.

Brady, J.W., The role of dynamics and solvation in protein structure and function, in *Food Proteins,* Kinsella, J.E. and Soucie, W.G., Eds., American Oil Chemists Society, Champaign, IL, 1989, chap. 1.

Brady, J.W. and Ha, S.N., Molecular dynamic simulations of the aqueous solvation of sugars, in *Water Relationships of Foods,* Duckworth, R.B., Ed., Academic Press, London, 1975, 3.

Braginsky, L.M. and Belevitskaya, M.A., Kinetics of droplets breakup in agitated vessels, in *Liquid–Liquid Systems,* Kulov, N.N., Ed., Nova Science, Commack, NY, 1996, chap. 1,

Brakenhoff, G.J., van der Voort, H.T.M., van Spronsen, E.A., and Nanninga, N., Three-dimensional imaging of biological structures by high resolution confocal laser scanning microscopy, *Scanning Microscopy,* **2**, 33, 1988.

Brash, J.L. and Horbett, T.A., *Proteins at Interfaces,* American Chemical Society, Washington, D.C., 1987.

Bremer, L.G.B., Fractal Aggregation in Relation to Formation and Properties of Particle Gels, Ph.D. thesis, Wageningen Agricultural University, Wageningen, Netherlands, 1992.

Bremer, L.G.B., Bijsterbosch, B.H., Walstra, P., and Van Vliet, T., Formation, properties and fractal structure of particle gels, *Advances in Colloid and Interface Science, 46*, 117, 1993.

Brennan, J.G., Butters, J.R., Cowell, N.D., and Lilly, A.E.V., *Food Engineering Operations*, Applied Science Publishers, London, 1981.

Britten, M. and Giroux, H.J., Coalescence index of protein-stabilized emulsions, *Journal of Food Science, 56*, 792, 1991.

Brooker, B.E., Imaging food systems by confocal laser scanning microscopy, in *New Physicochemical Techniques for the Characterization of Complex Food Systems*, Dickinson, E., Ed., Blackie Academic and Professional, London, 1995, chap. 2.

Brossard, C., Rousseau, F., and Dumont, J.P., Flavour release and flavour perception in oil-in-water emulsions: Is the link so close? in *Flavour Science: Recent Developments*, Tayloer, A.J. and Mottram, D.S., Eds., Royal Society of Chemistry, Cambridge, 1996, 375.

Bucheim, W. and Dejmek, P., Milk and dairy-type emulsions, in *Food Emulsions*, 2nd ed., Larsson, K. and Friberg, S., Eds., Marcel Dekker, New York, 1990, chap. 6.

Buffler, C.R., Advances in dielectric characterization of foods, in *Characterization of Food: Emerging Methods*, Gaonkar, A.G., Ed., Elsevier, Amsterdam, 1995, chap. 10.

Bujannunez, M.C. and Dickinson, E., Brownian dynamics simulation of a multisubunit deformable particle in simple shear-flow, *Journal of the Chemical Society, Faraday Transactions, 90*, 2737, 1994.

Buttery, R.G., Ling, L.C., and Guadagni, D.G., Volatilities of aldehydes, ketones and esters in dilute water solution, *Journal of Agricultural and Food Chemistry, 17*, 385, 1969.

Buttery, R.G., Bomden, J.L., Guadagni, D.G., and Ling, L.C., Some considerations of the volatilities of organic flavor compounds in foods, *Journal of Agricultural and Food Chemistry, 19*, 1045, 1971.

Buttery, R.G., Guadagni, D.G., and Ling, L.C., Flavor compounds: Volatilities in vegetable oil and oil–water mixtures. Estimation of odor thresholds, *Journal of Agricultural and Food Chemistry, 21*, 198, 1973.

Callaghan, P.T., Jolley, K.W., and Humphrey, R.S., Diffusion of fat and water in cheese as studied by pulsed field gradient nuclear magnetic resonance, *Journal of Colloid and Interface Science, 93*, 521, 1983.

Campanella, O.H., Dorward, N.M., and Singh, H., A study of the rheological properties of concentrated food emulsions, *Journal of Food Engineering, 25*, 427, 1995.

Campbell, I.J., The role of fat crystals in emulsion stability, in *Food Colloids*, Bee, R.D., Richmond, P., and Mingins, J., Eds., Royal Society of Chemistry, Cambridge, 1991, 272.

Campbell, I., Norton, I., and Morley, W., Factors controlling the phase inversion of oil-in-water emulsions, *Netherlands Milk and Dairy Journal, 50*, 167, 1996.

Carasso, M.L., Rowlands, W.N., and Kennedy, R.A., Electroacoustic determination of droplet size and zeta potential in concentrated intravenous fat emulsion, *Journal of Colloid and Interface Science, 174*, 405, 1995.

Carnie, S.L., Chan, D.Y.C., and Stankovich, J., Computation of forces between spherical colloidal particles: Non-linear Poisson–Boltzmann theory, *Journal of Colloid and Interface Science, 165*, 116, 1994.

Carr, J.M., Sufferling, K., and Poppe, J., Hydrocolloids and their use in the confectionery industry, *Food Technology, 49*(7), 41, 1995.

Carr, J., Baloga, D., Guinard, J.-X., Lawter, L., Marty, C., and Squire, C., The effect of gelling agent type and concentration on flavor release in model systems, in *Flavor–Food Interactions*, McGorrin, R.J. and Leland, J.V., Eds., American Chemical Society, Washington, D.C., 1996, chap. 9.

Cebula, D.J., McClements, D.J., Povey, M.J.W., and Smith, P.R., Neutron diffraction studies of liquid and crystalline trilaurin, *Journal of the American Oil Chemists Society, 69*, 130, 1992.

Cesero, A., The role of conformation on the thermodynamics and rheology of aqueous solutions of carbohydrate polymers, in *Water in Foods: Fundamental Aspects and Their Significance in Relation to the Processing of Foods*, Fito, P., Mullet, A., and McKenna, B., Eds., Elsevier Applied Science, London, 1994, 27.

Chang, C.M., Powries, W.D., and Fennema, O., Electron microscopy of mayonnaise, *Canadian Institute of Food Science and Technology Journal, 5*, 134, 1972.

Chapman, D., *The Structure of Lipids,* Methuen, London, 1965.

Charalambous, G. and Doxastakis, G., *Food Emulsifiers: Chemistry, Technology, Functional Properties and Applications,* Elsevier, Amsterdam, 1989.

Chen, J. and Anadarajah, A., Van der Waals attraction between spherical particles, *Journal of Colloid and Interface Science,* **180**, 519, 1996.

Chen, J. and Dickinson, E., Time-dependent competitive adsorption of milk proteins and surfactants in oil-in-water emulsions, *Journal of the Science of Food and Agriculture,* **62**, 283, 1993.

Chen, J., Dickinson, E., and Iveson, G., Interfacial interactions, competitive adsorption and emulsion stability, *Food Microstructure,* **12**, 135, 1993.

Chow, C.K., *Fatty Acids in Foods and Their Health Implication,* Marcel Dekker, New York, 1992.

Christenson, H.K., Fang, J., Ninham, B.W., and Parker, J.L., Effect of divalent electrolyte on the hydrophobic attraction, *Journal of Physical Chemistry,* **94**, 8004, 1990.

Christian, S.D. and Scamehorn, J.F., *Solubilization in Surfactant Aggregates,* Marcel Dekker, New York, 1995.

Chrysam, M.M., Margarine and spreads, in *Bailey's Industrial Oil and Fat Products,* Vol. 3, 5th ed., Hui, Y.H., Ed., John Wiley, New York, 1996, chap. 2.

Claesson, P.M., Experimental evidence for repulsive and attractive forces not accounted for by conventional DLVO theory, *Progress in Colloid and Polymer Science,* **74**, 48, 1987.

Claesson, P.M. and Christenson, H.K., Very long range attractive forces between uncharged hydrocarbon and fluorocarbon surfaces in water, *Journal of Physical Chemistry,* **92**, 1650, 1988.

Claesson, P.J., Blombert, E., Froberg, J.C., Nylander, T., and Arnebrant, T., Protein interactions at solid surfaces, *Advances in Colloid and Interface Science,* **57**, 161, 1995.

Claesson, P.M., Ederth, T., Bergeron, V., and Rutland, M.W., Techniques for measuring surface forces, *Advances in Colloid and Interface Science,* **67**, 119, 1996.

Clark, A.H., The application of network theory to food systems, in *Food Structure and Behaviour,* Blanshard, J.M.V. and Lillford, P.J., Eds., Academic Press, London, 1987, chap. 2.

Clark, A.H. and Lee-Tuffnell, C.D., Gelation of globular proteins, in *Functional Properties of Food Macromolecules,* Mitchell, J.R. and Ledward, D.A., Eds., Elsevier Applied Science, London, 1986, chap. 5.

Clausse, D., Dielectric properties of emulsions and related systems, in *Encyclopedia of Emulsion Technology,* Vol. 1, Blecher, P., Ed., Marcel Dekker, New York, 1983, chap. 9.

Clause, D., Research techniques utilizing emulsions, in *Encyclopedia of Emulsion Technology,* Vol. 2, Becher, P., Ed., Marcel Dekker, New York, 1985, chap. 2.

Clydesdale, F.M., Methods and measurements of food color, in *Theory, Determination and Control of Physical Properties of Food Materials,* Rha, C., Ed., D. Reidel Publishing, Boston, 1975, chap. 14.

Clydesdale, F.M. and Francis, F.J., *Food Colorimetry: Theory and Applications,* AVI Publishing, Westport, CT, 1975.

Corredig, M. and Dalgleish, D.G., A differential microcalorimetric study of whey proteins and their behaviour in oil-in-water emulsions, *Colloids and Surfaces B,* **4**, 411, 1995.

Coultate, T.P., *Food: The Chemistry of Its Components,* 3rd ed., Royal Society of Chemistry, Cambridge, 1996.

Couper, A., Surface tension and its measurement, in *Physical Methods of Chemistry,* Vol. IXA, Rossiter, B.W. and Baetzold, R.C., Eds., John Wiley & Sons, New York, 1993, chap. 1.

Coupland, J.N. and McClements, D.J., Lipid oxidation in food emulsions, *Trends in Food Science and Technology,* **7**, 83, 1996.

Coupland, J.N. and McClements, D.J., Droplet size determination in food emulsions: Comparison of ultrasonic and light scattering techniques, *Journal of Food Engineering,* submitted, 1998.

Coupland, J.N., Dickinson, E., McClements, D.J., Povey, M.J.W, and Rancourt de Mimmerand, C., Crystallization of simple paraffins and monoacid saturated triacylglycerols dispersed as an oil phase in water, in *Food Colloids and Polymers: Structure and Dynamics,* Dickinson, E. and Walstra, P., Eds., Royal Society of Chemistry, Cambridge, 1993.

Coupland, J.N., Zhu, Z., Wan, H., McClements, D.J., Nawar, W.W., and Chinachotti, P., Droplet composition affects the rate of oxidation of emulsified ethyl linoleate, *Journal of the American Oil Chemists Society,* **73**, 795, 1996.

Coupland, J.N., Brathwaite, D., Fairley, P., and McClements, D.J., Effect of ethanol on the solubilization of hydrocarbon emulsion droplets in nonionic surfactant micelles, *Journal of Colloid and Interface Science,* **190**, 71, 1997.

Courthaudon, J.-L., Dickinson, E., and Dalgleish, D.G., Competitive adsorption of β-casein and nonionic surfactants in oil-in-water emulsions, *Journal of Colloid and Interface Science,* **145**, 390, 1991a.

Courthaudon, J.-L., Dickinson, E., Matsumura, Y., and Clark, D.C., Competitive adsorption of β-lactoglobulin and polyoxyethylene sorbitan monostearate 20 at the oil–water interface, *Colloids and Surfaces,* **56**, 293, 1991b.

Courthaudon, J.-L., Dickinson, E., Matsumura, Y., and Williams, A., Influence of emulsifier on the competitive adsorption of whey proteins in emulsions, *Food Structure,* **10**, 109, 1991c.

Courthaudon, J.-L., Dickinson, E., and Christie, W.W., Competitive adsorption of lecithin and β-casein in oil-in-water emulsions, *Journal of Agriculture and Food Chemistry,* **39**, 1365, 1991d.

Cramer, C.J. and Tuhlar, D.G., *Structure and Reactivity in Aqueous Solution: Characterization of Chemical and Biological Systems,* American Chemical Society, Washington, D.C., 1994.

Crank, J., *The Mathematics of Diffusion,* Oxford University Press, London, 1975.

Curle, N. and Davies, H.J., *Modern Fluid Dynamics,* Vol. 1, D van Nostrand, London, 1968.

Dalgleish, D.G., Protein stabilized emulsions and their properties, in *Water and Food Quality,* Hardman, T.M., Ed., Elsevier Applied Science, London, 1989, chap. 6.

Dalgleish, D.G., Structure and properties of adsorbed layers in emulsions containing milk proteins, in *Food Macromolecules and Colloids,* Dickinson, E. and Lorient, D., Eds., Royal Society of Chemistry, Cambridge, 1995, 23.

Dalgleish, D.G., Food emulsions, in *Emulsions and Emulsion Stability,* Sjoblom, J., Ed., Marcel Dekker, New York, 1996a.

Dalgleish, D.G., Conformations and structures of milk proteins adsorbed to oil–water interfaces, *Food Research International,* **29**, 541, 1996b.

Dalgleish, D.G. and Hallet, F.R., Dynamic light scattering — Applications to food systems, *Food Research International,* **28**, 181, 1995.

Damodaran, S., Interrelationship of molecular and functional properties of food proteins, in *Food Proteins,* Kinsella, J.E. and Soucie, W.G., Eds., American Chemical Society, Champaign, IL, 1989, chap. 3.

Damodaran, S., Interfaces, protein films and foams, in *Advances in Food and Nutrition Research,* Vol. 34, Kinsella, J.E., Ed., Academic Press, San Diego, 1990, 1.

Damodaran, S., Structure–function relationships of food proteins, in *Protein Functionality in Food Systems,* Hettiarchchy, N. and Ziegler, G.R., Eds., Marcel Dekker, New York, 1994, 1.

Damodaran, S., Amino acids, peptides and proteins, in *Food Chemistry,* 3rd ed., Fennema, O.R., Ed., Marcel Dekker, New York, 1996, 321.

Damodaran, S., Food proteins: An overview, in *Food Proteins and Their Applications,* Marcel Dekker, New York, 1997, chap. 1.

Darling, D.F., Recent advances in the destabilization of dairy emulsions, *Journal of Dairy Research,* **49**, 695, 1982.

Das, A.K. and Ghosh, P.K., Concentrated emulsions — Investigation of polydispersity and droplet distortion and their effect on volume fraction and interfacial area, *Langmuir,* **6**, 1668, 1990.

Das, K.P. and Kinsella, J.E., Stability of food emulsions: Physicochemical role of protein and non-protein emulsifiers, *Advances in Food and Nutrition Research,* **34**, 81, 1990.

Davis, E.A., Thermal analysis, in *Introduction to the Chemical Analysis of Foods,* Nielsen, S.S., Ed., Bartlett and Jones, Boston, 1994a, chap. 34.

Davis, H.T., Factors determining emulsion type: Hydrophile–lipophile balance and beyond, *Colloids and Surfaces A,* **91**, 9, 1994b.

Davis, R.H., Velocities of sedimenting particles in suspensions, in *Sedimentation of Small Particles in a Viscous Liquid,* Tory, E.M., Ed., Computational Mechanics Publications, Southampton, U.K., 1996, chap. 6.

Davis, R.H., Schonberg, J.A., and Rallison, J.M., The lubrication force between two viscous drops, *Physics of Fluids A,* **1**, 77, 1989.

Dea, I.C.M., Polysaccharide conformation in solutions and gels, in *Food Carbohydrates,* Lineback, D.R. and Inglett, G.E., Eds., AVI Publishing, Westport, CT, 1982, chap. 22.

Deffenbaugh, L.B., Carbohydrate/emulsifier interactions, in *Food Emulsifiers and Their Applications,* Hasenhuettl, G.L. and Hartel, R.W., Eds., Chapman and Hall, New York, 1997, chap. 4.

Demetriades, K., Coupland, J.N., and McClements, D.J., Physical properties of whey protein stabilized emulsions as related to pH and NaCl, *Journal of Food Science,* **62**, 342, 1997a.

Demetriades, K., Coupland, J.N., and McClements, D.J., Physicochemical properties of whey protein stabilized emulsions as affected by heating and ionic strength, *Journal of Food Science,* **62**, 462, 1997b.

Derjaguin, B.V., *Theory of Stability of Colloids and Thin Films,* Consultants Bureau, New York, 1989.

Derjaguin, B.V., Churaev, N.V., and Muller, V.M., *Surface Forces,* Consultants Bureau, New York, 1987.

de Roos, A.L. and Walstra, P., Loss of enzyme activity due to adsorption onto emulsion droplets, *Colloids and Surfaces B,* **6**, 201, 1996.

Desrosier, N.W., *Elements of Food Technology,* AVI Publishing, Westport, CT, 1977, chap. 13.

de Vries, A.J., Effect of particle aggregation on the rheological behaviour of disperse systems, in *Rheology of Emulsions,* Sherman, P., Ed., Pergamon Press, New York, 1963.

de Wit, J.N. and van Kessel, T., Effects of ionic strength on the solubility of whey protein products. A colloid chemical approach, *Food Hydrocolloids,* **10**, 143, 1996.

Dickinson, E., A model of a concentrated dispersion exhibiting bridging flocculation and depletion flocculation, *Journal of Colloid and Interface Science,* **132**, 274, 1989.

Dickinson, E., *Introduction to Food Colloids,* Oxford University Press, Oxford, 1992.

Dickinson, E., Protein–polysaccharide interactions in food colloids, in *Food Colloids and Polymers: Stability and Mechanical Properties,* Dickinson, E. and Walstra, P., Royal Society of Chemistry, Cambridge, 1993, 77.

Dickinson, E., Colloidal aspects of beverages, *Food Chemistry,* **51**, 343, 1994.

Dickinson, E., *New Physico-Chemical Techniques for the Characterization of Complex Food Systems,* Chapman & Hall, London, 1995a.

Dickinson, E., Recent trends in food colloids research, in *Food Macromolecules and Colloids,* Dickinson, E. and Lorient, D., Eds., Royal Society of Chemistry, Cambridge, 1995b, 19.

Dickinson, E. and Euston, S.R., Computer-simulation of bridging flocculation, *Journal of the Chemical Society, Faraday Transactions,* **87**, 2193, 1991.

Dickinson, E. and Golding, M., Rheology of sodium caseinate stabilized oil-in-water emulsions, *Journal of Colloid and Interface Science,* **191**, 166, 1997a.

Dickinson, E. and Golding, M., Depletion flocculation of emulsions containing unadsorbed sodium caseinate, *Food Hydrocolloids,* **11**, 13, 1997b.

Dickinson, E. and Hong, S.-K., Surface coverage of β-lactoglobulin at the oil–water interface: Influence of protein heat treatment and various emulsifiers, *Journal of Agricultural and Food Chemistry,* **42**, 1602, 1994.

Dickinson, E. and Hong, S.T., Interfacial and stability properties of emulsions: Influence of protein heat treatment and emulsifiers, in *Food Macromolecules and Colloids,* Dickinson, E. and Lorient, D., Eds., Royal Society of Chemistry, Cambridge, 1995a, 269.

Dickinson, E. and Hong, S.T., Influence of water-soluble nonionic emulsifier on the rheology of heat-set protein-stabilized emulsion gels, *Journal of Agricultural and Food Chemistry,* **43**, 2560, 1995b.

Dickinson, E. and Iveson, G., Adsorbed films of β-lactoglobulin + lecithin at the hydrocarbon–water and triglyceride–water interfaces, *Food Hydrocolloids,* **6**, 553, 1993.

Dickinson, E. and Matsumura, Y., Time-dependent polymerization of β-lactoglobulin through disulphide bonds at the oil–water interface in emulsions, *International Journal of Biology and Macromolecules,* **13**, 26, 1991.

Dickinson, E. and McClements, D.J., *Advances in Food Colloids,* Chapman & Hall, London, 1995.

Dickinson, E. and Stainsby, G., *Colloids in Foods,* Applied Science Publishers, London, 1982.

Dickinson, E. and Stainsby, G, Emulsion stability, in *Advances in Food Emulsions and Foams,* Dickinson, E. and Stainsby, G., Eds., Elsevier Applied Science, London, 1988, chap. 1.

Dickinson, E. and Tanai, S., Temperature dependence of the competitive displacement of protein from the emulsion droplet surface by surfactants, *Food Hydrocolloids,* **6**, 163, 1992.

Dickinson, E. and Tong, S.-T., Interfacial and stability properties of emulsions: Influence of protein heat treatment and emulsifiers, in *Food Macromolecules and Colloids,* Dickinson, E. and Lorient, D., Eds., Royal Society of Chemistry, Cambridge, 1995, 269.

Dickinson, E. and Williams, A., Orthokinetic coalescence of protein-stabilized emulsions, *Colloids and Surfaces A*, **88**, 314, 1994.

Dickinson, E. and Yamamoto Y., Viscoelastic properties of heat-set whey protein stabilized emulsion gels with added lecithin, *Journal of Food Science,* **61**, 811, 1996.

Dickinson, E., Murray, B.S., and Stainsby, G., Coalescence stability of emulsion sized droplets at a planar oil–water interface and the relationship to protein film surface rheology, *Journal of the Chemical Society, Faraday Transactions 1*, **84**, 871, 1988.

Dickinson, E., Goller, M.I., McClements, D.J., and Povey, M.J.W., Ultrasonic monitoring of crystallization in oil-in-water emulsions, *Journal of the Chemical Society, Faraday Transactions,* **86**, 1147, 1990.

Dickinson, E., McClements, D.J., and Povey, M.J.W., Ultrasonic investigation of the particle size dependence of crystallization in n-hexadecane-in-water emulsions, *Journal of Colloid and Interface Science,* **142**, 103, 1991.

Dickinson, E., Hunt, J.A., and Horne, D.S., Calcium induced flocculation of emulsions containing adsorbed β-casein or phosvitin, *Food Hydrocolloids,* **6**, 359, 1992.

Dickinson, E., Owusu, R.K., Tan, S., and Williams, A., Oil-soluble surfactants have little effect on competitive adsorption of α-lactalbumin and β-lactoglobulin in emulsions, *Journal of Food Science*, **58**, 295, 1993a.

Dickinson, E., Owusu, R.K., and Williams, A., Orthokinetic destabilization of a protein stabilized emulsion by a water-soluble surfactant, *Journal of the Chemical Society, Faraday Transactions,* **89**, 865, 1993b.

Dickinson, E., Iveson, G., and Tanai, S., Competitive adsorption in protein stabilized emulsions containing oil-soluble and water-soluble surfactants, in *Food Colloids and Polymers: Stability and Mechanical Properties,* Dickinson, E. and Walstra, P., Eds., Royal Society of Chemistry, Cambridge, 1993c, 312.

Dickinson, E., Goller, M.I., and Wedlock, D.J., Creaming and rheology of emulsions containing polysaccharide and non-ionic or anionic surfactants, *Colloids and Surfaces A,* **75**, 195, 1993d.

Dickinson, E., Goller, M.I., and Wedlock, D.J., Osmotic pressure, creaming and rheology of emulsions containing non-ionic polysaccharide, *Journal of Colloid and Interface Science,* **172**, 192, 1995.

Dickinson, E., Hong, S.T., and Yamanoto, Y., Rheology of heat-set emulsion gels containing β-lactoglobulin and small-molecule surfactants, *Netherlands Milk and Dairy Journal,* **50**, 199, 1996.

Dickinson, E., Golding, M., and Povey, M.J.W., Rheology and dynamics of water-in-oil emulsions containing sodium caseinate, *Journal of Colloid and Interface Science,* **185**, 515, 1997.

Dill, K.A., Dominant forces in protein folding, *Biochemistry,* **29**, 7133, 1990.

Doi, E., Gels and gelling of globular proteins, *Trends in Food Science and Technology,* **4**, 1, 1993.

Ducker, W.A. and Pashley, R.M., Forces between mica surfaces in the presence of rod-shaped divalent counterions, *Langmuir,* **8**, 109, 1992.

Duckworth, R.B., *Water Relationships of Foods,* Academic Press, London, 1975.

Dukhin, A.S. and Goetz, P.J., Acoustic and electroacoustic spectroscopy, *Langmuir,* **12**, 4336, 1996.

Dukhin, S. and Sjoblom, J., Kinetics of Brownian and gravitational coagulation in dilute emulsions, in *Emulsions and Emulsion Stability,* Sjoblom, J., Ed., Marcel Dekker, New York, 1996.

Dukhin, S.S., Krietchamer, G., and Miller R., *Dynamics of Adsorption at Liquid Interfaces: Theory, Experiment and Application,* Elsevier, Amsterdam, 1995.

Eads, T.M., Molecular origins of structure and functionality in foods, *Trends in Food Science and Technology,* **5**, 147, 1994.

Eagland, D., Nucleic acids, peptides and proteins, in *Water: A Comprehensive Treatise,* Vol. 4, Franks, F., Ed., Plenum Press, New York, 1975, chap. 5.

Eastoe, J., Small-angle neutron scattering and neutron reflection, in *New Physicochemical Techniques for the Characterization of Complex Food Systems,* Dickinson, E., Ed., Blackie Academic and Professional, London, 1995, chap. 12.

Edwards, S.F., Lillford, P.J., and Blanshard, J.M.V., Gels and networks in practice and theory, in *Food Structure and Behaviour,* Blanshard, J.M.V. and Lillford, P.J., Eds., Academic Press, London, 1987, chap. 1.

Eisenberg, D. and Kauzmann, W., *The Structure and Properties of Water,* Oxford University Press, Oxford, 1969.

Elworthy, P.H., Florence, A.T., and Macfarlane, C.B., *Solubilization by Surface-Active Agents and Its Application in Chemistry and the Biological Sciences,* Chapman and Hall, London, 1968.

Epstein, P.S. and Carhart, R.R., The absorption of sound in suspensions and emulsions, *Journal of the Acoustical Society of America,* **25**, 553, 1953.

Esselink, K., Hilbers, P.A.J., van Os, N.M., Smit, B., and Karaborni, S., Molecular dynamics simulations of mode oil/water/surfactant systems, *Colloids and Surfaces A,* **91**, 155, 1994.

Evans, D.F. and Wennerstrom, H., *The Colloidal Domain: Where Physics, Chemistry, Biology and Technology Meet,* VCH Publishers, New York, 1994.

Everett, D.H., *Basic Principles of Colloid Science,* Royal Society of Chemistry, Cambridge, 1988.

Evison, J., Dickinson, E., Owusu Apenten, R.K., and Williams, A., Formulation and properties of protein stabilized water-in-oil-in-water multiple emulsions, in *Food Macromolecules and Colloids,* Dickinson, E. and Lorient, D., Eds., Royal Society of Chemistry, Cambridge, 1995, 235.

Farinato, R.S. and Rowell, R.L., Optical properties of emulsions, in *Encyclopedia of Emulsion Technology,* Vol. 1, Becher, P., Marcel Dekker, New York, 1983, chap. 8.

Fellows, P., *Food Processing Technology: Principles and Practice,* VCH Publishers, Weinheim, Germany, 1988, chap. 3 and 4.

Fennema, O.R., Water and protein hydration, in *Food Proteins,* AVI Publishing, Westport, CT, 1977, chap. 3.

Fennema, O.R., Ed., *Food Chemistry,* 3rd ed., Marcel Dekker, New York, 1996a.

Fennema, O.R., in *Food Chemistry,* 3rd ed., Fennema, O.R., Ed., Marcel Dekker, New York, 1996b, chap. 2.

Fennema, O.R. and Tannenbaum, S.R., Introduction to food chemistry, in *Food Chemistry,* 3rd ed., Fennema, O.R., Ed., Marcel Dekker, New York 1996, 1.

Fisher, L.R. and Parker, N.S., How do food emulsion stabilizers work? *CSIRO Food Research Quarterly,* **45**, 33, 1985.

Fito, P., Mullet, A., and McKenna, B., *Water in Foods: Fundamental Aspects and Their Significance in Relation to the Processing of Foods,* Elsevier Applied Science, London, 1994.

Formo, M.W., Physical properties of fats and fatty acids, in *Bailey's Industrial Oil and Fat Products,* Vol. 1, 5th ed., Swern, D., Ed., John Wiley & Sons, New York, 1979, chap. 3.

Fox, P.F. and Condon, J.J., *Food Proteins,* Applied Science Publishers, London, 1982.

Francis, F.J., Basic concepts of colorimetry, in *Theory, Determination and Control of Physical Properties of Food Materials,* Rha, C., Ed., D. Reidel Publishing, Boston, 1975, chap. 14.

Francis, F.J., Colorimetric properties of foods, in *Engineering Properties of Foods,* 2nd ed., Rao, M.A. and Rizvi, S.S.H., Eds., Marcel Dekker, New York, 1995, chap. 10.

Francis, F.J. and Clydesdale, F.M., *Food Colorimetry: Theory and Applications,* AVI Publishing, Westport, CT, 1975.

Frankel, E.N., Recent advances in lipid oxidation, *Journal of Food Science and Agriculture,* **54**, 495, 1991.

Frankel, E.N., Evaluation of antioxidant activity of rosemary extracts, carnosol and carnosic acid in bulk vegetable oils and fish oil and their emulsions, *Journal of the Science of Food and Agriculture,* **72**, 201, 1996.

Frankel, E.N., Huang, S.W., Kanner, J., and German, J.B., Interfacial phenomena in the evaluation of antioxidants: Bulk oils versus emulsions, *Agriculture and Food Chemistry,* **42**, 1054, 1994.

Franklin, F., *College Physics,* 4th ed., Harcourt Brace Jovanovich, New York, 1977.

Franks, F., *Water: A Comprehensive Treatise,* Vol. 1–7, Plenum Press, New York, 1972–82.

Franks, F., The solvent properties of water, in *Water: A Comprehensive Treatise,* Vol. 2, Franks, F., Ed., Plenum Press, New York, 1973, chap. 1.

Franks, F., The hydrophobic interaction, in *Water: A Comprehensive Treatise,* Vol. 4, Franks, F., Ed., Plenum Press, New York, 1975a, chap. 1.

Franks, F., Water, ice and solutions of simple molecules, in *Water Relationships of Foods,* Duckworth, R.B., Ed., Academic Press, London, 1975b, 3.

Franks, F., Water and aqueous solutions: Recent advances, in *Properties of Water in Foods in Relation to Quality and Stability,* Simatos, D. and Multon, J.L., Eds., Martinus Nijhoff Publishers, Dordrecht, 1985a, 1.

Franks, F., Complex aqueous systems at subzero temperatures, in *Properties of Water in Foods in Relation to Quality and Stability,* Simatos, D. and Multon, J.L., Eds., Martinus Nijhoff Publishers, Dordrecht, 1985b, 497.

Franks, F., Hydration phenomena: An update and implications for the food processing industry, in *Water Relationships in Foods: Advances in the 1980's and Trends for the 1990's,* Levine, H. and Slade, L., Eds., Plenum Press, New York, 1991, 1.

Franzen, K.L. and Kinsella, J.E., Physicochemical aspects of food flavouring, *Chemistry and Industry,* **21**, 505, 1975.

Friberg, S., Emulsion stability, in *Food Emulsions,* 3rd ed., Friberg, S. and Larsson, K., Eds., Marcel Dekker, New York, 1997, chap. 1.

Friberg, S.E. and El-Nokaly, M., Multilayer emulsions, in *Physical Properties of Foods,* Peleg, M. and Bagley, E.B., Eds., AVI Publishing, Westport, CT, 1983, chap. 5.

Friberg, S. and Larsson, K., Eds., *Food Emulsions,* 3rd ed., Marcel Dekker, New York, 1997.

Friedman, M., *The Chemistry and Biochemistry of the Sulfhydryl Group in Amino Acids, Peptides and Proteins,* Pergamon Press, Oxford, 1973.

Galema, S.A. and Hoiland, H., Stereochemical aspects of hydration of carbohydrates in aqueous solutions: Density and ultrasonic measurements, *Journal of Physical Chemistry,* **95**, 5321, 1991.

Gaonkar, A., *Characterization of Foods: Emerging Methods,* Elsevier, Amsterdam, 1995.

Garrouch, A.A., Lababidi, H.M.S., and Gharbi, R.B., Dielectric dispersion of dilute suspensions of colloid particles: Practical applications, *Journal of Physical Chemistry,* **100**, 16996, 1996.

Garside, J., General principles of crystallization, in *Food Structure and Behaviour*, Blanshard, J.M.V. and Lillford, P., Eds., Academic Press, London, 1987, chap. 3.

Garti, N. and Sato, K., *Crystallization and Polymorphism of Fats and Fatty Acids,* Marcel Dekker, New York, 1988.

Gelin, B.R., *Molecular Modeling of Polymer Structures and Properties,* Hanser Publishers, Munich, 1994.

Glicksman, M., Background and classification, in *Food Hydrocolloids,* Vol. I, Glicksman, M., Ed., CRC Press, Boca Raton, FL, 1982a, chap. 1.

Glicksman, M., Functional properties, in *Food Hydrocolloids,* Vol, I, Glicksman, M., Ed., CRC Press, Boca Raton, FL, 1982b, chap. 3.

Glicksman, M., Food applications of gums, in *Food Carbohydrates,* Lineback, D.R. and Inglett, G.E., Eds., AVI Publishing, Westport, CT, 1982c, chap. 15.

Goff, H.D., Interactions and contributions of stabilizers and emulsifiers to development of structure in ice-cream, in *Food Colloids and Polymers: Stability and Mechanical Properties,* Dickinson, E. and Walstra, P., Eds., Royal Society of Chemistry, Cambridge, 1993, 71.

Goff, H.D., Liboff, M., Jordon, W.K., and Kinsella, J.E., The effect of polysorbate on the fat emulsion of ice-cream mix: Evidence from transmission electron microscopy studies, *Food Microstructure,* **6**, 193, 1987.

Goodwin, J.W. and Ottewill, R.H., Properties of concentrated colloidal dispersions, *Journal of the Chemical Society, Faraday Transactions,* **87**, 357, 1991.

Gopal, E.S.R., Principles of emulsion formation, in *Emulsion Science*, Sherman, P., Ed., Academic Press, London, 1968, chap. 1.

Gregory, J., The calculation of Hamaker constants, *Advances in Colloid and Interface Science,* **2**, 396, 1969.

Gregory, J., Approximate expressions for retarded van der Waals interaction, *Journal of Colloid Science*, **83**, 138, 1981.

Griffin, W.G. and Griffin, M.C.A., The attenuation of ultrasound in aqueous suspensions of casein micelles from bovine milk, *Journal of the Acoustical Society of America,* **87**, 2541, 1990.

Grover, G.S. and Bike, S.G., Monitoring of flocculation in-situ in sterically stabilized silica dispersions using rheological techniques, *Langmuir,* **11**, 1807, 1995.

Guichard, E., Interactions between pectins and flavor compounds in strawberry jam, in *Flavor–Food Interactions,* McGorrin, R.J. and Leland, J.V., Eds., American Chemical Society, Washington, D.C., 1996, chap. 11.

Gunstone, F.D. and Norris, F.A., *Lipids in Foods: Chemistry, Biochemistry and Technology,* Pergamon Press, Oxford, 1983.

Gurr, E., *Encyclopedia of Microscopic Stains,* Williams and Wilkins, Baltimore, 1961.

Guyot, C., Bonnafont, C., Lesschaeve, I., Issanchou, S., Voilley, A., and Spinnler, H.E., Effect of fat content on odor intensity of three aroma compounds in model emulsions: δ-Decalactone, diacetyl and butyric acid, *Journal of Agricultural and Food Chemistry,* **44**, 2341, 1996.

Hallet, F.R., Particle size analysis by dynamic light scattering, *Food Research International,* **27**, 195, 1994.

Halliwell, B. and Gutteridge, J.M.C., *Free Radicals in Biology and Medicine,* 2nd ed., Clarendon Press, London, 1991.

Hamilton, R.J. and Bhati, A., *Recent Advances in Chemistry and Technology of Fats and Oils,* Elsevier Applied Science, London, 1987.

Hansen, A.P. and Booker, D.C., Flavor interactions with casein and whey protein, in *Flavor–Food Interactions,* McGorrin, R.J. and Leland, J.V., Eds., American Chemical Society, Washington, D.C., 1996, chap. 7.

Harper, W.J. and Hall, C.W., *Dairy Technology and Engineering,* AVI Publishing, Westport, CT, 1976.

Harrison, M., Hills, B.P., Bakker, J., and Clothier, T., Mathematical models of flavor release from liquid emulsions, *Journal of Food Science,* **62**, 653, 1997.

Harvey, B.A., Druaux, C., and Voilley, A., Effect of protein on the retention and transfer of aroma compounds at the lipid water interface, in *Food Macromolecules and Colloids,* Dickinson, E. and Lorient, D., Eds., Royal Society of Chemistry, Cambridge, 1995, 154.

Harwalker, V.R. and Ma, C.Y., *Thermal Analysis of Foods,* Elsevier Science, New York, 1990.

Hasenhuettl, G.L., Overview of food emulsifiers, in *Food Emulsifiers and Their Applications,* Hasenhuettl, G.L. and Hartel, R.W., Eds., Chapman and Hall, New York, 1997.

Hasenhuettl, G.L. and Hartel, R.W., Eds., *Food Emulsifiers and Their Applications,* Chapman and Hall, New York, 1997.

Hasted, J.B., Liquid water: Dielectric properties, in *Water: A Comprehensive Treatise,* Vol. 1, Franks, F., Ed., Plenum Press, New York, 1972, chap. 7.

Hato, M., Maruta, M., and Yoshida, T., Surface forces between protein A adsorbed mica surfaces, *Colloids and Surface A,* **109**, 345, 1996.

Hau, M.Y.M., Gray, D.A., and Taylor, A.J., Binding of volatiles to starch, in *Flavor–Food Interactions,* McGorrin, R.J. and Leland, J.V., Eds., American Chemical Society, Washington, D.C., 1996, chap. 10.

Hauser, H., Lipids, in *Water: A Comprehensive Treatise,* Vol. 4, Franks, F., Ed., Plenum Press, New York, 1975, chap. 4.

Heertje, I. and Paques, M., Advances in electron microscopy, in *New Physicochemical Techniques for the Characterization of Complex Food Systems,* Dickinson, E., Ed., Blackie Academic and Professional, London, 1995, chap. 1.

Hemar, Y., Herrmann, N., Lemarechal, P., Hocquart, R., and Lequeux, F., Effective medium model for ultrasonic attenuation due to the thermo-elastic effect in concentrated emulsions, *Journal of Physique II,* **7**, 637, 1997.

Hermansson, A.M., Gel structure of food biopolymers, in *Food Structure — Its Creation and Evaluation,* Blanshard, J.M.V. and Mitchell, J.R., Eds., Butterworths, London, 1988, chap. 3.

Hernandez, E. and Baker, R.A., Turbidity of beverages with citrus oil clouding agent, *Journal of Food Science,* **56**, 1024, 1991.

Hernandez, E., Baker, R.A., and Crandall, P.G., Model for evaluating the turbidity in cloudy beverages, *Journal of Food Science,* **56**, 747, 1991.

Hernqvist, L., On the structure of triglycerides in the liquid state and fat crystallinization, *Fette Seifen Anstrichmittel,* **86**, 297, 1984.

Hernqvist, L., Polymorphism of triglycerides: A crystallographic review, *Food Structure,* **9**, 39, 1990.

Hiemenz, P.C., *Principles of Colloid and Surface Chemistry,* Dekker, New York, 1986.

Hollingsworth, P., Food research: Cooperation is the key, *Food Technology,* **49**(2), 67, 1995.

Holt, W.J.C. and Chan, D.Y.C., Pair interactions between heterogeneous spheres, *Langmuir,* **13**, 1577, 1997.

Horne, D.S., Light scattering studies of colloid stability and gelation, in *New Physicochemical Techniques for the Characterization of Complex Food Systems,* Dickinson, E., Ed., Blackie Academic and Professional, London, 1995, chap. 11.

Huang, S.W., Frankel, E.N., and German, J.B., Antioxidant activity of α- and γ-tocopherols in bulk oils and oil-in-water emulsions, *Journal of Agricultural and Food Chemistry*, **42**, 2108, 1994.

Huang, S.W., Frankel, E.N., Schwarz, K., Aeschbach, R., and German, J.B., Antioxidant activity of carnosic acid and methyl carnosate in bulk oils and oil-in-water emulsions, *Journal of Agricultural and Food Chemistry*, **44**, 2951, 1996a.

Huang, S.W., Frankel, E.N., Schwarz, K., and German, J.B., Effect of pH on antioxidant activity of alpha tocopherol and trolox in oil-in-water emulsions, *Journal of Agricultural and Food Chemistry*, **44**, 2496, 1996b.

Huang, S.W., Hopia, A., Schwarz, K., Frankel, E.N., and German, J.B., Antioxidant activity of alpha tocopherol and trolox in different lipid substrates — Bulk oils versus oil-in-water emulsions, *Journal of Agricultural and Food Chemistry*, **44**, 444, 1996c.

Huang, S.W., Frankel, E.N., Aeschabach, R., and German, J.B., Partition of selected antioxidants in corn oil-in-water model systems, *Journal of Agricultural and Food Chemistry*. **45**, 1991, 1997.

Hunt, J.A. and Dalgleish, D.G., Effect of pH on the stability and surface composition of emulsions made with whey protein isolate, *Journal of Food Science*, **59**, 2131, 1994.

Hunt, J.A. and Dalgleish, D.G., Heat stability of oil-in-water emulsions containing milk proteins: Effect of ionic strength and pH, *Journal of Food Science*, **60**, 1120, 1995.

Hunt, J.A. and Dalgleish, D.G., The effect of the presence of KCl on the adsorption behavior of whey-protein and caseinate in oil-in-water emulsions, *Food Hydrocolloids*, **10**, 159, 1996.

Hunter, R.J., *Foundations of Colloid Science*, Vol. 1, Oxford University Press, Oxford, 1986.

Hunter, R.J., *Foundations of Colloid Science*, Vol. 2, Oxford University Press, Oxford, 1989.

Hunter, R.J., *Introduction to Modern Colloid Science*, Oxford University Press, Oxford, 1993.

Hunter, R.J., The electroacoustic characterization of colloidal suspensions, in *Handbook on Ultrasonic and Dielectric Characterization Techniques for Suspended Particles*, Hackley, V.A. and Texter, J., Eds., American Chemical Society, Westerville, OH, 1998.

Hussin, A.B.B.H. and Povey, M.J.W., A study of dilation and acoustic propagation in solidifying fats and oils: Experimental, *Journal of the American Oil Chemists Society*, **61**, 560, 1984.

Hutchings, J.B., *Food Colour and Appearance*, Blackie Academic and Professional, London, 1994.

Isaacs, E.E., Huang, M., and Babchin, A.J., Electroacoustic method for monitoring the coalescence of water-in-oil emulsions, *Colloids and Surfaces*, **46**, 177, 1990.

Isengard, H.D., Rapid water determination in food stuffs, *Trends in Food Science and Technology*, **6**, 155, 1995.

Israelachvili, J.N., *Intermolecular and Surface Forces*, Academic Press, London, 1992.

Israelachvili, J.N., The science and application of emulsions — An overview, *Colloids and Surfaces*, **91**, 1, 1994.

Israelachvili, J.N. and Berman A., Irreversibility, energy dissipation and time effects in intermolecular and surface interactions, *Israel Journal of Chemistry*, **35**, 85, 1995.

Israelachvili, J.N. and Pashly, R.M., Measurement of the hydrophobic interaction between two hydrophobic surfaces in aqueous electrolyte solutions, *Journal of Colloid Science*, **98**, 500, 1984.

Israelachvili, J.N. and Wennerstrom, H., Role of hydration and water-structure in biological and colloidal interactions, *Nature*, **379**, 219, 1996.

Jackel, V.K., Uber die Funktionen des Schuzkolloids, *Kolloid-Zeitschrift und Zeitshrift fur Polymere*, **197**, 143, 1964.

Janssen, J.J.M., Boon, A., and Agterof, W.G.M., Influence of dynamic interfacial properties on droplet breakup in simple shear flocs, *Fluid Mechanics and Transport Phenomena*, **40**, 1929, 1994.

Jaynes, E.N., Applications in the food industry. II, in *Encyclopedia of Emulsion Technology*, Vol. 2, Becher, P., Ed., Marcel Dekker, New York, 1983, chap. 6.

Jenkins, P. and Snowden, M., Depletion flocculation in colloidal dispersions, *Advances in Colloid and Interface Science*, **68**, 57, 1996.

Jokela, P., Fletcher, P.D.I., Aveyard, R., and Lu, J.R., The use of computerized microscopic image analysis to determine droplet size distributions, *Journal of Colloid and Interface Science*, **134**, 417, 1990.

Jost, R., Baechler, R., and Masson, G., Heat gelation of oil-in-water emulsions stabilized by whey protein, *Journal of Food Science*, **51**, 440, 1986.

Jungermann, E., Fat based surface-active agents, in *Bailey's Industrial Oil and Fat Products*, Vol. 1, Swern, D., Ed., John Wiley & Sons, New York, 1979, chap. 9.

Kabalnov, A.S. and Shchukin, E.D., Ostwald ripening theory: Applications to fluorocarbon emulsion stability, *Advances in Colloid and Interface Science,* **38**, 69, 1992.

Kabalnov, A. and Weers, J., Macroemulsion stability within the Winsor III region: Theory versus experiment, *Langmuir,* **12**, 1931, 1996.

Kabalnov, A. and Wennerstrom, H., Macroemulsion stability: The orientated wedge theory revisited, *Langmuir,* **12**, 276, 1996.

Kabalnov, A., Weers, J., Arlauskas, R., and Tarara, T., Phospholipids as emulsion stabilizers. 1. Interfacial tension, *Langmuir,* **11**, 2966, 1995.

Kalab, M., Electron microscopy of milk products: A review of techniques, *Scanning Electron Microscopy,* **3**, 453, 1981.

Kalab, M., Allan-Wojtas, P., and Miller, S.S., Microscopy and other imaging techniques in food structure analysis, *Trends in Food Science and Technology,* **6**, 180, 1995.

Kallay, N., Hlady, V., Jednacak-Biscan, J., and Milonjic, S., Techniques for the study of adsorption from solution, in *Physical Methods of Chemistry,* Vol. IXA, Rossiter, B.W. and Baetzold, R.C., Eds., John Wiley & Sons, New York, 1993, chap. 2.

Kandori, K., Applications of microporous glass membranes: Membrane emulsification, in *Food Processing: Recent Developments,* Gaonkar, A.G., Ed., Elsevier Science, Amsterdam, 1995.

Karbstein, H. and Schubert, H., Developments in the continuous mechanical production of oil-in-water macro-emulsions, *Chemical Engineering and Processing,* **34**, 205, 1995.

Karplus, M. and Porter, R.N., *Atoms and Molecules: An Introduction for Students of Physical Chemistry,* Benjamin-Cummins, Menlo Park, CA, 1970.

Kato, A., Fujishige, T., Matsudomi, N., and Kogayashi, K., Determination of emulsifying properties of some proteins by conductivity measurements, *Journal of Food Science,* **50**, 56, 1985.

Katz, F., Technology trends, *Food Technology,* **51**(6), 46, 1997.

Kawano, S., Progress in application of NIR and FT-IR in food characterization, in *Characterization of Foods: Emerging Techniques,* Gaonkar, A.G., Ed., Elsevier, Amsterdam, 1995, chap. 8.

Keikens, F., Optimization of electrical conductance measurements for the quantification and prediction of phase separation in O/W emulsions containing hydroxpropylmethylcelluloses as emulsifying agents, *International Journal of Pharmaceutics,* **146**, 239, 1997.

Kellens, M., Meeussen, W., and Reynaers, H., Study of the polymorphism and the crystallization kinetics of tripalmitin, *Journal of the American Oil Chemists Society,* **69**, 906, 1992.

Kerker, M., *The Scattering of Light and Other Electromagnetic Radiation,* Academic Press, New York, 1969.

Kern, C.W. and Karplus, M., The water molecule, in *Water: A Comprehensive Treatise,* Vol. 1, Franks, F., Ed., Plenum Press, New York, 1972, chap. 2.

Kessler, H.G., *Food Engineering and Dairy Technology,* Verlag A., Kessler Publishing, Freising, Germany, 1981.

Kinsella, J.E., Relationships between structure and functional properties of food proteins, in *Food Proteins,* Fox, P.F. and Condon, J.J., Eds., Applied Science Publishers, London, 1982, 51.

Kinsella, J.E., in *Flavor Chemistry of Lipid Foods,* Min, D.B. and Smouse, T.H., American Oil Chemists Society, Champaign, IL, 1989, 376.

Kinsella, J.E. and Whitehead, D.M., Proteins in whey: Chemical, physical and functional properties, *Advances in Food and Nutrition Research,* **33**, 343, 1989.

Kirby, A.R., Gunning, A.P., and Morris, V.J., Atomic force microscopy in food research: A new technique comes of age, *Trends in Food Science and Technology,* **6**, 359, 1995.

Kitakara, A. and Watanabe, A., *Electrical Phenomenon at Interfaces: Fundamentals, Measurements and Applications,* Marcel Dekker, New York, 1984.

Klemaszeski, J., Haque, Z., and Kinsella, J.E., An electronic imaging system for determining the droplet size and dynamic breakdown of protein stabilized emulsions, *Journal of Food Science,* **54**, 440, 1989.

Kokini, J.L., The physical basis of liquid food texture and texture–taste interactions, *Journal of Food Engineering,* **6**, 51, 1987.

Kokini, J.L., Eads, T., and Ludescher, R.D., Research needs on the molecular basis for food functionality, *Food Technology,* **47**(3), 36S, 1993.

Kresheck, G.C., Surfactants, in *Water: A Comprehensive Treatise,* Vol. 4, Franks, F., Ed., Plenum Press, New York, 1975, chap. 2.

Krog, M.J., Riisom, T.H., and Larsson, K., Applications in the food industry. I, in *Encyclopedia of Emulsion Technology,* Vol. 2, Becker, P., Ed., Marcel Dekker, New York, 1983, chap. 5.

Kumosinski, T.F., Brown, E.M., and Farrell, H.M., Molecular modeling in food research: Technology and techniques, *Trends in Food Science and Technology,* **2**, 110, 1991a.

Kumosinski, T.F., Brown, E.M., and Farrell, H.M., Molecular modeling in food research: Applications, *Trends in Food Science and Technology,* **2**, 190, 1991b.

Lachaise, J., Mendiboure, B., Dicharry, C., Marion, G., and Salager, J.L., Simulation of the overemulsification phenomenon in turbulent stirring, *Colloids and Surfaces A,* **110**, 1, 1996.

Land, D.G., Perspectives on the effects of interactions on flavour perception: An overview, in *Flavor– Food Interactions,* McGorrin, R.J. and Leland, J.V., Eds., American Chemical Society, Washington, D.C., 1996.

Land, J. and Zana, R., Chemical relaxation methods, in *Surfactant Solutions: New Methods of Investigation,* Zana, R., Ed., Marcel Dekker, New York, 1987.

Landy, P., Courthaudon, J.L., Dubois, C., and Voilley, A., Effect of interface in model food emulsions on the volatility of aroma compounds, *Journal of Agricultural and Food Chemistry,* **44**, 526, 1996.

Langourieux, S. and Crouzet, J., Protein–aroma interactions, in *Food Macromolecules and Colloids,* Dickinson, E. and Lorient, D., Eds., Royal Society of Chemistry, Cambridge, 1995, 123.

Lapasin, R. and Pricl, S., *Rheology of Industrial Polysaccharides: Theory and Applications,* Blackie Academic and Professional, London, 1995.

Larison, K.D., *Handbook of Fluorescent Probes and Research Chemicals,* Molecular Probes, Eugene, OR, 1992.

Launay, B., Doublier, J.L., and Cuvelier, G., Flow properties of aqueous solutions of polysaccharides, in *Functional Properties of Food Macromolecules,* Mitchell, J.R. and Ledward, D.A., Eds., Elsevier, London, 1986, chap. 1.

Lawson, H.W., *Food Oils and Fats: Technology, Utilization and Nutrition,* Chapman and Hall, New York, 1995.

Ledward, D.A., Gelation of gelatin, in *Functional Properties of Food Macromolecules,* Mitchell, J.R. and Ledward, D.A., Eds., Elsevier Applied Science, London, 1986, chap. 4.

Lee, S.Y. and Morr, C.V., Fixation and staining milkfat globules in cream for transmission and scanning electron microscopy, *Journal of Food Science,* **57**, 887, 1992.

Lees, L.H. and Pandolfe, W.D., Homogenizers, in *Encyclopedia of Food Engineering,* Hall, C.W., Farrall, A.W., and Rippen, A.L., Eds., AVI Publishing, Westport, CT, 1986, 467.

Lehnert, S., Tarabishi, H., and Leuenberger, H., Investigation of thermal phase inversion in emulsions, *Colloids and Surfaces A,* **91**, 227, 1994.

Lehninger, A.L., Nelson, D.L., and Cox, M.M., *Principles of Biochemistry,* 2nd ed., Worth Publishers, New York, 1993.

Levine, H. and Slade, L., *Water Relationships in Foods: Advances in the 1980's and Trends for the 1990's,* Plenum Press, New York, 1991.

Leward, D.A., Gelation of gelatin, in *Functional Properties of Food Macromolecules*, Mitchell, J.R. and Ledward, D.A., Eds., Elsevier Applied Science, London, 1986, chap. 4.

Li, X., Cox, J.C., and Flumerfelt, R.W., Determination of emulsion size distribution by NMR restricted diffusion measurement, *AIChE Journal,* **38**, 1671, 1992.

Lickiss, P.D. and McGrath, V.E., Breaking the sound barrier, *Chemistry in Britain,* **32**, 47, 1996.

Lindman, B., Carlsson, A., Gerdes, S., Karlstrom, G., Picullel, L., Thalberg, K., and Zhang, K., Polysaccharide–surfactant systems: Interactions, phase diagrams and novel gels, in *Food Colloids and Polymers: Stability and Mechanical Properties,* Dickinson, E. and Walstra, P., Eds., Royal Society of Chemistry, Cambridge, 1993, 113.

Linfield, W.M., *Anionic Surfactants: Parts I and II,* Marcel Dekker, New York, 1976.

Lips, A., Campbell, I.J., and Pelan, E.G., Aggregation mechanisms in food colloids and the role of biopolymers, in *Food Polymers, Gels and Colloids,* Dickinson, E., Ed., Royal Society of Chemistry, Cambridge, 1991, 1.

Lips, A., Westbury, T., Hart, P.M., Evans, I.D., and Campbell, I.J., Aggregation mechanisms in food colloids and the role of biopolymers, in *Food Colloids and Polymers: Stability and Mechanical Properties,* Dickinson, E. and Walstra, P., Eds., Royal Society of Chemistry, Cambridge, 1993, 31.

Lissant, K.J., *Demulsification: Industrial Applications,* Marcel Dekker, New York, 1983.

Liu, S. and Masliyah, J.H., Rheology of suspensions, in *Suspensions: Fundamentals and Applications in the Petroleum Industry,* Schramm, L.L., Ed., American Chemical Society, Washington, D.C., 1996, chap. 3.

Lloyd, N.E., Determination of surface-average particle diameter of colored emulsions by reflectance, and application to emulsion stability studies, *Journal of Colloid and Interface Science,* 14, 441, 1959.

Loncin, M. and Merson, R.L., *Food Engineering: Principles and Selected Applications,* Academic Press, New York, 1979.

Lovsey, S.W., *Theory of Neutron Scattering from Condensed Matter,* Clarendon Press, Oxford, 1984.

Lucassen-Reyanders, E.H. and Kuijpers, K.A., The role of interfacial properties in emulsification, *Colloids and Surfaces,* 65, 175, 1992.

Luckham, P.F. and Costello, B.A. de L., Recent developments in the measurement of interparticle forces, *Advances in Colloid and Interface Science,* 44, 193, 1993.

Ludescher, R.D., Molecular dynamics of food proteins: Experimental techniques and observations, *Trends in Food Science and Technology,* 1, 145, 1990.

Luyten, H., Jonkman, M., Kloek, W., and van Vliet, T., Creaming behaviour of dispersed particles in dilute xanthan solutions, in *Food Colloids and Polymers: Stability and Mechanical Properties,* Dickinson, E. and Walstra, P., Eds., Royal Society of Chemistry, Cambridge, 1993, 224.

Ma, Y., Varadan, V.K., and Varadan, V.V., Comments on ultrasonic propagation in suspensions, *Journal of the Acoustical Society of America,* 87, 2779, 1990.

MacGregor, E.A. and Greenwood, C.T., *Polymers in Nature,* John Wiley, New York, 1980.

MacKay, R.A., Solubilization, in *Nonionic Surfactants: Physical Chemistry,* Schick, M.J., Ed., Marcel Dekker, New York, 1987, 297.

Macosko, C.W., *Rheology: Principles, Measurements and Applications,* VCH Publishers, New York, 1994.

Magdassi, S., *Surface Activity of Proteins: Chemical and Physicochemical Modifications,* Marcel Dekker, New York, 1996.

Magdassi, S. and Kamyshny, A., Surface activity and functional properties of proteins, in *Surface Activity of Proteins: Chemical and Physicochemical Modifications,* Magdassi, S., Ed., Marcel Dekker, New York, 1996, chap. 1.

Mahanty, J. and Ninham, B.W., *Dispersion Forces,* Academic Press, New York, 1976.

Malmsten, M., Lindstrom, A.-L., and Warnheim, T., Ellipsometry studies of interfacial film formation in emulsion systems, *Journal of Colloid and Interface Science,* 173, 297, 1995.

Marra, J., Direct measurements of attractive van der Waals and adhesion forces between uncharged lipid bilayers in aqueous solutions, *Journal of Colloid and Interface Science,* 109, 11, 1986.

Matsubara, T. and Texter, J., In situ voltammetric determinations of solute distribution coefficients in emulsions, *Journal of Colloid and Interface Science,* 112, 421, 1986.

Matthew, J.B., Electrostatic effects in proteins, *Annual Review in Biophysics and Biophysical Chemistry,* 14, 387, 1985.

Mayhill, P.G. and Newstead, D.F., The effect of milkfat fractions and emulsifier type on creaming in normal-solids UHT recombined milk, *Milchwissenschaft,* 47, 75, 1992.

McCarthy, W.W., Ultrasonic emulsification, *Drug and Cosmetic Industry,* 94(6), 821, 1964.

McClements, D.J., The Use of Ultrasonics for Characterizing Fats and Emulsions, Ph.D. thesis, University of Leeds, United Kingdom, 1989.

McClements, D.J., Ultrasonic characterization of emulsions and suspensions, *Advances in Colloid and Interface Science,* 37, 33, 1991.

McClements, D.J., Ultrasonic determination of depletion flocculation in oil-in-water emulsions containing a non-ionic surfactant, *Colloids and Surfaces A,* 90, 25, 1994.

McClements, D.J., Principles of ultrasonic droplet size determination, *Langmuir,* 12, 3454, 1996.

McClements, D.J. and Dungan, S.R., Factors that affect the rate of oil exchange between oil-in-water emulsion droplets stabilized by a non-ionic surfactant: Droplet size, surfactant concentration and ionic strength, *Journal of Physical Chemistry,* 97, 7304, 1993.

McClements, D.J. and Dungan, S.R., Effect of colloidal interactions on the rate of interdroplet heterogeneous nucleation in oil-in-water emulsions, *Journal of Colloid and Interface Science,* 186, 17, 1997.

McClements, D.J. and Fairley, P., Ultrasonic pulse echo reflectometer, *Ultrasonics*, **29**, 58, 1991.

McClements, D.J. and Fairley, P., Frequency scanning ultrasonic pulse echo reflectometer, *Ultrasonics*, **30**, 403, 1992.

McClements, D.J. and Keogh, M.K., Physical properties of cold-setting gels prepared from heat-denatured whey protein isolate, *Journal of Food Chemistry and Agriculture*, **69**, 7, 1995.

McClements, D.J. and Povey, M.J.W., Solid fat content determination using ultrasonic velocity measurements, *International Journal of Food Science and Technology*, **22**, 491, 1987.

McClements, D.J. and Povey, M.J.W., Comparison of pulsed NMR and ultrasonic velocity techniques for determining solid fat contents, *International Journal of Food Science and Technology*, **23**, 159, 1988.

McClements, D.J. and Povey, M.J.W., Scattering of ultrasound by emulsions, *Journal of Physics D: Applied Physics*, **22**, 38, 1989.

McClements, D.J., Povey, M.J.W., Jury, M., and Betsansis, E., Ultrasonic characterization of a food emulsion, *Ultrasonics*, **28**, 266, 1990.

McClements, D.J., Dungan, S.R., German, J.B., Simoneau, C., and Kinsella, J.E., Droplet size and emulsifier type affect crystallization and melting of hydrocarbon-in-water emulsions, *Journal of Food Science*, **58**, 1148, 1993a.

McClements, D.J., Dungan, S.R., German, J.B., and Kinsella, J.E., Factors which affect oil exchange between oil-in-water emulsion droplets stabilized by whey protein isolate: Protein concentration, droplet size and ethanol, *Colloids and Surfaces A*, **81**, 203, 1993b.

McClements, D.J., Monahan, F.J., and Kinsella, J.E., Effect of emulsion droplets on the rheology of whey protein isolate gels, *Journal of Texture Studies*, **24**, 411, 1993c.

McClements, D.J., Monahan, F.J., and Kinsella, J.E., Disulfide bond formation affects the stability of whey protein stabilized emulsion, *Journal of Food Science*, **58**, 1036, 1993d.

McClements, D.J., Dickinson, E., Dungan, S.R., Kinsella, J.E., Ma, J.E., and Povey, M.J.W., Effect of emulsifiers type on the crystallization kinetics of oil-in-water emulsions containing a mixture of solid and liquid droplets, *Journal of Colloid and Interface Science*, **160**, 293, 1993e.

McClements, D.J., Povey, M.J.W., and Dickinson, E., Absorption and velocity dispersion due to crystallization and melting of emulsion droplets, *Ultrasonics*, **31**, 443, 1993f.

McClements, D.J., Han, S.W., and Dungan, S.R., Interdroplet heterogeneous nucleation of supercooled liquid droplets by solid droplets in oil-in-water emulsions, *Journal of the American Oil Chemists Society*, **71**, 1385, 1994.

McClements, D.J., Chantrapornchai, W., and Clydesdale, F., Prediction of food emulsion color using light scattering theory, *Journal of Food Science*, submitted, 1998a.

McClements, D.J., Hemar, Y., and Herrmann, N., Incorporation of thermal overlap effects into multiple scattering theory, *Journal of the Acoustical Society of America*, submitted, 1998b.

McNulty, P.B., Flavor release — Elusive and dynamic, in *Food Structure and Behaviour*, Blanshard, J.M.V. and Lillford, P., Eds., Academic Press, London, 1987.

McNulty, P.B. and Karel, M., Factors affecting flavour release and uptake in O/W emulsions. II. Stirred cell studies, *Journal of Food Technology*, **8**, 319, 1973.

Mei, L., McClements, D.J., Wu, J., and Decker, E.A., Iron-catalyzed lipid oxidation in emulsions as affected by surfactant, pH and NaCl, *Food Chemistry*, **61**, 307, 1998.

Melik, D.H. and Fogler, H.S., Fundamentals of colloidal stability in quiescent media, in *Encyclopedia of Emulsion Technology*, Vol. 3, Becher, P., Marcel Dekker, New York, 1988, chap. 1.

Menon, W.B. and Wasan, D.T., Demulsification, in *Encyclopedia of Emulsion Technology*, Vol. 2, Becher, P., Ed., Marcel Dekker, New York, 1985, chap. 1.

Mewis, J. and Macosko, C.W., Suspension rheology, in *Rheology: Principles, Measurements and Applications*, Macosko, C.W., Ed., VCH Publishers, New York, 1994, chap. 10.

Miklavic, S.J. and Ninham, B.W., Competition for adsorption sites by hydrated ions, *Journal of Colloid and Interface Science*, **134**, 305, 1990.

Mikula, R.J., Emulsion characterization, in *Emulsions: Fundamentals and Applications in the Petroleum Industry*, Schramm, L.L., Ed., American Chemical Society, Washington, D.C., 1992, chap. 3.

Miles, M.J. and McMaster, T.J., Scanning probe microscopy of food-related systems, in *New Physicochemical Techniques for the Characterization of Complex Food Systems*, Dickinson, E., Ed., Blackie Academic and Professional, London, 1995, chap. 3.

Miles, C.A., Fursey, G.A.J., and Jones, C.D.J., Ultrasonic estimation of solid/liquid ratios in fats, oils and adipose tissue, *Journal of the Science of Food and Agriculture,* **36,** 215, 1985.

Miles, C.A., Shore, A., and Langley, K.R., Attenuation of ultrasound in milks and creams, *Ultrasonics,* **28,** 394, 1990.

Miller, C.A. and Neogi, P., *Interfacial Phenomena,* Marcel Dekker, New York, 1985.

Milling, A., Mulvaney, P., and Larson, E., Direct measurement of repulsive van der Waals interactions using an atomic force microscope, *Journal of Colloid and Interface Science,* **180,** 460, 1996.

Min, D.B., Crude fat analysis, in *Introduction to the Chemical Analysis of Foods,* Nielsen, S.S., Ed., Bartlett and Jones, Boston, 1994, chap. 12.

Mishchuk, N.A., Sjoblom, J., and Dukhin, S.S., Influence of retardation and screening on van der Waals attractive forces on reverse coagulation of emulsions in the secondary minimum, *Colloid Journal,* **57,** 785, 1995.

Mishchuk, N.A., Sjoblom, J., and Dukhin, S.S., The effect of retardation and screening of van der Waals attractive forces on the breaking of doublet of drops during sedimentation, *Colloid Journal,* **58,** 210, 1996.

Mitchell, J.R. and Ledward, D.A., *Functional Properties of Food Macromolecules,* Elsevier Applied Science, London, 1986.

Mittal, K.L., *Solution Chemistry of Surfactants,* Vol. 1, Plenum Press, New York, 1979.

Monahan, F.J., McClements, D.J., and German, J.B., Disulfide-mediated polymerization reactions and physical properties of heated WPI-stabilized emulsions. *Journal of Food Science,* **61,** 504, 1996.

Moran, D.P.J., Fats in spreadable products, in *Fats in Food Products,* Blackie Academic and Professional, London, 1994.

Moran, D.P.J. and Rajah, K.K., *Fats in Food Products,* Blackie Academic and Professional, London, 1994.

Morris, V.J., Gelation of polysaccharides, in *Functional Properties of Food Macromolecules*, Mitchell, J.R. and Ledward, D.A., Eds., Elsevier Applied Science, London, 1986, chap. 3.

Mukerjee, P., Solubilization in aqueous micellar systems, in *Solution Chemistry of Surfactants,* Vol. 1, Mittal, K.L., Ed., Plenum Press, London, 1979.

Mulder, H. and Walstra, P., *The Milk Fat Globule: Emulsion Science as Applied to Milk Products and Comparable Foods,* Pudoc, Wageningen, Netherlands, 1974.

Mullin, J.W., *Crystallization,* 3rd ed., Butterworth-Heinmann, Oxford, 1993.

Mulvihill, D.M. and Kinsella, J.E., Gelation characteristics of whey proteins and β-lactoglobulin, *Food Technology,* **41,** 102, 1987.

Mulvihill, D.M. and Kinsella, J.E., Gelation of β-lactoglobulin: Effects of sodium chloride and calcium chloride on the rheological and structural properties of gels, *Journal of Food Science,* **53,** 231, 1988.

Murray, B.S. and Dickinson, E., Interfacial rheology and the dynamic properties of adsorbed films of food proteins and surfactants, *Food Science and Technology International,* **2,** 131, 1996.

Murrell, J.N. and Boucher, E.A., *Properties of Liquids and Solutions,* John Wiley & Sons, Chichester, U.K., 1982.

Murrell, J.N. and Jenkins, A.D., *Properties of Liquids and Solutions,* 2nd ed., John Wiley & Sons, Chichester, U.K., 1994.

Myers, D., *Surfactant Science Technology,* VCH Publishers, Weinheim, Germany, 1988.

Nakai, S. and Li-Chan, E., *Hydrophobic Interactions in Food Systems,* CRC Press, Boca Raton, FL, 1988.

Nakamura, H., Roles of electrostatic interactions in proteins, *Quarterly Review of Biophysics,* **29,** 1, 1996.

Nawar, W.W., Lipids, in *Food Chemistry,* 3rd ed., Fennema, O.R., Ed., Marcel Dekker, New York, 1996, 225.

Nielsen, S.S., Introduction to food analysis, in *Introduction to the Chemical Analysis of Foods,* Nielsen, S.S., Ed., Jones and Bartlett, Boston, 1994, chap. 1.

Ninham, B.W. and Yaminsky, V., Ion binding and ion specificity: The Hofmeister effect and Onsager and Lifshitz theories, *Langmuir,* **13,** 2097, 1997.

O'Brien, R.W., Cannon, D.W., and Rowlands, W.N., Electroacoustic determination of particle size and zeta potential, *Journal of Colloid and Interface Science,* **173,** 406, 1995.

O'Donnell, C.D., Controlling the fat: What's new, what's to come, *Prepared Foods,* p. 65, July 1995.

Oh, S.G., Jobalia, M., and Shah, D.O., The effect of micellar lifetime on the droplet size in emulsions, *Journal of Colloid and Interface Science,* **155**, 511, 1993.

Okshima, H., Electrostatic interaction between two spherical colloidal particles, *Advances in Colloid and Interface Science,* **53**, 77, 1994.

O'Neill, T.E., Flavor binding to food proteins: An overview, in *Flavor–Food Interactions,* McGorrin, R.J. and Leland, J.V., Eds., American Chemical Society, Washington, D.C., 1996, chap. 6.

Oretega-Vinuesa, J.L., Marin-Rodriguez, A., and Hidalgo-Alvarez, R., Colloidal stability of polymer colloids with different interfacial properties: Mechanisms, *Journal of Colloid and Interface Science,* **184**, 259, 1996.

Orr, C., Determination of particle size, in *Encyclopedia of Emulsion Technology,* Vol. 3, Becher, P., Ed., Marcel Dekker, New York, 1988, chap. 3.

Overbosch, P., Afterof, W.G.M., and Haring, P.G.M., Flavor release in the mouth, *Food Reviews International,* **7**, 137, 1991.

Ozilgen, S., Simoneau, C., German, J.B., McCarthy, M.J., and Reid, D.S., Crystallization kinetics of emulsified triglycerides, *Journal of the Science of Food and Agriculture,* **61**, 101, 1993.

Pacek, A.W., Moore, I.P.T., Nienow, A.W., and Calabrese, R.V., Video technique for measuring dynamics of liquid–liquid dispersion during phase inversion, *AIChE Journal,* **40**, 1940, 1994.

Pailthorpe, B.A. and Russel, W.B., Retarded van der Waals interaction between spheres, *Journal of Colloid and Interface Science,* **89**, 563, 1982.

Pal, R., Rheology of polymer thickened emulsions, *Journal of Rheology,* **36**, 1245, 1992.

Pal, R., Techniques for measuring the composition (oil and water content) of emulsions — A state of the art review, *Colloids and Surfaces A,* **84**, 141, 1994.

Pal, R., Rheology of emulsions containing polymeric liquids, in *Encyclopedia of Emulsion Technology,* Vol. 4, Becher, P., Ed., Marcel Dekker, New York, 1996, chap. 3.

Pal, R., Yan, Y., and Masliyah, J.H., Rheology of emulsions, in *Emulsions: Fundamentals and Applications in the Petroleum Industry,* Schramm, L.L., Ed., American Chemical Society, Washington, D.C., 1992, chap. 4.

Pandolfe, W.D., Homogenizers, in *Encyclopedia of Food Science and Technology,* John Wiley & Sons, New York, 1991, 1413.

Pandolfe, W.D., Effect of premix condition, surfactant concentration and oil level on the formation of oil-in-water emulsions by homogenization, *Journal of Dispersion Science and Technology,* **16**, 633, 1995.

Pandolfe, W.D. and Masucci, S.F., An instrument for rapid spectroturbimetric analysis, *American Laboratory,* p. 40, August 1984.

Parsegian, V.A., Reconciliation of van der Waals force measurements between phosphatidylcholine bilayers in water and between bilayer coated mica surfaces, *Langmuir,* **9**, 3625, 1993.

Partmann, W., The effects of freezing and thawing on food quality, in *Water Relationships of Foods,* Duckworth, R.B., Ed., Academic Press, London, 1975, 505.

Pashley, R.M., McGuiggan, P.M., Ninham, B.W., and Evans, D.F., Attractive forces between uncharged hydrophobic surfaces: Direct measurements in aqueous solutions, *Science,* p. 1088, 1985.

Patterson, D., *Pigments,* Elsevier, Amsterdam, 1967.

Paulaitis, M.E., Garde, S., and Ashbaugh, H.S., The hydrophobic effect, *Current Opinion in Colloid and Interface Science,* **1**, 376, 1996.

Pearce, K.N. and Kinsella, J.E., Emulsifying properties of proteins: Evaluation of a turbidimetric technique, *Journal of Agricultural and Food Chemistry,* **26**, 716, 1978.

Peleg, M., Fractals and foods, *Critical Reviews in Food Science and Nutrition,* **33**, 149, 1993.

Penner, M.H., Basic principles of spectroscopy, in *Introduction to the Chemical Analysis of Foods,* Nielsen, S.S., Ed., Jones and Bartlett, London, 1994a, chap. 22.

Penner, M.H., Ultraviolet, visible and fluorescence spectroscopy, in *Introduction to the Chemical Analysis of Foods,* Nielsen, S.S., Ed., Jones and Bartlett, London, 1994b, chap. 23.

Pettitt, D.J., Waybe, J.E.B., Nantz, J.R., and Shoemaker, C.F., Rheological properties of solutions and emulsions stabilized with xanthan gum and propylene glycol alginate, *Journal of Food Science,* **60**, 528, 1995.

Phipps, L.W., Heterogeneous and homogeneous nucleation in supercooled triglycerides and n-paraffins, *Transactions of the Faraday Society,* **60**, 1873, 1964.

Phipps, L.W., *The High Pressure Dairy Homogenizer*, The National Institute for Research in Dairying, Reading, England, 1985.

Pike, O.A., Fat characterization, in *Introduction to the Chemical Analysis of Foods*, Nielsen, S.S., Ed., Jones and Bartlett, London, 1994, chap. 13.

Pinfield, V.J., Dickinson, E., and Povey, M.J.W., Modeling of concentration profiles and ultrasound velocity profiles in a creaming emulsion: Importance of scattering effects, *Journal of Colloid and Interface Science,* **166**, 363, 1994.

Pinfield, V.J., Dickinson, E., and Povey, M.J.W., Modeling of combined creaming and flocculation in emulsions, *Journal of Colloid and Interface Science,* **186**, 80, 1997.

Pomeranz, Y. and Meloan, C.E., *Food Analysis,* 3rd ed., Chapman and Hall, New York, 1994.

Pothakamury, U.E. and Barbosa-Canovas, G.V., Fundamental aspects of controlled release in foods, *Trends in Food Science and Technology,* **6**, 397, 1995.

Povey, M.J.W., Ultrasound studies of shelf-life and crystallization, in *New Physicochemical Techniques for the Characterization of Complex Food Systems*, Dickinson, E., Ed., Blackie Academic and Professional, London, 1995, chap. 9.

Povey, M.J.W., *Ultrasonic Techniques for Fluids Characterization,* Academic Press, San Diego, 1997.

Rabinovich, Y.A.I. and Derjaguin, B.V., Interaction of hydrophobized filaments in aqueous electrolyte solutions, *Colloids and Surfaces,* **30**, 243, 1988.

Race, S.W., Improved product quality through viscosity measurement, *Food Technology,* **45**, 86, 1991.

Rao, M.A., Rheological properties of fluid foods, in *Engineering Properties of Foods,* 2nd ed., Rao, M.A. and Rizvi, S.S.H., Eds., Marcel Dekker, New York, 1995, chap. 1.

Rao, V.N.M., Delaney, R.A.M., and Skinner, G.E., Rheological properties of solid foods, in *Engineering Properties of Foods,* 2nd ed., Rao, M.A. and Rizvi, S.S.H., Eds., Marcel Dekker, New York, 1995, chap. 2.

Reddy, S.R. and Fogler, H.S., Emulsion stability: Determination from turbidity, *Journal of Colloid and Interface Science,* **79**, 101, 1981.

Reichardt, C., *Solvents and Solvent Effects in Organic Chemistry,* 2nd ed., VCH Publishers, Weinheim, Germany, 1988.

Reiner, E.S. and Radke, C.J., Double layer interactions between charge regulated surfaces, *Advances in Colloid and Interface Science,* **47**, 59, 1993.

Rha, C.K. and Pradipasena, P., Viscosity of proteins, in *Functional Properties of Food Macromolecules,* Mitchell, J.R. and Ledward, D.A., Eds., Elsevier, London, 1986, chap. 2.

Richmond, J.M., *Cationic Surfactants: Organic Chemistry,* Marcel Dekker, New York, 1990.

Rizvi, S.S.H., Singh, R.K., Hotchkiss, J.H., Heldman, D.R., and Leung, H.K., Research needs in food engineering, processing and packaging, *Food Technology,* **47**(3), 268, 1993.

Roberts, D.D. and Acree, T.E., Retronasal flavor release in oil and water model systems with an evaluation of volatility predictors, in *Flavor–Food Interactions,* McGorrin, R.J. and Leland, J.V., Eds., American Chemical Society, Washington, D.C., 1996, chap. 16.

Roberts, D.D., Elmore, J.S., Langley, K.R., and Bakker, J., Effects of sucrose, guar gum and carboxymethylcellulose on the release of volatile flavor compounds under dynamic conditions, *Journal of Agricultural and Food Chemistry,* **44**, 1321, 1996.

Robin, O., Britten, M., and Paquin, P., Influence of the disperse phase distribution on the electrical conductivity of liquid O/W model and dairy emulsions, *Journal of Colloid and Interface Science,* **167**, 401, 1994.

Robinson, G.W., Zhu, S.-B., Singh, S., and Evans, M.W., *Water in Biology, Chemistry and Physics: Experimental Overviews and Computational Methodologies,* World Scientific, Singapore, 1996.

Rogers, N.K., in *Prediction of Protein Structure and the Principles of Protein Conformation,* Fasman, G.D., Ed., Plenum Press, New York, 1989, 359.

Roozen, J.P., Frankel, E.N., and Kinsellan, J.E., Enzymic and autooxidation of lipids in low fat foods: Model of linoleic acid emulsion in emulsified hexadecane, *Food Chemistry,* **50**, 33, 1994a.

Roozen, J.P., Frankel, E.N., and Kinsellan, J.E., Enzymic and autooxidation of lipids in low fat foods: Model of linoleic acid emulsion in emulsified triolein and vegetable oils, *Food Chemistry,* **50**, 39, 1994b.

Rosen, M.J., *Surfactants and Interfacial Phenomenon,* Wiley-Interscience, New York, 1978.

Roth, C.M. and Lenhoff, A.M., Improved parametric representation of water dielectric data for Lifshitz theory calculations, *Journal of Colloid and Interface Science,* **179**, 637, 1996.

Sader, J.E., Carnie, S.L., and Chan, D.Y.C., Accurate analytic formulas for the double-layer interaction between spheres, *Journal of Colloid and Interface Science,* **171**, 4, 1995.

Salager, J.L., Phase transformation and emulsion inversion on the basis of the catastrophe theory, in *Encyclopedia of Emulsion Technology,* Vol. 3, Becher, P., Marcel Dekker, New York, 1988, chap. 1.

Salvador, D., Bakker, J., Langley, K.R., Potjewijd, R., Martin, A., and Elmore, S., Flavor release of diacetyl from water, sunflower oil and emulsions in model systems, *Food Quality Preference,* **5**, 103, 1994.

Sand, R., Structure and conformation of hydrocolloids, in *Food Hydrocolloids,* Vol. I, Glicksman, M., Ed., CRC Press, Boca Raton, FL, 1982, chap. 2.

Sato, K., Crystallization of fats and fatty acids, in *Crystallization and Polymorphism of Fats and Fatty Acids,* Garti, N. and Sato, K., Eds., Marcel Dekker, New York, 1988, 227.

Schenkel, J.H. and Kitchner, J.A., A test of the Derjaguin–Verwey–Overbeek theory with a colloidal suspension, *Transactions of the Faraday Society,* **56**, 161, 1960.

Schmidt, R.K., Tasaki, K., and Brady, J.W., Computer modeling studies of the interaction of water with carbohydrates, *Journal of Food Engineering,* **22**, 43, 1994.

Schubert, H. and Armbruster, H., Principles of formation and stability of emulsions, *International Chemical Engineering,* **32**, 14, 1992.

Schultz, K.W., Day, E.A., and Sinnbuber, R.O., *Lipids and Their Oxidation,* AVI Publishing, Westport, CT, 1962.

Sears, F.W. and Salinger, G.L., *Thermodynamics, Kinetic Theory and Statistical Thermodynamics,* Addison-Wesley, Reading, MA, 1975.

Seebergh, J.E. and Berg, J.C., Depletion flocculation of aqueous, electrostatically stabilized latex dispersions, *Langmuir,* **10**, 454, 1994.

Semenova, M.G., Factors determining the character of biopolymer–biopolymer interactions in multi-component aqueous solutions modeling food systems, in *Macromolecular Interactions in Food Technology,* Parris, N., Kato, A., Creamer, L.K., and Pearce, J., Eds., American Chemical Society, Washington, D.C., 1996, chap. 3.

Shaw, D.J., *Introduction to Colloid and Surface Chemistry,* 3rd ed., Butterworths, London, 1980.

Sherman, P., Ed., *Emulsion Science,* Academic Press, London, 1968a.

Sherman, P., General properties of emulsions and their constituents, in *Emulsion Science,* Sherman, P., Ed., Academic Press, London, 1968b, chap. 3.

Sherman, P., Rheology of emulsions, in *Emulsion Science,* Sherman, P., Ed., Academic Press, London, 1968c, chap. 4.

Sherman, P., *Industrial Rheology with Particular Reference to Foods, Pharmaceuticals and Cosmetics,* Academic Press, London, 1970.

Sherman, P., Rheology of emulsions, in *Encyclopedia of Emulsion Technology, Basic Theory,* Vol. 1, Becher, P., Ed., Marcel Dekker, New York, 1982, chap. 7.

Sherman, P., A critique of some methods proposed for evaluating the emulsifying capacity and emulsion stabilizing performance of vegetable proteins, *Italian Journal of Food Science,* **1**, 3, 1995.

Shinoda, K. and Friberg, S., *Emulsions and Solubilization,* John Wiley & Sons, New York, 1986.

Shinoda, K. and Kunieda, H., Phase properties of emulsions: PIT and HLB, in *Encyclopedia of Emulsion Technology,* Vol. 1, Becher, P., Ed, Marcel Dekker, New York, 1983.

Shoemaker, C.F., Nantz, J., Bonnans, S., and Noble, A.C., Rheological characterization of dairy products, *Food Technology,* **46**, 98, 1992.

Siano, S.A., Applications development with radio-frequency dielectric spectroscopy: Emulsions, suspensions and biomass, in *Handbook on Ultrasonic and Dielectric Characterization Techniques for Suspended Particles,* Hackley, V.A. and Texter, J., Eds., American Chemical Society, Westerville, OH, 1998.

Simatos, D. and Multon, J.L., *Properties of Water in Foods in Relation to Quality and Stability,* Martinus Nijhoff Publishers, Dordrecht, Netherlands, 1985.

Simic, M.G., Jovanovic, S.V., and Niki, E., Mechanisms of lipid oxidative processes and their inhibition, in *Lipid Oxidation in Food,* St. Angelo, A.J., Ed., American Chemical Society, Washington, D.C., 1992, chap. 2.

Simoneau, C., McCarthy, M.J., Kauten, R.J., and German, J.B., Crystallization dynamics in model emulsions from magnetic resonance imaging, *Journal of the American Oil Chemists Society,* **68**, 481, 1991.

Simoneau, C., McCarthy, M.J., Reid, D.S., and German, J.B., Influence of triglyceride composition on crystallization kinetics of model emulsions, *Journal of Food Engineering,* **19**, 365, 1993.

Sjoblom, J., Fordedal, H., and Skodvin, T., Flocculation and coalescence in emulsions studied by dielectric spectroscopy, in *Emulsion and Emulsion Stability,* Sjoblom, J., Ed., Marcel Dekker, New York, 1996, chap. 8.

Skoda, W. and van den Tempel, M., Crystallization of emulsified triglycerides, *Journal of Colloid Science,* **18**, 568, 1963.

Skodvin, T., Sjoblom, J., Saeten, J.O., Warnhim, T., and Gestblom, B., A time domain dielectric spectroscopy study of some model emulsions and liquid margarines, *Colloids and Surfaces A,* **83**, 75, 1994.

Skoog, D.A., West, D.M., and Holler, F.J., *Analytical Chemistry: An Introduction,* 6th ed., Saunders College Publishing, Philadelphia, 1994.

Sloan, A.E., Top 10 trends to watch and work on, *Food Technology,* **48**(7), 89, 1994.

Sloan, A.E., America's appetite '96: The top 10 trends to watch and work on, *Food Technology,* **50**(7), 55, 1996.

Smart, M.G., Fulcher, R.G., and Pechak, D.G., Recent developments in the microstructural characterization of foods, in *Characterization of Foods: Emerging Techniques,* Gaonkar, A.G., Ed., Elsevier, Amsterdam, 1995, chap. 11.

Smith, A.L. and Mitchell, D.P., The centrifuge technique in the study of emulsion stability, in *Theory and Practice of Emulsion Technology,* Smith, A.L., Ed., Academic Press, London, 1976, chap. 4.

Smith, N.J. and Williams, P.A., Depletion flocculation of polystyrene lattices by water-soluble polymers, *Journal of the Chemical Society, Faraday Transactions,* **91**, 1483, 1995.

Smolin, L.A. and Grosvenor, M.B., *Nutrition: Science and Applications,* Saunders College Publishing, Fort Worth, TX, 1994.

Soderberg, I., Hernqvist, L., and Bucheim, W., Milk fat crystallization in natural milk fat globules, *Milchwissenschaft,* **44**, 403, 1989.

Soderman, O. and Bailnov, B., NMR self-diffusion studies of emulsions, in *Emulsions and Emulsion Stability,* Sjoblom, J., Ed., Marcel Dekker, New York, 1996, chap. 8.

Soderman, O., Lonnqvist, I., and Bailnov, B., NMR self-diffusion studies of emulsion systems. Droplet sizes and microstructure of the continuous phase, in *Emulsions — A Fundamental and Practical Approach,* Sjoblom, J., Ed., Kluwer, Dordrecht, Netherlands, 1992, 239.

Sonntag, N.O.V., Structure and composition of fats and oils, in *Bailey's Industrial Oil and Fat Products,* Vol. 1, Swern, D., Ed., John Wiley & Sons, New York, 1979a, chap. 1.

Sonntag, N.O.V., Reactions of fats and fatty acids, in *Bailey's Industrial Oil and Fat Products,* Vol. 1, Swern, D., Ed., John Wiley & Sons, New York, 1979b, chap. 2.

Sonntag, N.O.V., Sources, utilization and classification of oils and fats, in *Bailey's Industrial Oil and Fat Products,* Vol. 1, Swern, D., Ed., John Wiley & Sons, New York, 1979c, chap. 5.

Sperry, P.R., A simple quantitative model for the volume restriction flocculation of latex by water-soluble polymers, *Journal of Colloid Science,* **87**, 375, 1982.

Spicer, P.T. and Pratsinis, S.E., Shear-induced flocculation — The evolution of floc structure and the shape of the size distribution at steady-state, *Water Research,* **30**, 1049, 1996.

Sridhar, S. and McDonald, M.J., Protein hydration investigations with high-frequency dielectric spectroscopy, *Journal of Physical Chemistry,* **98**, 6644, 1994.

Stang, M., Karbstein, H., and Schubert, H., Adsorption kinetics of emulsifiers at oil–water interfaces and their effect on mechanical emulsification, *Chemical Engineering and Processing,* **33**, 307, 1994.

St. Angelo, A., brief introduction to food emulsions and emulsifiers, in *Food Emulsifiers: Chemistry, Technology, Functional Properties and Applications,* Charalambous, G. and Doxastakis, G., Eds., Elsevier, Amsterdam, 1989.

St. Angelo, A.J., *Lipid Oxidation in Food,* American Chemical Society, Washington, D.C., 1992.

Stauffer, C.E., Emulsifiers for the food industry, in *Bailey's Industrial Oil and Fat Products,* Vol. 3, 5th ed., Hui, Y.H., Ed., John Wiley, New York, 1996, chap. 12.

Stephan, A.M., *Food Polysaccharides and Their Applications,* Marcel Dekker, New York, 1995.

Stoll, S. and Buffle, J., Computer-simulation of bridging flocculation processes — The role of colloid to polymer concentration ratio on aggregation kinetics, *Journal of Colloid and Interface Science,* **180**, 548, 1996.

Stone, H.A., Dynamics of drop deformation and breakup in viscous fluids, *Annual Review of Fluid Mechanics,* **26**, 65, 1994.

Stothart, P.H., Developments in the application of small angle neutron scattering to food systems, in *Characterization of Foods: Emerging Techniques,* Gaonkar, A.G., Ed., Elsevier, Amsterdam, 1995, chap. 9.

Strawbridge, K.B., Ray, E., Hallett, F.R., Tosh, S.M., and Dalgleish, D.G., Measurement of the particle size distribution in milk homogenized by a microfluidizer — Estimation of populations of particles with radii less than 100 nm, *Journal of Colloid and Interface Science,* **171**, 392, 1995.

Suggett, A., Water–carbohydrate interactions, in *Water Relationships of Foods,* Duckworth, R.B., Ed,, Academic Press, London, 1975a, 23.

Suggett, A., Polysaccharides, in *Water: A Comprehensive Treatise,* Vol. 4, Franks, F., Ed., Plenum Press, New York, 1975b, chap. 6.

Swaisgood, H.E., Characteristics of milk, in *Food Chemistry,* 3rd ed., Fennema, G.R., Ed., Marcel Dekker, New York, 1996, chap. 14.

Swift, C.E., Lockett, C., and Fryer, P.J., Comminuted meat emulsions — The capacity of meat for emulsifying fat, *Food Technology,* **15**, 469, 1961.

Tadros, T.F., Fundamental principles of emulsion rheology and their applications, *Colloids and Surfaces,* **91**, 30, 1994.

Tadros, Th.F., Correlation of viscoelastic properties of stable and flocculated suspensions with their interparticle interactions, *Advances in Colloid and Interface Science,* **68**, 97, 1996.

Tadros, T.F. and Vincent, B., Liquid/liquid interfaces, *in Encyclopedia of Emulsion Technology,* Vol. 1, Becher, P., Ed., Marcel Dekker, New York, 1983.

Taisne, L., Walstra, P., and Cabane, B., Transfer of oil between emulsion droplets, *Journal of Colloid and Interface Science,* **184**, 378, 1996.

Tan, C.-T., Beverage emulsions, in *Food Emulsions,* 2nd ed., Larsson, K. and Friberg, S.E., Eds., Marcel Dekker, New York, 1990, chap. 10.

Tanford, C., *The Hydrophobic Effect,* Wiley, New York, 1980.

Taylor, P., Ostwald ripening in emulsions, *Colloids and Surfaces A,* **99**, 175, 1995.

Taylor, A.J. and Linforth, R.S.T., Flavour release in the mouth, *Trends in Food Science and Technology,* **7**, 444, 1996.

Texter, J., Beverly, T., Templar, S.R., and Matsubara, T., Partitioning of para-phenylenediamines in oil-in-water emulsions, *Journal of Colloid and Interface Science,* **120**, 389, 1987.

Thomson, D.M.H., The meaning of flavor, in *Developments in Food Flavours,* Birch, G.G. and Lindley, M.G., Eds., Elsevier Applied Science, London, 1986, 1.

Timms, R.E., Crystallization of fats, *Chemistry and Industry,* p. 342, May 1991.

Timms, R.E., Crystallization of fats, in *Developments in Oils and Fats,* Hamilton, R.J., Ed., Blackie Academic and Professional, London, 1995, chap. 8.

Tinoco, I., Sauer, K., and Wang, J.C., *Physical Chemistry: Principles and Applications in Biological Sciences,* 2nd ed., Prentice-Hall, Englewood Cliffs, NJ, 1985.

Tolstoguzov, V., Structure–property relationships in foods, in *Macromolecular Interactions in Food Technology,* Parris, N., Kato, A., Creamer, L.K., and Pearce, J., Eds., American Chemical Society, Washington, D.C., 1996, chap. 1.

Tornberg, E. and Hermannson, A.M., Functional characteristics of protein stabilized emulsions. Effect of processing, *Journal of Food Science,* **42**, 468, 1977.

Tory, E.M., *Sedimentation of Small Particles in a Viscous Liquid,* Computational Mechanics Publications, Southampton, U.K., 1996.

Trainer, M.N., Freud, P.J., and Leonardo, E.M., High-concentration submicron particle size distribution by dynamic light scattering, *American Laboratory,* p. 34, July 1992.

Tung, M.A. and Jones, L.J., Microstructure of mayonnaise and salad dressing, in *Studies of Food Microstructure,* Holcomb, D.N. and Kalab, M., Eds., Scanning Electron Microscopy, AMF O'Hare, IL, 1981, 231.

Tung, M.A. and Paulson, A.T., Rheological concepts for probing ingredient interactions in food systems, in *Ingredient Interactions: Effects on Food Quality,* Gaonkar, A., Ed., Marcel Dekker, New York, 1995.

Tunon, I., Silla, E., and Pascual-Ahuir, J.L., Molecular surface area and hydrophobic effect, *Protein Engineering,* **5**, 715, 1992.

Turnbull, D. and Cormia, D., Kinetics of crystal nucleation in some normal alkane liquids, *Journal of Chemical Physics,* **34**, 820, 1961.

Uriev, N.B., Structure, rheology and stability of concentrated disperse systems under dynamic conditions, *Colloids and Surfaces A,* **87**, 1, 1994.

Vaessen, G.E.J. and Stein, H.N., The applicability of catastrophe theory to emulsion phase inversion, *Journal of Colloid and Interface Science,* **176**, 378, 1995.

van Boekel, M.A.J.S., Estimation of solid liquid ratios in bulk fats and emulsions by pulsed NMR, *Journal of the American Oil Chemists Society,* **58**, 768, 1981.

van Boekel, M.A.J.S. and Walstra, P., Stability of oil-in-water emulsions with crystals in the disperse phase, *Colloids and Surfaces,* **3**, 109, 1981.

van de Hulst, H.C., *Light Scattering by Small Particles,* John Wiley & Sons, New York, 1957.

van den Enden, J.C., Waddington, D., van Aalst, H., van Kralingen, C.G., and Packer, K.J., Rapid determination of water droplet size distributions by PFG-NMR, *Journal of Colloid and Interface Science,* **140**, 105, 1990.

van Gunsteren, W.F., The role of computer simulation techniques in protein engineering, *Protein Engineering,* **2**, 5, 1988.

van Holde, K.E., Effects of amino acid composition and microenvironment on protein structure, in *Food Proteins,* Whitaker, J.R. and Tannenbaum, S.R., Eds., AVI Publishing, Westport, CT, 1977, chap. 1.

van Vliet, T., Mechanical properties of concentrated food gels, in *Food Macromolecules and Colloids,* Dickinson, E. and Lorient, D., Eds., Royal Society of Chemistry, Cambridge, 1995, 447.

van Vliet, T. and Walstra, P., Weak particle networks, in *Food Colloids,* Bee, R.D., Richmond, P., and Mingins, J., Eds., Royal Society of Chemistry, Cambridge, 1989, 206.

Vaughan, J.G., Food emulsions, in *Food Science and Technology. A Series of Monographs. Food Microscopy*, Vaughan, J.G., Ed., Academic Press, London, 1979.

Vodovotz, Y., Vittadini, E., Coupland, J., McClements, D.J., and Chinachoti, P., Bridging the gap: Use of confocal microscopy in food research, *Food Technology,* **50**, 74, 1996.

Vold, M.J., The effect of adsorption on the van der Waals interaction of spherical colloidal particles, *Journal of Colloid Science,* **16**, 1, 1961.

Waddington, D., Some applications of wide line NMR in the oils and fats industry, in *Fats and Oils: Chemistry and Technology,* Hamilton, R.J. and Bhati, A., Eds., Applied Science Publishers, London, 1980, chap. 2.

Walstra, P., On the crystallization habit in fat globules, *Netherlands Milk and Dairy Journal,* **21**, 166, 1967.

Walstra, P., Estimating globule-size distribution of oil-in-water emulsions by spectroturbidimetry, *Journal of Colloid and Interface Science,* **27**, 493, 1968.

Walstra, P., Formation of emulsions, in *Encyclopedia of Emulsion Technology,* Vol. 1, Becher, P., Ed., Marcel Dekker, New York, 1983, chap. 2.

Walstra, P., Fat crystallization, in *Food Structure and Behaviour,* Blanshard, J.M.V. and Lillford, P., Eds., Academic Press, London, 1987, chap. 5.

Walstra, P., Introduction to aggregation phenomenon in food colloids, in *Food Colloids and Polymers: Stability and Mechanical Properties,* Dickinson, E. and Walstra, P., Eds., Royal Society of Chemistry, Cambridge, 1993a, 3.

Walstra, P., Principles of emulsion formation, *Chemical Engineering Science,* **48**, 333, 1993b.

Walstra, P., Emulsion stability, in *Encyclopedia of Emulsion Technology,* Vol. 4, Becher, P., Ed., Marcel Dekker, New York, 1996a, chap. 1.

Walstra, P., Disperse systems: Basic considerations, in *Food Chemistry,* 3rd ed., Fennema, O.R., Ed., Marcel Dekker, New York, 1996b, chap. 3.

Walstra, P. and van Beresteyn, E.C.H., Crystallization of milk fat in the emulsified state, *Netherlands Milk and Dairy Journal,* **29**, 35, 1975.

Wedzicha, B.L., Distribution of low-molecular weight food additives in dispersed systems, in *Advances in Food Emulsions,* Dickinson, E. and Stainsby, G., Eds., Elsevier, London, 1988, chap. 10.

Wedzicha, B.L., Zeb, A., and Ahmed, S., Reactivity of food preservatives in dispersed systems, in *Food Polymers, Gels and Colloids,* Dickinson, E., Ed., Royal Society of Chemistry, Cambridge, 1991, 180.

Wehling, R.L., Infrared spectroscopy, in *Introduction to the Chemical Analysis of Foods,* Nielsen, S.S., Ed., Bartlett and Jones, Boston, 1994, chap. 24.

Wei, Y.-Z., Kumbharkhane, A.C., Sadeghi, M., Sage, J.T., Tiam, W.D., Champion, P.M., Reiner, E.S., and Radke, C.J., Double layer interactions between charge-regulated colloidal systems: Pair potentials for spherical particles bearing ionogenic surface groups, *Advances in Colloid and Interface Science,* **47**, 59, 1993.

Wei, Y.-Z., Kumbharkhane, A.C., Sadeghi, M., Sage, J.T., Tiam, W.D., Champion, P.M., Radke, C.J., and McDonald, M.J., Protein hydration investigations with high-frequency dielectric spectroscopy, *Journal of Physical Chemistry,* **98**, 6644, 1994.

Weiss, T.J., *Food Oils and Their Uses,* 2nd ed., AVI Publishing, Westport, CT, 1983.

Weiss, J., Coupland, J.N., and McClements, D.J., Solubilization of hydrocarbon droplets suspended in a non-ionic surfactant solution, *Journal of Physical Chemistry,* **100**, 1066, 1996.

Whitaker, J.R., Denaturation and renaturation of proteins, in *Food Proteins,* Whitaker, J.R. and Tannenbaum, S.R., Eds., AVI Publishing, Westport, CT, 1977, chap. 2.

Whitaker, J.R. and Tannenbaum, S.R., Eds., *Food Proteins,* AVI Publishing, Westport, CT, 1977.

Whorlow, R.W., *Rheological Techniques,* 2nd ed., Ellis Horwood, New York, 1992.

Williams, A.A., Modern sensory analysis and the flavour industry, in *Developments in Food Flavours,* Birch, G.G. and Lindley, M.G., Eds., Elsevier Applied Science, London, 1986, 1.

Williams, A., Janssen, J.J.M., and Prins, A., Behaviour of droplets in simple shear flow in the presence of a protein emulsifier, *Colloids and Surfaces A,* **125**, 189, 1997.

Wilson, R.H., Recent developments in infrared spectroscopy and microscopy, in *New Physicochemical Techniques for the Characterization of Complex Food Systems,* Dickinson, E., Ed., Blackie Academic and Professional, London, 1995, chap. 8.

Wunderlich, B., *Thermal Analysis,* Academic Press, New York, 1990.

Xang, H.J. and Xu, G.D., The effect of particle refractive index on size measurement, *Powder Technology,* **70**, 189, 1992.

Xiong, Y.L. and Kinsella, J.E., Influence of fat globule membrane composition and fat type on the rheological properties of milk based composition gels. 1. Methodology, *Milchwissenschaft,* **46**, 150, 1991.

Xiong, Y.L., Aguilera, J.M., and Kinsella, J.E., Emulsified milkfat effects on rheology of acid-induced milk gels, *Journal of Food Science,* **56**, 920, 1991.

Yaminsky, V.V., Ninham, B.W., Christenson, H.K., and Pashley, R.M., Adsorption forces between hydrophobic monolayers, *Langmuir,* **12**, 1936, 1996a.

Yaminsky, V.V., Jones, C., Yaminsky, F., and Ninham, B.W., Onset of hydrophobic attraction at low surfactant concentrations, *Langmuir,* **12**, 3531, 1996b.

Yost, R.A. and Kinsella, J.E., Properties of acidic whey protein gels containing emulsified butterfat, *Journal of Food Science,* **57**, 158, 1993.

Young, S.L., Sarda, X., and Rosenberg, M., Microencapsulation properties of whey proteins. 1. Microencapsulation of anhydrous milk fat, *Journal of Dairy Science,* **76**, 2868, 1993a.

Young, S.L., Sarda, X., and Rosenberg, M., Microencapsulation properties of whey proteins. 2. Combination of whey proteins with carbohydrates, *Journal of Dairy Science,* **76**, 2878, 1993b.

Zhang, X. and Davis, H.D., The rate of collisions due to Brownian or gravitational motion on small droplets, *Journal of Fluid Mechanics,* **230**, 479, 1991.

Zhang, H.J. and Xu, G.D., The effect of particle refractive index on size measurement, *Powder Technology,* **70**, 189, 1992.

Ziegler, G.R. and Foegedding, E.A., The gelation of proteins, in *Advances in Food and Nutrition Research,* Vol. 34, Kinsella, J.E., Ed., Academic Press, San Diego, 1990, 203.

Zielinski, R.J., Synthesis and composition of food-grade emulsifiers, in *Food Emulsifiers and Their Applications,* Hasenhuettl, G.L. and Hartel, R.W., Eds., Chapman and Hall, New York, 1997, chap. 2.

Index

A

Absorbance, 287
Absorption, 291, 292
    in concentrated emulsions, 290–291
    defined, 286
    kinetics of, 286–287
    thermodynamics of, 317
Absorption coefficient, 290
Absorption frequency, 43, 44
Absorption spectrum, 287, 291
Accelerated creaming test, 198
Activation energy
    for crystallization, 88
    for emulsion stability, 188, 189
    for nucleation, 90, 92
Acylglycerols, 84
Adsorbed ions, 49
Adsorbed layer
    creaming and, 194–195
    rheology of, 155
    thickness, 313
    van der Waals interactions and, 46–47, 48
Adsorption, 54
    at an interface, 129
    biopolymers, 119
    competitive, 143–146, 155, 298
    creaming and, 194–195
    droplet charge and, 13
    free energy of, 54
    ions, 48
    kinetics, 140–142, 148, 149, 155, 156, 298
        emulsifier, 152, 170
        emulsifier to interface, 130–131
    negative, 130
    surfactant molecules, 36
Adsorption efficiency, 144
Adsorption equilibrium constant, 53
Aggregates, 261–264
Aggregation, 24, 65, 143, 259, 273, 313, see also
        specific topics
    biopolymers and, 112–116, 120
    coalescence, 41, 42, 221–222
    creaming and, 193
    critical concentration, 76
    droplet, colloidal interactions and, 39–42, 48, 59
    electrical charge and, 13
    emulsion rheology and, 257–259
    flocculation, 41
    hydrophobic interactions and, 117
    kinetics, 3, 4
    protein, 36, 103, 118, 180
    stability to, 49
    surfactants and, 107
Alcohols, 69, 100, 180, 228, 270
Aldehydes, 270, 272
Alkanes, 272

Amino acids, 111, 112, 113, 114, 199, 145
Amphiphilic molecules, 5
Angular scattering, 309
Antioxidants, 85, 233, 268
Apparent modulus, 237
Apparent viscosity, nonideal liquid, 240, 241, 242
Appearance, 1, 267, 285–294, 310, see also specific
        topics
    biopolymers and, 111
    color, 291–292
    droplet concentration and, 6
    droplet crystallinity and, 14
    droplet size and, 178
    electrostatic interactions and, 49
    flocculation and, 209
    gravitational separation and, 189
    homogeneity, 292
    interaction of light waves with emulsions, 285–291
    lipids and, 85, 86
    opacity, 291
    particle size distribution and, 7
    phase inversion and, 229
Aqueous phase, 2–3, 5, 6
    electrical conductivity, 326
    flavor partitioning and, 268, 269, 270, 271, 273,
        274, 275, 276
    flavor release and, 278, 279, 280, 282
    ionic strength and, 48
    pH and ionic strength and, 13, 48
Aqueous solution, 97–98, see also Water
    biopolymers in, 23, 116–119
    molecular interactions, 23
    viscosity of, 120–121
Archimedes principle, 324
Aroma, 1, 267, 268, 279, 280–283
    droplet crystallinity and, 14
    ingredient partitioning and, 5
    lipids and, 85, 94–95
Association colloids, 104
Atomic force microscopy, 157, 306–307
Attenuation coefficient, 318, 319, 320, 334
Autocorrelation function, 312, 313

B

Babcock method, 323
Bancroft's rule, 107
Bending modulus, 72
Benzene, 103
Beverages, 1, 235, 288, 289, see also specific
        beverages
Bilayers, 104, 105, 109
Bimodal distribution, 11, 12
Binding, 281–282
    irreversible, 277, 281
    reversible, 276–277, 281
Binding coefficient, 272
Bingham plastic, 244

nonideal liquid, 239–243
oil, 87
phase inversion and, 232
water, 87
Viscosity ratio, 180, 181
Vitamins, 84, 85, 97, 268
Volatile components, 267
Volatile compounds, release of, 280–283
Volatility
   electrostatic interactions and, 13
   water and, 97
Volume frequency, 7, 8
Volume–surface mean diameter, 9

W

Wall slip, 253
Water, 84, 95
   bulk physicochemical properties, 87, 97
   electrical conductivity, 325
   electrostatic interactions, 29
   hydrogen bonding, 35
   hydrophobic interactions, 35, 36
   interactions, 25
      of biopolymers with, 116–119
      with dipolar solutes, 98, 100–101
      with ionic solutes, 98–100
      with nonpolar solutes, 98, 101–104
   interface with oil, see Oil–water interface
   molecular structure and organization, 23, 95–97
   physicochemical properties, 22–23, 44
   solubility, of biopolymers, 116–119
   viscosity of, 239
Water content, proximate analysis, 323
Water-holding capacity, 35, 122
Water-in-oil emulsion, 4, 217
   analysis by electron microscopy, 302, 306
   coalescence, 214
   creaming, 194

crystallization, 94
defined, 2
droplet crystallinity and, 14, 332, 333
flavor partitioning, 274
flavor release, 282–283
formation of, 162
gravitational separation, 189
phase inversion, 229–232
rheology, 264, 265
stability, 197
surfactants and, 107, 108, 109
viscosity, 255
Weak flocculation, 41
Weak nuclear interactions, 17
Weber number, 165, 166, 167
Weighting agents, 196
Wetting, 136–138, 143
Whey protein, 78–79, 204, 207, 264
Whipped cream, 185
Whipped topping, 222, 225
Whipping, 161
Wilhelmy plate method, 146, 148–149, 155

X

X-ray diffraction, 226
X-ray scattering, 156, 157
Xanthan, 33, 111, 208

Y

Yield stress, 121, 194, 212, 243–244, 263
Yogurt, 1, 122
Young equation, 137

Z

Zero interaction regime, 63
Zeta potential, 322, 334–337
Zetasizer®, 335